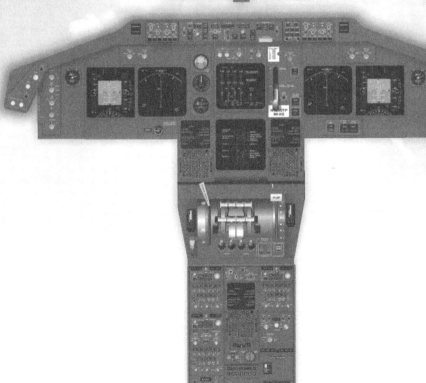

747-400 SIMULATOR TECHNIQUES ...

© MIKE RAY 2014
published by UNIVERSITY of TEMECULA PRESS

... and PROCEDURES FOR STUDY and REVIEW

CAPTAIN MIKE RAY'S

747-400

PILOT HANDBOOK

SIMULATOR AND CHECKRIDE

PROCEDURES MANUAL

This document is crammed with lotsa stuff that was stolen, copied, some of it made up, collected, and even a whole bunch of personal experiences from a pile of stuff hoarded over a career and put together in this one place for your enjoyment and edification.

© MIKE RAY 2014
WWW.UTEM.COM

MY LAWYER WANTS TO SAY SOMETHING (probably worthless) ...

THE LEGAL STUFF

To be read aloud as rapidly as possible!

Reproduction or use of editorial or pictorial content in any manner is prohibited without express permission. No part of this document may be reproduced, copied, adapted, or transmitted in any form or by any means without written permission. The author (me) and the publisher (me again) makes no representations or warranties with respect to the contents hereof and specifically disclaims any implied warranties of merchantability or fitness for any particular purpose, even the simulator portion of the check-ride but certainly not actually flying the jet. No patent liability is assumed with respect to the use of the information contained herein. Further, we reserve the right to revise, change, or alter this publication and to make changes from time to time in the contents hereof without any obligation to notify anybody of such changes or revisions. While every precaution (OK there is probably some precaution I forgot) has been taken in the preparation of this book, the publisher assumes no (what is there about "NO" that is ambiguous?) responsibilities for errors or omissions. Neither is any liability assumed for damages (such as a busted check-ride) resulting from the use of the information contained herein.
This manual is sold "as-is" and you are assuming the entire risk as to its performance for your needs and purposes. We are not liable for any damages, direct or indirect, resulting from any error, inclusion, exclusion, inference, misunderstanding, screw-up, goober, outright lie or mistake.
Of course, what dolt would not realize that Boeing is a registered trademark of Boeing Aircraft Corporation and the use of the term in no way implies that they endorse, support, or even know or care about this manual. Anybody else we may have mentioned in this publication, either by using their registered name or trademark or in some way referencing them obliquely, in no possible way infers or claims that they endorse this product. Duh!
Gosh, I hope this covers my six.

QUICK SUMMARY OF THIS STATEMENT:

DON'T SUE ME FOR BUSTING YOUR CHECK-RIDE OVER SOMETHING THAT I SAID IN THIS DOCUMENT.

© MIKE RAY 2016

Published by
UNIVERSITY of TEMECULA PRESS

160930
v12

CREATED IN THE UNITED STATES of AMERICA
VISIT the WEBSITE:
www.utem.com

DISCLAIMERS

THIS IS REALLY SERIOUS STUFF

Of course, I wouldn't deliberately put anything in a book like this that I didn't think was pretty close to being the truth and in pretty good agreement with every jot and tittle of the documents from people who are in charge; but, hey, I am only an ex-airline pilot. My intentions are that everything written here is to be considered subordinate to and subject to correction by reference to the plethora of "**OFFICIAL**" publications and documentation available from the Company, the FAA, Boeing Airplane Company, your Mother-in-law, or anybody else that has jurisdiction or oversight in these matters.

Those guys represent the *"right stuff,"* not me! **PERIOD!**

REMINDER!

This material is written for and intended to be used in the SIMULATOR ONLY. It does not imply or suggest that there is any carry over value to the operation of the "REAL" airplane. Of course, current company SOPs and FAA mandated procedures and operational guidelines ALWAYS supercede this information. Remember, this material is STRICTLY for study and review in preparation for operating a SIMULATOR. PERIOD!

The FABULOUS BOEING 747-400

What a marvelous and wonderful treasure we have in these unbelievably fantastic sky-machines. Who would have believed just a generation ago that these truly incredible airplanes would even exist. Who would have made the prediction that there would be literally thousands of these remarkable jet airplanes filling the stratosphere. Routinely flying at incredible altitudes high above the earth, heights that were unimaginable just a few years ago; and yet, doing so safely and comfortably and at speeds that were beyond human conceptualization in those days. In turn this has given human beings the undreamed of mobility and unfettered ability to traverse the surface of our earth at will, and do so often and relatively inexpensively. We are indeed blessed to be living in this age of such miraculous technologies.

... And it is this specific airplane, the iconic Boeing 747-400 which has made a major contribution to the fulfillment of man's aching desire to fly. This modest document celebrates this unique and historic airplane that has flown billions of air-miles as it continues to ceaselessly circumnavigate our earthly home.

some probably totally worthless
COMMENTS

Mike Ray

What arrogant moron would somehow think in his wildest imaginations that he in any way had the authority or knowledge to write a manual about how to pass a check-ride on the Fabulous Boeing 747-400 Monster-Jet. Only an egotistical ex-airline pilot, with time on his hands, would wade into that forbidden territory, a no-man's land that even the engineers and test pilots would not venture to tread. Fearlessly, it is I (the fool that I am) that is willing to put down my meager and limited (sometimes outright wrong) knowledge into a book that preserves my ignorance for all posterity. So... this is a problem since I am not an artist or a writer or even a very good communicator; however, since there is a lot of good technology out there, I can make the reader think that I know what I am doing. Just go with me on this, I'll try to make it as "professional looking" as I can.

If past experience is any indication, there are a whole lot of you out there that are a lot smarter than I am, ...and are more than willing to tell me so ... but, bring it on. I welcome your comments, vituperating, whining, complaining, yelling and screaming.

The whole idea here is that a bunch of ordinary, garden variety airline pilots can get together and pool their knowledge in some place where we can all benefit from each others comments and experiences. If I hear from you about how we can make this book better or how to change it to make it "right-er," then I will do it. My plan is to make a series of "short" printing runs, attempting to keep the book up to date and continually improving along the way ... but, **what is desperately needed is input from the troopers out in the trenches**.

Let's keep those cards and letters coming in.

the boring BIO

Tall guy, bald head, mustache ... worked for United Airlines for over 32 years. He is, thankfully, retired now. He has been extremely happy married forever to his beautiful teenage sweetheart. He has two gorgeous daughters, three angelic granddaughters and a terrific grandson.
He is a fulfilled and happy man. Life is good!

Website: www.utem.com
e-mail: mikeray@utem.com

MIKE RAY
CAPTAIN, UAL ret.

CONTACT ME!

FAX: 951-698-3676

ACKNOWLEDGEMENTS

THESE GUYS HELPED ME!

Of course, I wouldn't want to tell you the names of all the people who helped me out here, because then this would turn into some kinda written grammy awards thing. You know, "I'd like to thank God, my Mother, my plumber, Uncle Jack, etc. But there really were people who spent a lot of their own precious time looking over the raw material and actually got involved in the preparation of this manual.

First, there wouldn't be a 747-400 manual if Captain Jim Marshall hadn't kept up the pressure. This guy was always asking me when it would be ready and how was I doing.

Dr. Jack Rubino, who is a long-long time friend and probably has more knowledge and training in jet airplanes than most full time line pilots. I don't know what I would have done without his willingness to be a part of this process.

Capt. Bill Dobbs who took time to spend with me on all those details that I couldn't understand.

My readers and proofers who spent their layovers and potential TV watching time to pore over the boring drafts and present their corrections and additions.

Capt. Jim Marshall,
Dr. Jack Rubino
Capt. Bill Dobbs,
Capt. Ernie Yoshimoto,
Capt. Don Weber,
Capt. Rob Toro.

If you see these guys around, tell them thank you for me.

My Editor, Teri Lee VanNyhuis who waded through material of which she had no knowledge. Keeping me straight on my punctuation and spelling.

My printer, and longtime friend, David Marlow for all his assistance in making ideas become reality.

My wonderful, understanding wife, Midge, for all her assistance in making this "manual" business operate smoothly.

Some personal thoughts about...

FLYING THE FABULOUS
BOEING 747-400

There is a more than just a small amount of arrogance associated with the whole concept that a mere human being can assume that they can actually manipulate the controls and "fly" what is arguably the one of the largest and most complicated machines ever to achieve airliner status. However, it was my personal great fortune to have been among the chosen few to actually have been a real, honest-to-goodness 747-400 airline Captain, fortuitously granted the privilege of being in charge of this incredible piece of human engineering. Without a doubt, even for an airplane so large, it has the most pilot friendly flight deck I ever strapped on. She has the most powerful set of throttles I have ever pushed. The comfortable yoke has the silky smooth feel of a expensive race car and the airplane responds like a young gazelle. In short, actually flying this marvelous and wonderful airplane is a bucket list experience.

    The downside of the whole -400 operation; however, was the long range mission. While take-off is routinely made at incredibly heavy gross weights; the lengthy flight plans require flying for hours in cruise using automated flight controls and navigation aids. Once the airplane burns down literally tons of fuel and gets to destination ... many hours have passed. As a result, the pilots only get to hand fly and land the light and extremely responsive airplane once or twice a month. Even though the heavy airplane (some models up to 1.2 million pounds gross) handles nicely and has plenty of power margin for engine failure; it literally dances when it is at landing weights. A light -400 will leap forward and upward when the thrust levers are even lightly pushed forward. Truly amazing. And with all those tires and articulated trucks on the undercarriage, it is almost difficult to make a bad landing.

I have a fellow 747-400 pilot friend who used to say, "You love the airplane; but you hate the mission."

The AWESOME PRIVILEGE

It should be intuitively obvious that what this book is going to attempt to do in the following few skimpy pages is virtually impossible. In fact, it is probably pure folly to even attempt to describe and understand how this magnificent 747-400 machine operates using any written format ... she simply must be experienced to be appreciated. We are among the few that have been awarded the privilege to get to know about this airplane as intimately as we are about to. Getting to fly her is a whole different, challenging and rewarding experience by any standard, and this wonderful airplane is as delightful to fly as she is beautiful to look at. Come on, let's start our relationship with our new mistress and get to know her better.

The Beautiful Boeing 747-400

T.O.C.
TABLE of CONTENTS

Intro Stuff	1 - 12
Flight Deck Layout	13
Big Five Instruments	30
CDU (Computer Display Unit)	32
MCP (Mode Control Panel)	36
PFD (Primary Flight Display)	38
ND (Navigation Display)	48
ECU	50
FLOWS	53
Pre-Flight Section	59
Flows (S-L-U-G-O)	61
"S"	62
Exterior Inspection	83
Captain's FMCS Set-up	84
F/O FMCS Verification	96
"L!!" Captain's Cockpit Preparation	102
"U"	108
"G"	122
"O"	128
Speeds	137
"S"	138
WX Radar	147
F/O Cockpit Set-up	148
Final Cockpit Preparation	159
Fuel Stuff	170
Fuel Panel Set-up	174
Overweight Landing	178
Fuel Jettison	180
Takeoff Brief	184
Captain's 5 things	186
Push-Back	193
Engine Start	198
After Start	206
Before Taxi out	209
Taxi Out	210
Before Take-Off	216
WIND	218
EOSID	219
TOGA	222
Cleared into Position	223
Take-Off	228
5 T/O Scenarios	232
ABORTS	234
Hot Brakes	238
V1 Cut	239
V2 Cut	246
ENG FIRE on T/O	247
Normal Take-Off Profile	248
ATO (After T/O)	256
VNAV	260
Holding	262
AIRBAG	267
CDAP Profiles	280
Visual Approach	292
SLAM-DUNK	296
NON-ILS Approaches	302
ILS Approaches	318
GO-Around (Missed Approach)	326
Landing	333
Taxi-in and Parking	343
QRC (Emergency Checklists)	357
Limits and Specifications	387

DISCLAIMER

A note about the differences you will encounter between this manual and your specific airline's SOPs, airplane differences, and unique simulation details.

*T*he layout of the cockpit, the way the material is presented, and the verbiage used may, in fact, differ from airline to airline and from flight simulation to flight simulation. It is a fact that this is a problem for Mr. Boeing and he would much prefer it if everyone had the exact same cockpit set-up. This doesn't happen for several reasons.

One reason is that the "experts" at the various airlines have differing views as to what is important and what is not. As incredible as it may seem, these "customers" are willing to pay considerable sums of money to Mr. Boeing to move and alter the basic design to accommodate their individual desires. They will demand that Mr. Boeing rewire and move basic items from the Boeing original design for, what I consider usually to be petty considerations.

Then what happens when airline "A" buys a used airplane from airline "B" is that the differences become glaring and annoying when a pilot flies the different airplanes. What you will notice, if you are working for a large airline, is that eventually that airline will acquire a "fleet" of airplanes with many differing and unique cockpit and system layouts.

So, when you read this book, you will try and compare the systems controls and layout with what you are familiar with ... and you will notice that there are probably lots of those differences. This DOES NOT negate the value of this manual. These differences to be expected.

It should be noted, that when the simulations or manuals (such as this one) are developed, the writers and computer geniuses seek to display a "common" or representative cockpit layout that is useful to the most situations.

That is what I have tried to do in this presentation. The basic procedures and airplane layout presented here is from my experience at a major airline. I consider it to be a good representation and will fulfill the need for a very basic set of procedures that should be applicable to whatever airplane or simulation you are flying.

Here is what I suggest you do when you notice that there are some significant differences between what you are looking at in this book and what you are seeing in your specific personal airline/simulation ...

USE THE INFORMATION WHERE YOU CAN.
It never hurts to take another point of view and understand it.

INTRODUCING the FABULOUS 747-400 FLIGHT DECK

WOW!! SURE ARE A LOTTA LITTLE TV SCREENS.

The beautiful flight deck for the most wonderful airplane ever conceived by mankind. I think I can safely say that there has never been a more pilot oriented jet than this gigantic flying airliner.

The first time a pilot climbs up the stairs and onto the flight-deck, the cockpit seems incredibly complex. It was my experience, however, that it becomes something intensely beautiful and extremely functional, and the more you use it, the more you get used to it. While there is a fairly steep learning curve, it eventually becomes a part of you. As a pilot, you will find that those Boeing master craftsmen who created it had magic in their minds.

So, in order to achieve the level of understanding and familiarity with it, we must start with the basics. To help you start to begin to assimilate the diverse collection of strange and unknown dials and gauges, I have broken the 747-400 universe into 10 separate domains:

> UPPER OVERHEAD PANEL
> CIRCUIT BREAKERS
> MAIN OVERHEAD PANEL
> MCP PANEL
> LEFT FORWARD PANEL
> CENTER FORWARD PANEL
> RIGHT FORWARD PANEL
> CDU PANEL
> THROTTLE QUADRANT
> LOWER CONSOLE

I know, it seems impossible at first, that a mere human being could come to comprehend and understand the totality of the -400 electronics suite. It seems far too complex to grasp; but I assure you as time goes by, and more and more of the operational understanding penetrates your psyche, you will begin to "get it." I personally feel that up until about 100 hours, pilots will be operating on mostly "conditioning" and not true "learning." Then, magically, somewhere around 300 hours, they begin to operate the airplane with a more complete understanding of the systems and the way seemingly unrelated systems actually affect each other. They begin to "learn' and actually reach "habituation" in more and more areas.

**LIVE IT,
FEEL IT,
KNOW IT,
LOVE IT.**

However, there is an inherent desire on the part of pilots to reach a certain point in their learning where they cease striving to learn and then to stagnate at that "comfort zone." I suggest that the student pilot must ALWAYS remain a student, always pushing the knowledge envelope and continue to demand more of oneself and ones information bank.

To that end, let's begin our journey. Let's begin our romance and love affair with our beautiful and demanding mistress:
The fabulous Boeing 747-400.

Behold; She waits.

The 10 parts of **the 747-400 universe**

UPPER OVERHEAD PANEL

CIRCUIT BREAKER PANEL

It seems impossibly complicated when you first view the vast array of doo-hickeys, thing-a-ma-bobs, veeblefetzers, and what-not located in the cockpit suite of instrumentation and controls. To assist us in mentally separating the material into more bite sized chunks, let's divide the whole magilla into 10 distinct parts.

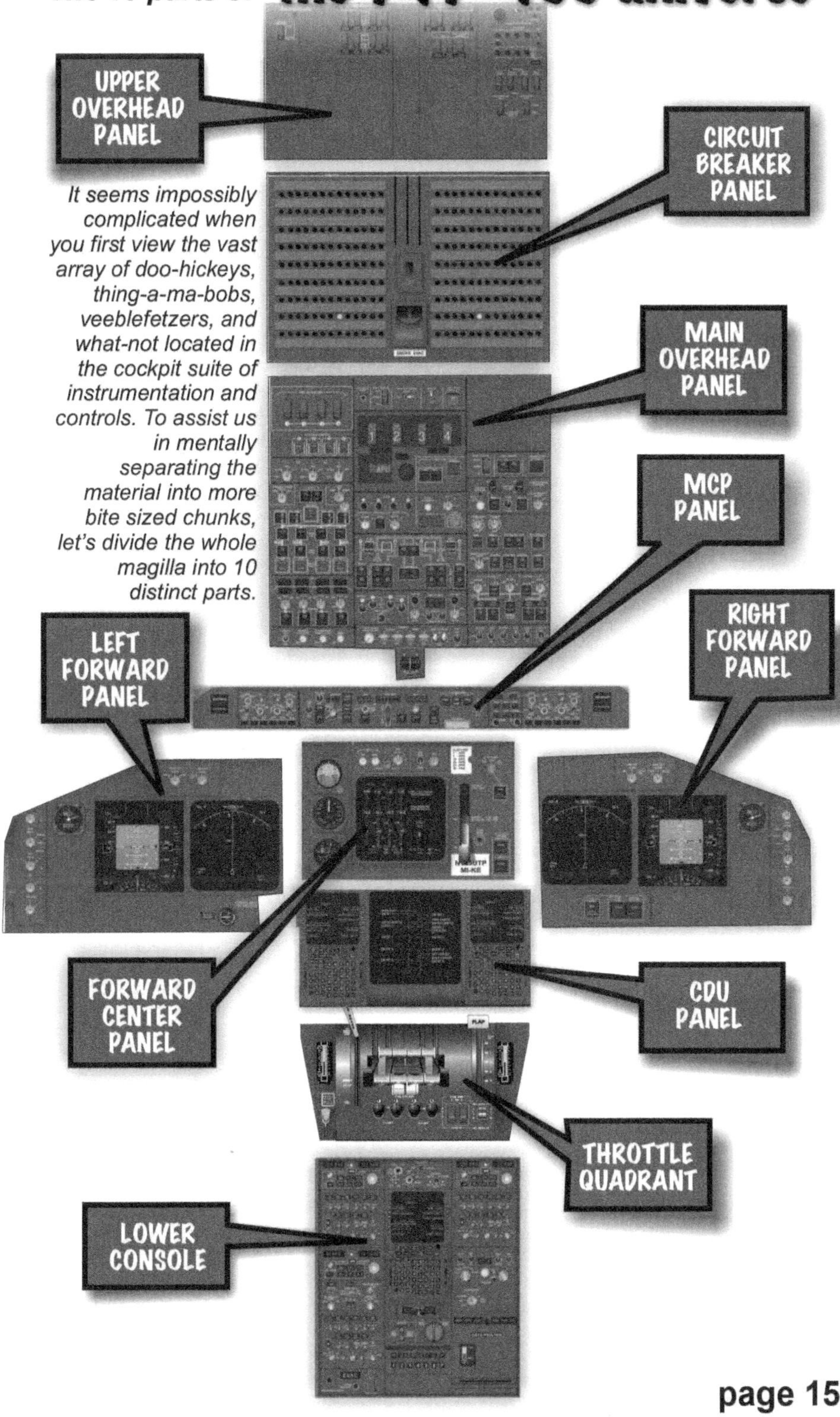

MAIN OVERHEAD PANEL

MCP PANEL

LEFT FORWARD PANEL

RIGHT FORWARD PANEL

FORWARD CENTER PANEL

CDU PANEL

THROTTLE QUADRANT

LOWER CONSOLE

page 15

UPPER OVERHEAD PANEL

- FLIGHT DECK ACCESS LIGHTS
- FLIGHT CONTROL HYD SHUTOFF VALVES
- COCKPIT VOICE RECORDER
- FIRE EXTINGUISHER SQUIB TEST PANEL
- ENGINE GENERATOR FIELD SWITCHES
- APU GENERATOR FIELD SWITCHES
- EEC MAINT PANEL
- GROUND TESTS SWITCH
- SPLIT SYSTEM BREAKER
- CENTER WING TANK SCAVENGE PUMP
- RESERVE 2 & 3 TRANSFER

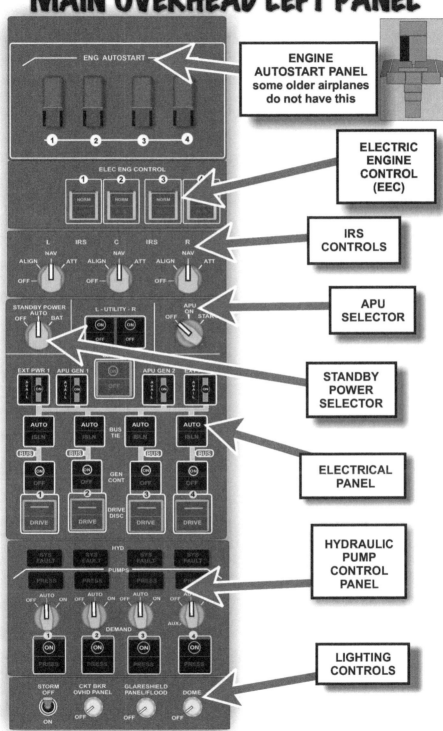

MAIN OVERHEAD CENTER PANEL

FIRE OVHT TEST.
EMERG LIGHTS.
OBS AUDIO SEL.
SERVC INTPHN.
FUEL XFR 1&4

FIRE EXTINGUISHER PANEL

CARGO FIRE EXTINGUISHER PANEL

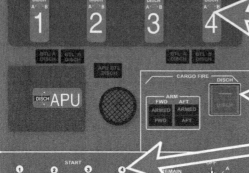

ENGINE START PANEL

FUEL JETTISON PANEL

FUEL TANK CONTROL PANEL

WINDOW HEAT CONTROLS

NACELLE ANTI-ICE
some older a/c may have this control panel.

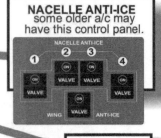

LIGHTING CONTROLS

"WET" COMPASS

page 18

© MIKE RAY 2014
published by UNIVERSITY of TEMECULA PRESS

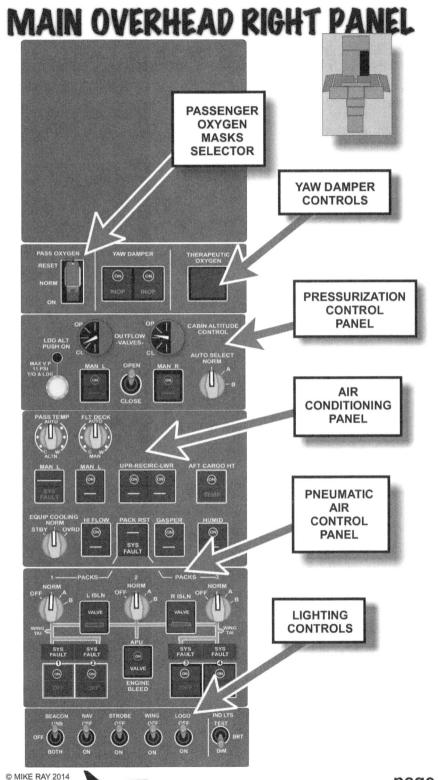

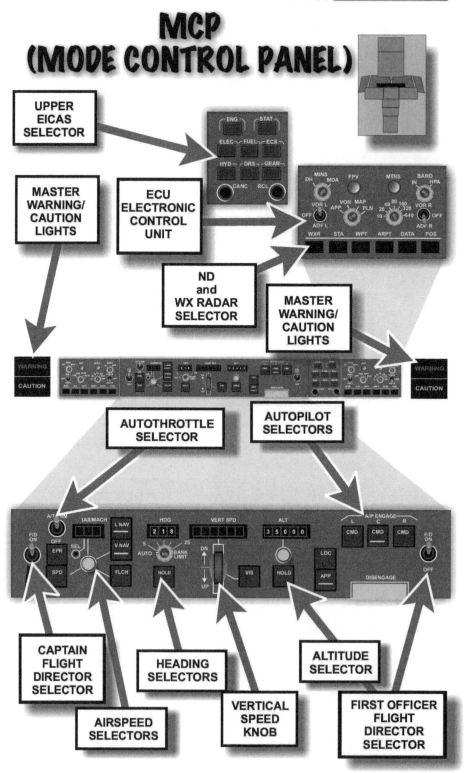

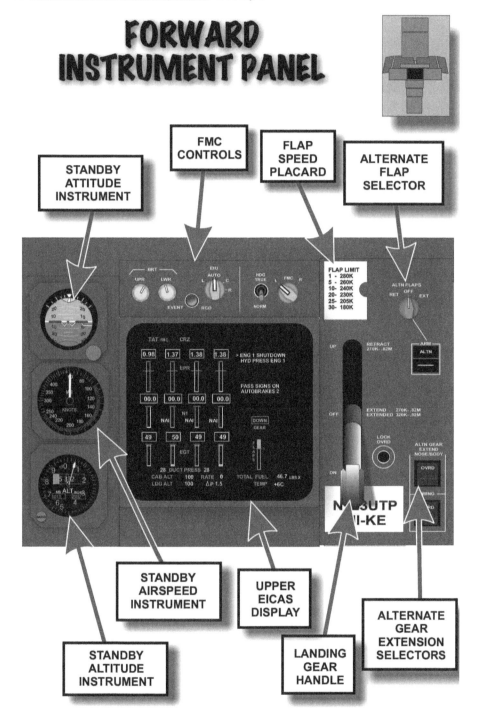

747-400 SIMULATOR TECHNIQUES ...

LEFT FORWARD INSTRUMENT PANEL (CAPTAIN'S)

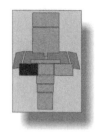

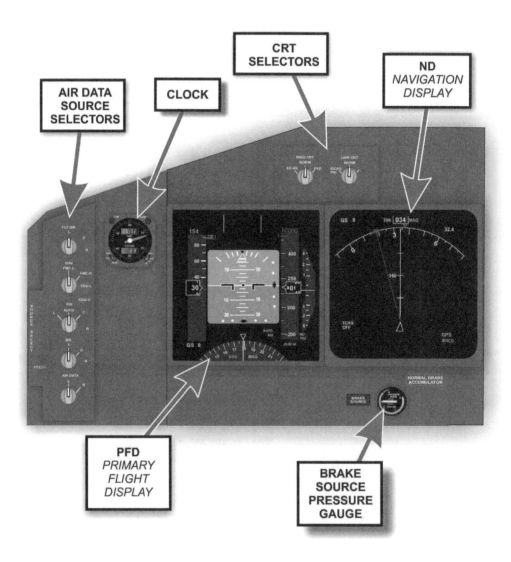

page 22

© MIKE RAY 2014
published by UNIVERSITY of TEMECULA PRESS

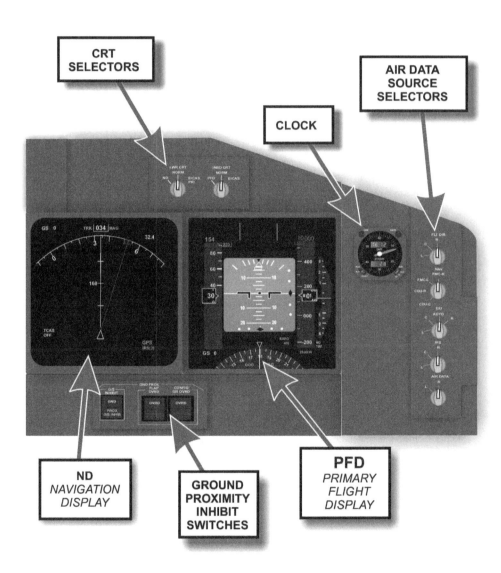

CDU / FMC PANEL

This is a MULTIFUNCTION DISPLAY, known on the 747-400 as the CDU (pronounced CEE-DEE-YEW) wnich stands for Control Display Unit. Sometimes the CDU is confused with the FMC (Flight Management Computer) also call the FMS on some airlines. These units are also referred to as the CDU/FMC.

It may not be evident to the initial reader, but in the cockpit there are three CDUs. Two of them are dedicated for pilot use in navigation and other immediate information and communications with the FMC for actually operating the airplne. Each pilot has their own CDU and since the both CDUs are connected, and they continuously communicate with the FMC in a COOPERATIVE manner, each pilot MUST BE careful to avoid placing CONFLICTING commands into the CDU.
It is usually consider good pilot technique for only one pilot to enter data into the CDU/FMC at a time.

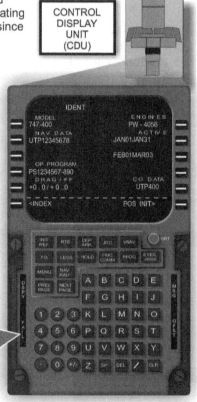

The third CDU is located on the lower console and is generally considered to be allocated specifically for communications with AIRINC, ATC, and the COMPANY.

The FMC contains TWO stored databases and ONE "top level" dynamic database. These are:
- NAV DATA BASE
- PERFORMANCE DATABASE
- ENGINE DATABASE

The FMC also contains TWO sets of NAV DATA. Each set contains dated that is valid for 28 days. Airline maintenance is responsible for updating the FMC databases.

If an FMC FAILS, the FAIL LIGHT on the associated CDU illuminates and the EICAS advisory panel wil display either L or R FMC FAIL.

The CDUs are designed to be operated independently and simultaneously, each unit displaying different messages appropriate to their use. However, data entries made on one CDU will be shared with BOTH FMCs to ensure that data entered by one pilot will be consistent between the two systems.

Nomenclature used in this document will refer to the the FMC that is displaying the FMC on its side to be the "ON-SIDE FMC". However, since the systems are data interconnected, either FMC can supply data to the other side of the cockpit by pilot selection. This seems difficult, but it is really intuitive. This system makes all flight management tasks and functions available as long as both CDUs and FMCs are operating

For aircraft equipped with a CENTER CDU; even though it CANNOT operate as an FMC interface unit, it uses the LEFT NAV SOURCE select switch to determine which FMC will be the source of radio tuning and flight plan data which supports its backup capabilities.

page 24

LOWER EICAS DISPLAY
called the SECONDARY DISPLAY

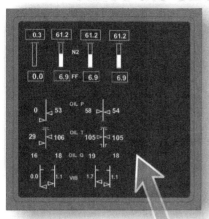

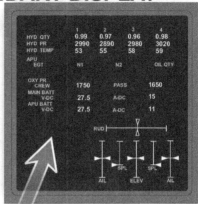

ENG DISPLAY
This display includes:
- N2
- FF (Fuel Flow)
- OIL PRESSURE
- OIL TEMPERATURE
- OIL QUANTITY
- VIBRATION

STAT DISPLAY
This display includes:
- HYDRAULIC QUANTITY
- HYDRAULIC PRESSURE
- HYDRAULIC TEMPERATURE
- APU STATUS (When running).
- OXYGEN PRESSURE
- MAIN BATTERY VOLTAGE
- APU BATTERY VOLTAGE

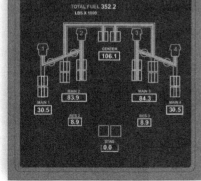

*The selector for the **SYSTEM SYNOPTICS** selector is located next to the **MCP** (Mode Control Panel).*

SYSTEM SYNOPTIC DISPLAYS
- ELEC (Electrical)
- FUEL (fuel)
- ECS (Environmental Control System)
- HYD (Hydraulic)
- DRS (Doors)
- GEAR (landing wheels)

(I have just displayed only one of the several System Synoptic displays that are available)

747-400 SIMULATOR TECHNIQUES ...

THROTTLE QUADRANT

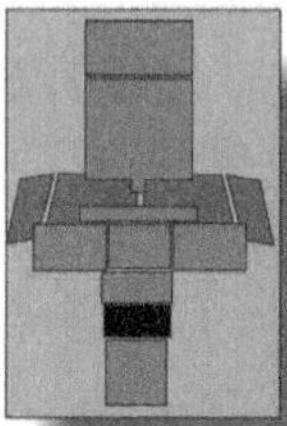

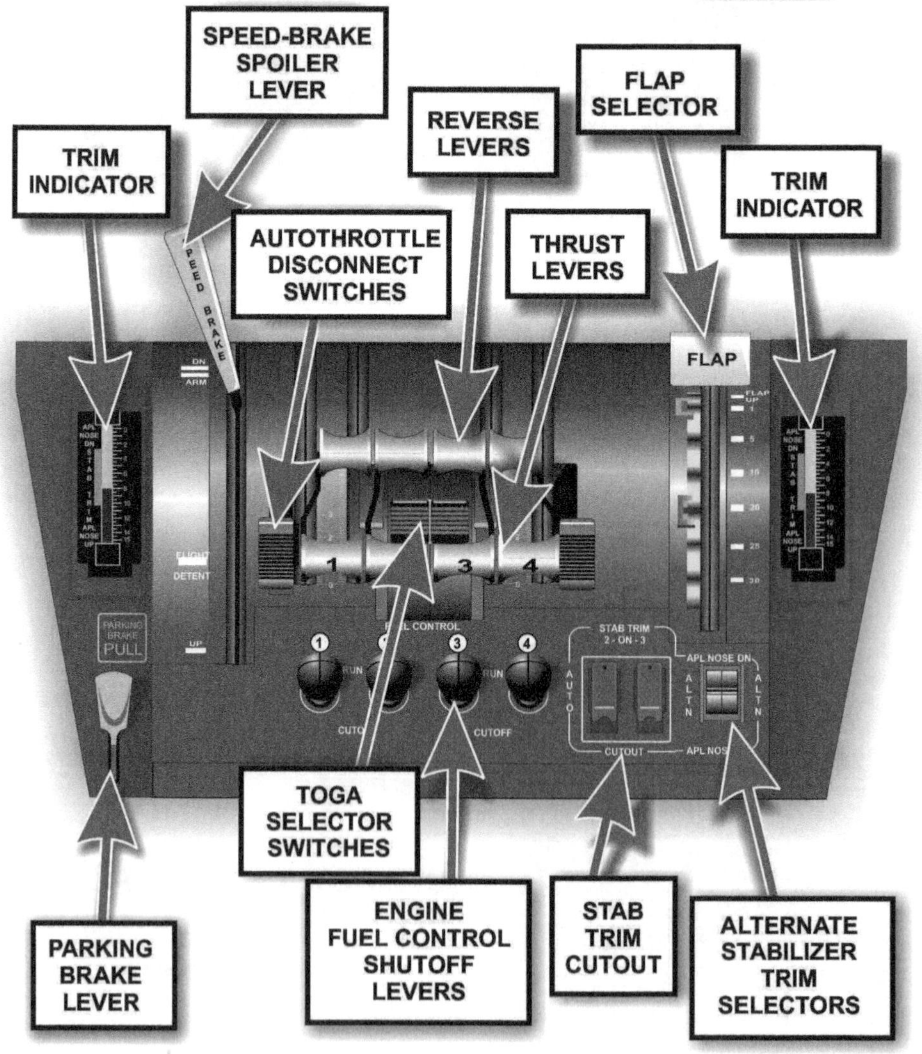

- SPEED-BRAKE SPOILER LEVER
- FLAP SELECTOR
- REVERSE LEVERS
- TRIM INDICATOR
- TRIM INDICATOR
- AUTOTHROTTLE DISCONNECT SWITCHES
- THRUST LEVERS
- TOGA SELECTOR SWITCHES
- ENGINE FUEL CONTROL SHUTOFF LEVERS
- STAB TRIM CUTOUT
- ALTERNATE STABILIZER TRIM SELECTORS
- PARKING BRAKE LEVER

page 26

© MIKE RAY 2014
published by UNIVERSITY of TEMECULA PRESS

LOWER CONSOLE (LEFT PANEL)

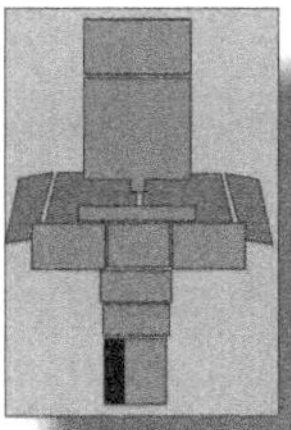

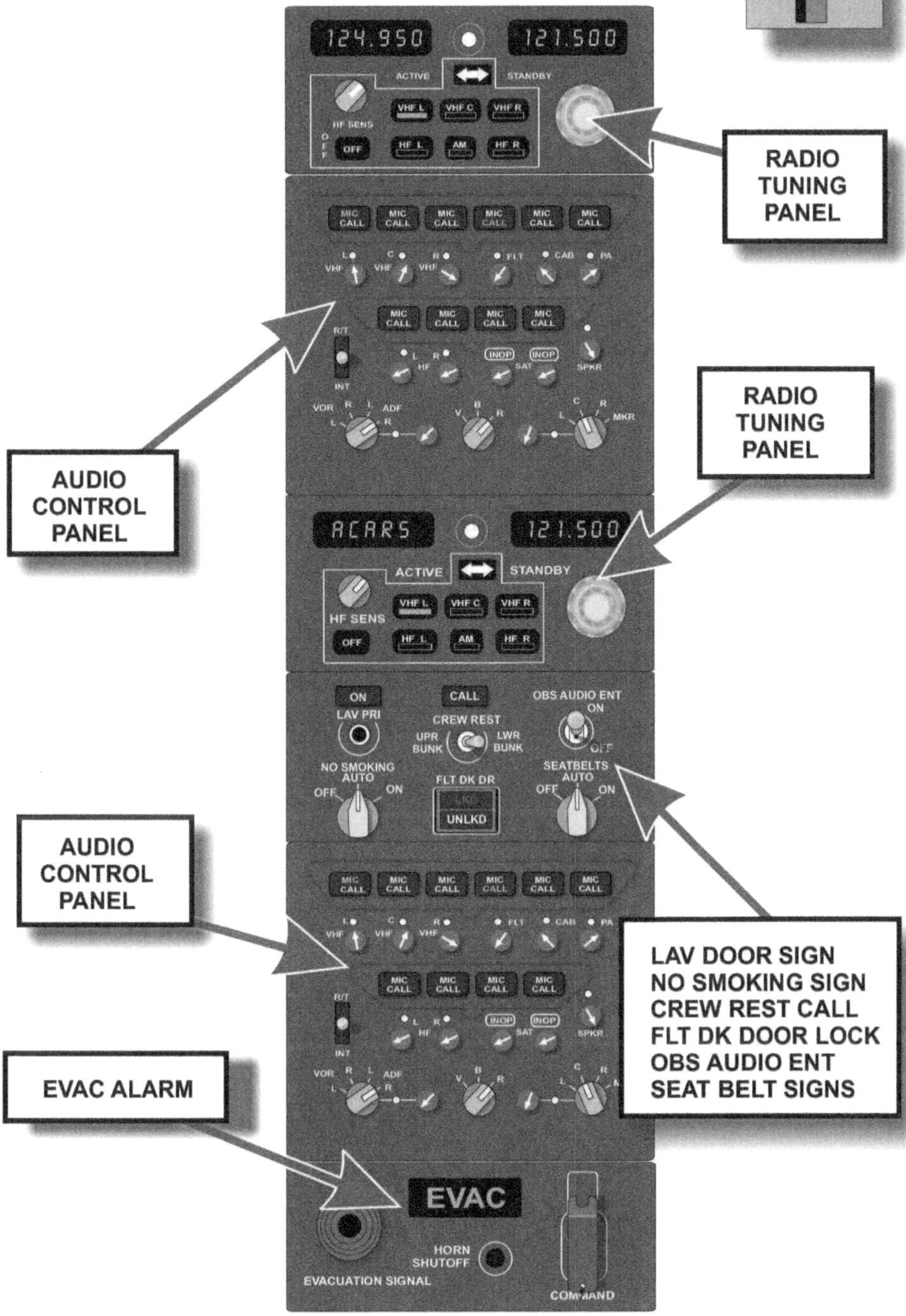

RADIO TUNING PANEL
AUDIO CONTROL PANEL
RADIO TUNING PANEL
AUDIO CONTROL PANEL
EVAC ALARM
LAV DOOR SIGN
NO SMOKING SIGN
CREW REST CALL
FLT DK DOOR LOCK
OBS AUDIO ENT
SEAT BELT SIGNS
124.950
121.500
ACARS
121.500
EVAC
EVACUATION SIGNAL
HORN SHUTOFF
COMMAND

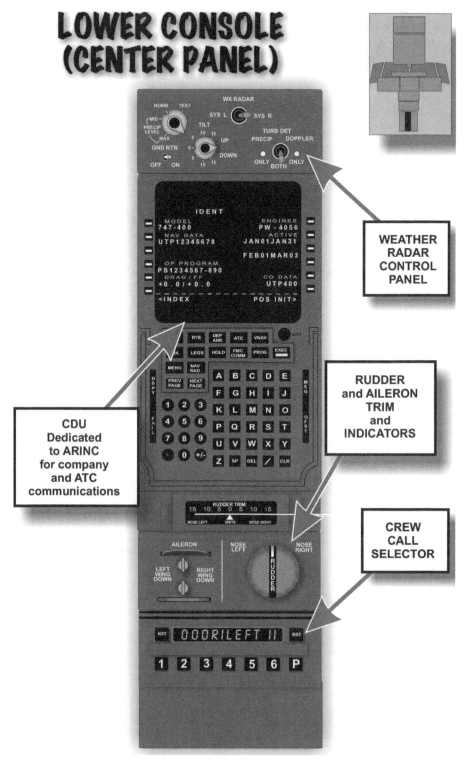

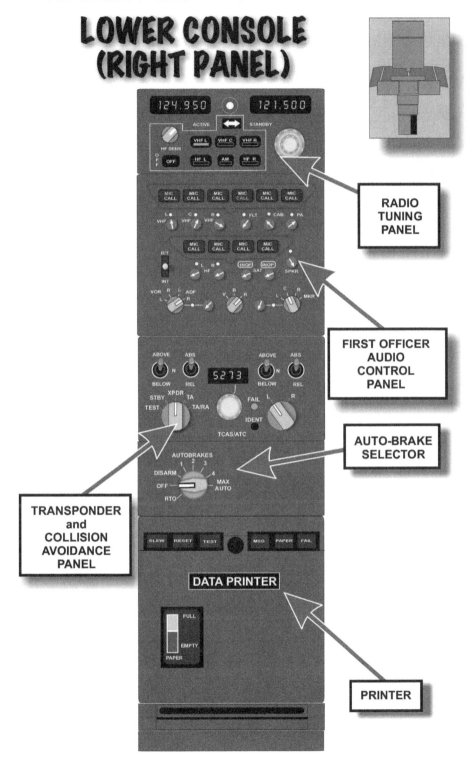

Welcome to...

the incredible
BOEING 747-400
"GLASS"

Some boring introductory comments

What is it that makes the Boeing 747-400 different from earlier Jurassic versions of this fabulous flying machine? The answer lies in the way that modern computerized technology has been integrated into the pilot-airplane interface. This fusion of technologies has completely altered the view of the relationship between the pilot and the operating systems of the airplane. While this Boeing masterpiece may be the most complex air machine ever to be operated in the public sector ... the onboard computers and control manipulation devices have come together to create an absolutely fantastic blending of man and machine. While initially, for the pilot who is new to "glass," the learning curve seems steep, as the whole concept becomes more familiar and the details more clearly perceived and understood, the sheer genius of the displays and the knobs and buttons becomes clear. This system; referred to by the aircrew as the "glass" (**EFIS - E**lectronic **F**light **I**nstrument **S**ystem), has proven to be far superior to the old "steam" gauges of the past.

How can a mere human operate the 747-400?

The pilot uses these five BASIC tools:

There are three DISPLAY tools:
The PFD (Primary Flight Display)
The ND (Navigation Display Unit)
The ECU (EFIS Control Unit)

There are two CONTROL tools:
The CDU (Control Display Unit)
The MCP (Mode Control Panel)

These 5 instruments send information to the **FMC** (**F**light **M**anagement **C**omputer) which also receives input from the **IRS**' (**I**nertial **R**eference **S**ystem) and then sends the appropriate signals to the control surface and the engines to accomplish the desired flight path.

Let's break these components out and look at them one at a time:

... and **PROCEDURES FOR STUDY and REVIEW** ONLY!

Initially the **TOTALLY BAFFLING** glass cockpit seems impossible, but it is actually "**SIMPLE**" ... once you know what makes it work. The Boeing system has a "common" design concept that is used on all their glass airplanes and here is a very brief description of that **EFIS** (Electronic Flight Instrument System) in it's **SIMPLIFIED** bare bones iteration.

The BIG FIVE of the "GLASS" flightdeck.

There are five **BASIC** parts to the system. Here are those fundamental "cockpit" parts that the pilot must learn to manipulate in order to "*fly the glass.*"

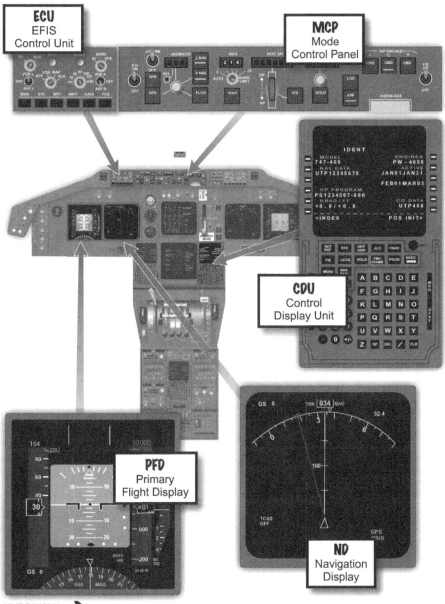

ECU — EFIS Control Unit
MCP — Mode Control Panel
CDU — Control Display Unit
PFD — Primary Flight Display
ND — Navigation Display

© MIKE RAY 2014
WWW.UTEM.COM

page 31

747-400 SIMULATOR TECHNIQUES ...

The CDU
Control Display Unit

This is the pilot's access port to the very heart of the computerized control mechanisms of the Boeing 747-400. It is the interface between the human and the Flight Management Computer (FMC). This is the device that we use to talk to the airplane and tell it what we want it to do.

Let's understand some of the very basic things about how to operate this simple unit.

Before we get to the content of the screens, let's understand how to manipulate the controls on the CDU itself. Here are 5 areas we will cover initially:

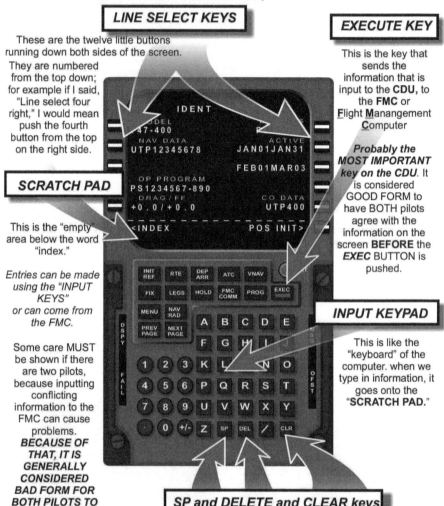

LINE SELECT KEYS

These are the twelve little buttons running down both sides of the screen. They are numbered from the top down; for example if I said, "Line select four right," I would mean push the fourth button from the top on the right side.

EXECUTE KEY

This is the key that sends the information that is input to the **CDU,** to the **FMC** or **F**light **M**anangement **C**omputer

Probably the MOST IMPORTANT key on the CDU. It is considered GOOD FORM to have BOTH pilots agree with the information on the screen **BEFORE** the **EXEC** BUTTON is pushed.

SCRATCH PAD

This is the "empty" area below the word "index."

Entries can be made using the "INPUT KEYS" or can come from the FMC.

Some care MUST be shown if there are two pilots, because inputting conflicting information to the FMC can cause problems. **BECAUSE OF THAT, IT IS GENERALLY CONSIDERED BAD FORM FOR BOTH PILOTS TO BE INPUTTING DATA INTO THE COMPUTER AT THE SAME TIME.**

INPUT KEYPAD

This is like the "keyboard" of the computer. when we type in information, it goes onto the **"SCRATCH PAD."**

SP and DELETE and CLEAR keys

DEL key places the word **DELETE** in the scratchpad, and when Line Selected will delete that selected line on the screen, IF the information is deletable. **CLR** key simply removes the last letter of the scratch pad entry (whole entry if held down long enough).
SP key enters a space in the scratchpad when using the **CDU** for **ACARS** or **SATCOM**.

... and **PROCEDURES FOR STUDY and REVIEW** ONLY!

The CDU

Simply do these **2** KEYSTROKES!

THE SIMPLE SECRET of getting into THE "GLASS"

There are TWO SIMPLE KEYSTROKES that will give you access to the whole GUTS of the FMC/CDU. ANYTIME you want to go into the very heart of the "MAGIC BOX,"

1 INIT REF button
upper-left corner of keypad

2 "INDEX" LINE SELECT-6L
lower-left corner of CRT

which causes the CDU "BOX" to reveal ...

"The 20"
CDU/FMC pages

8 CRT PAGES

PLUS

12 CDU KEYS

INIT REF
RTE
DEP ARR
ATC
VNAV
FIX
LEGS
HOLD
FMC COMM
PROG
MENU
NAV RAD

Each of these pages may have additional related pages so that there are a large total number of actual pages available for the pilot to use.

Getting familiar with all the options available and using the FMC to it's maximum potential requires numerous hours of actual hands on operation. In this book, we will only expose the reader to a very few of the operations for flight.

© MIKE RAY 2014
WWW.UTEM.COM

page 33

747-400 SIMULATOR TECHNIQUES ...

The CDU continued ...

Here is a simple sample problem to demonstrate how to enter data into the CDU using the keypad.

In our example, we wish to tell the computer where the starting airport is. For our example, let's assume that we are at Los Angeles International Airport.

THE KEYPAD TECHNIQUE

HOW TO ...
use the CDU to input data to FMC.

NOTE:
Generally speaking (there are exceptions, but):
1. **DASHED** lines indicate data entry is **OPTIONAL**.
2. **BOX** prompts indicate that data entry is **REQUIRED**.

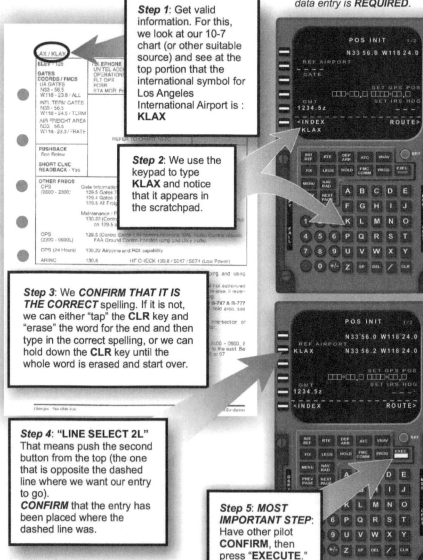

Step 1: Get valid information. For this, we look at our 10-7 chart (or other suitable source) and see at the top portion that the international symbol for Los Angeles International Airport is : **KLAX**

Step 2: We use the keypad to type **KLAX** and notice that it appears in the scratchpad.

Step 3: We **CONFIRM THAT IT IS THE CORRECT** spelling. If it is not, we can either "tap" the **CLR** key and "erase" the word for the end and then type in the correct spelling, or we can hold down the **CLR** key until the whole word is erased and start over.

Step 4: "LINE SELECT 2L"
That means push the second button from the top (the one that is opposite the dashed line where we want our entry to go).
CONFIRM that the entry has been placed where the dashed line was.

Step 5: **MOST IMPORTANT STEP**:
Have other pilot **CONFIRM**, then press "**EXECUTE**."

... and **PROCEDURES FOR STUDY and REVIEW** ONLY!

HOW TO ...
use the CDU to input data to FMC.

THE LINE SELECT KEY TECHNIQUE

Step 1: Get valid information. We can see that the "**LAST POS**" is very close to the "**AIRPORT**" longitude and latitude. A valid position check would show that the airplane last position was on the airport we have selected.
NOTE: However, It "could" be that the IRS itself was replaced with an IRS unit flown in from another station and may have not been updated. In that case the last position might NOT be the departure airport.

Step 2: LIne select the information you wish to place in the scratchpad. In our example, we have elected to tell the **IRS** to use our "**LAST POS**" for our "**SET IRS POS**" position.
NOTE: It is **NOT** required for the **IRS** to know the exact position of the airplane. It could have been towed from a remote parking area without power, for example.

Step 3: **CONFIRM THAT THE CORRECT INFORMATION** is placed in the scratchpad.

Step 4: **LINE SELECT 4R** ... that is, push the button abeam the boxes labelled "**SET IRS POS**" and observe the boxes replaced with the position selected. **EXECUTE** light should come on.

Step 5: **MOST IMPORTANT STEP**: Have other pilot **CONFIRM**, then press "**EXECUTE**" button. Light should go out. Information is now sent to the FMC.

747-400 SIMULATOR TECHNIQUES ...

We'll get back to programming the CDU in greater detail later on in the book.
The second part of the EFIS that we want to look at is:

The MCP
Mode Control Panel

You will quickly become used to integrating these various units into your flight management flows and your hands and eyes will be darting from one place to another in an appropriate manner. However, at this point, while you are still new to the systems, it will be useful to break out the various features of the MCP and analyze each one in cursory detail.

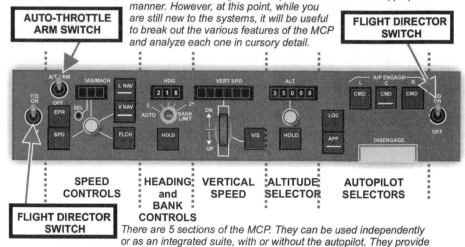

AUTO-THROTTLE ARM SWITCH

FLIGHT DIRECTOR SWITCH

SPEED CONTROLS

HEADING and BANK CONTROLS

VERTICAL SPEED

ALTITUDE SELECTOR

AUTOPILOT SELECTORS

FLIGHT DIRECTOR SWITCH

There are 5 sections of the MCP. They can be used independently or as an integrated suite, with or without the autopilot. They provide input to both the autopilot and the PFD and can be used to actually

SPEED CONTROLS

The 5 different speed selections each have their own venue of operation. We will discuss them more in detail later in the text.
The window can be set using the selector knob and is tied directly to the "SALMON BUG" on the AIRSPEED INDICATOR.

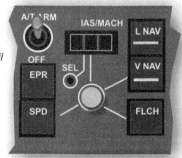

HEADING and BANK CONTROLS

The **HEADING KNOB** inserts the desired heading into the indicator on the MCP
... it also slews the heading on the ND (NAVIGATION DISPLAY) and represents (usually) the heading that the pilot desires the airplane to turn towards. If the airplane is operating in modes other than "heading select," the heading selector is de-activated.
There'll be more to say about this later.
BANK ANGLES are critical at altitude where excessive bank angles "may" compromise the airplane's stall margin during rough air operations. Further, during turns at lower altitudes when under the control of ATC, their expectation will be that the pilot will use 25 degree bank turns.
Once again, the bank angle limiter only works in the heading select mode.

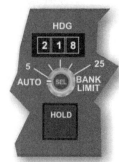

VERTICAL SPEED CONTROLS

Pushing the selector and "rolling" the thumb wheel will induce a vertical speed indication on the PFD. If on autopilot, the airplane will attempt to pitch up or down to meet the selected vertical speed.

If manually flown, the pitch bar on the PFD will indicate the appropriate pitch for the pilot to use.

If V/S is selected when FLCH or VNAV is engaged, then (if the AUTO-THROTTLES are engaged) the throttle will retard or increase so as to maintain the speed indicated on the SPEED SELECTOR.

If operating in ALT HOLD or VNAV PATH, you must select SPD mode for the AUTO-THROTTLES to target the airspeed.

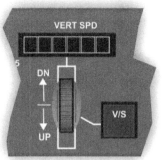

ALTITUDE SELECTOR

With a target altitude set in this window, and using the FLCH (referred to as "FLITCH"), the airplane will attempt to go to that altitude.
If the AUTO-THROTTLES are engaged, it will use whatever power is necessary up to "CLIMB POWER" or (if descending) retards the throttles to IDLE.

The FLCH speed is displayed in each PFD, and SPD window opens, and
FLCH SPD opens in FMA (PITCH WINDOW).

DETAIL: *If the aircraft thinks it can make the altitude change in less than two minutes, it does in the THR mode.*
If it thinks it will take more than two minutes, it goes to THR REF but does not display as an FMA.

AUTOPILOT SELECTORS

There are three (3) different autopilots on the airplane and they can be engaged once the airplane has climbed above 250 feet AGL (SOP dictates 800 feet AGL) after Take-off and are capable (if appropriate airport conditions and equipment exists) to fly rest of the whole flight, make the approach, land and roll-out without being dis-engaged. Truly amazing ... and really accurate.

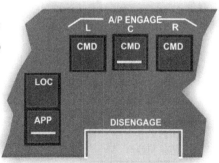

This comprises the **AFDS** or "AUTO FLIGHT DIRECTOR SYSTEM." The FMCs (FLIGHT MANAGEMENT COMPUTERS) automatically maintain pitch, roll, and thrust when both the autopilots and the auto-throttles are engaged.

747-400 SIMULATOR TECHNIQUES ...

The PFD
Primary Flight Display

Mr. Boeing decided that pilots needed only one instrument to stare at when they were flying. So, he created the PFD. It is virtually encrusted with information. Some of the information is non-critical ... some is absolutely essential. So, here is an attempt to show the various things about the PFD that we should know about.

Remember: This is only a cursory once over. Learning the PFD takes many many hours of useage.

The PFD is laid out in six (6) different sections:
- ATTITUDE INDICATOR
- FLIGHT MODE ANNUNCIATION
- AIRSPEED INDICATOR TAPE
- ALTITUDE INDICATOR
- HEADING INDICATOR
- VERTICAL SPEED INDICATOR

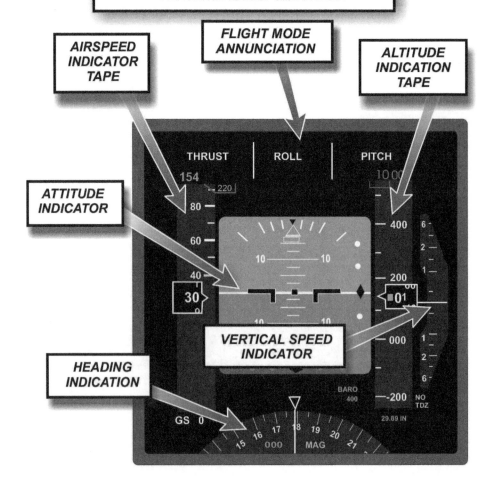

page 38

© MIKE RAY 2014
published by UNIVERSITY of TEMECULA PRESS

... and PROCEDURES FOR STUDY and REVIEW ONLY!

The FLIGHT MODE ANNUNCIATORS

At the top of the PFD are three boxes. These are the Flight Mode Annunciators.

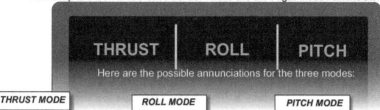

Here are the possible annunciations for the three modes:

THRUST MODE

THR REF - *Thrust is set to the "selected" limit on the EICAS.*
HOLD - *Throttles are disconnected and can be set manually.*
SPD - *Autothrottle commands the speed indicated on the PFD. That speed can be set by the SPEED KNOB on the MCP or by using the CDU.*
IDLE - *Displayed only when throttles are moving to idle. When they get to idle they will display HOLD.*
THR - *Applies thrust necessary to maintain requested vertical speed.*

ROLL MODE

ARMED (WHITE) indications:

LNAV - *Above 50 feet, will engage at "capture" point for active leg.*
LOC - *If within 120 degrees of track, indicates AFDS will capture.*
ROLLOUT - *appears below 1500 feet and engages at 5 feet RADALT.*
TO/GA - *On the ground: armed when a single Flight Director turned ON.*

ENGAGED (GREEN) indications:

TO/GA - *Maintains GROUND TRACK.*
LNAV - *If above 50 feet, within 2 ½ miles of active leg; will capture active leg displayed on ND.*
HDG SEL - *AFDS will turn to maintain heading selected on MCP.*
LOC - *AFDS will follow LOC course inbound.*
ROLLOUT - *Airplane will track runway centerline after touchdown.*
HDG HOLD - *Holds the present heading. If in turn, it will hold the heading AFTER rollout. Does not "correct back."*
ATT - *Holds existing bank angle IF greater than 5 degrees.*

PITCH MODE

ARMED (WHITE) indications:

VNAV - *Will engage above 400 feet. Will Follow vertical commands set in CDU.*
G/S - *Glide slope armed for capture.*
FLARE - *Appears below 1500 feet RADALT, engages at 50 feet.*
TO/GA - *On ground, a single Flight Director wil arm.*
In Air - Glide slope capture or FLAPS not up.

ENGAGED (GREEN) indications:

TO/GA - *On ground gives 8 degree pitch up signal.*
In flight; the AFDS commands the lesser of 15 degrees or below pitch limit whiskers. As the rate of climb increases, the indication transitions to AIRSPEED.
VNAV SPD - *Maintains FMC selected airspeed. If "SPEED INTERVENE" (that is selected by pushing the speed selector knob on the MCP) maintains the MCP selected speed.*
VNAV ALT - *Airplane is being held at the altitude selected in the MCP.*
VNAV PATH - *Airplane is following the path calculated by the FMC. See the "LEGS" indication on the CDU.*
FLCH SPD - *Maintains the MCP selected airspeed.*
ALT - *airplane is holding or capturing the altitude set on the MCP.*
V/S - *Maintains the selected vertical speed on the MCP.*
NOTE: Airplanes CAN fly away from selected altitude.
G/S - *Follows Glide Slope.*
FLARE - *Starts flare maneuver at 50 feet.*

page 39

The ATTITUDE INDICATOR

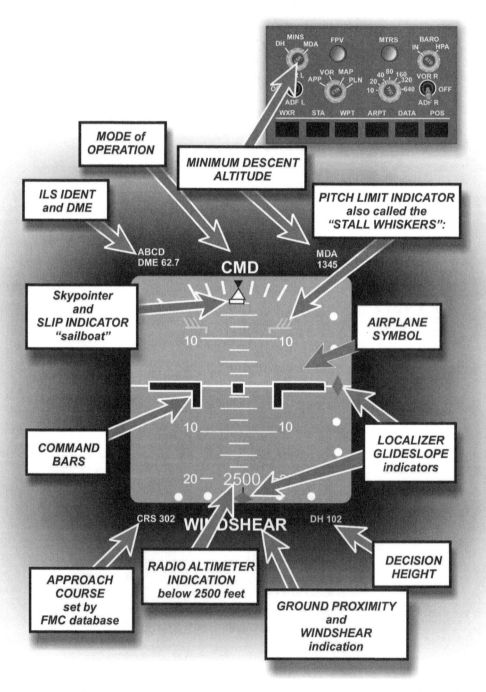

... and PROCEDURES FOR STUDY and REVIEW ONLY!

MODE of OPERATION:
This is a **VERY IMPORTANT** indication. If at any time you wish to know if the Autopilot is controlling the airplane... **LOOK HERE**:
CMD (green) means the autopilot is engaged.
FD (green) means ONLY the Flight Directors are engaged.
During the auto-coupled approach, you wil see either:
LAND 3 (green)
LAND 2 (green)
NO AUTOLAND (amber)

MINIMUM DESCENT ALTITUDE and DECISION HEIGHT:

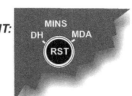

An "MDA" or "DH" is set on the PFD using the EFIS panel.
The DH/MDA selector has three knobs:
OUTER knob selects either DH or MDA,
MIDDLE knob rotates to select the appropriate value,
INNER knob is a push to reset the DH alert.

PITCH LIMIT INDICATOR:
Only displayed IF the flaps are extended. Indicates the PITCH at which the STALL SHAKER will activate. It calculates the pitch based on existing flight conditions.

AIRPLANE SYMBOL:
This symbol is fixed in the instrument and does not move. Some airlines use the "FLY-BAR" indication. I have flown both and can attest to the fact that the fixed bars are far superior.

LOCALIZER and GLIDESLOPE indicators:
ONLY displayed when ILS in use.
The "diamonds" respond to the airplane's position relative to the ground generated signal.
When within 2 1/3 dots of center, diamond turns black.
At low altitudes, indicators flash when excessive deviation.
IF (God forbid) the airplane gets to a low altitude with LNAV engaged, but LOC armed but not captured; the LOCALIZER scale changes to amber and the indicator flashes. This is your "**LOCALIZER NOT CAPTURED**" signal. **HELLO!!!**

GROUND PROXIMITY and WINDSHEAR indication:
To be discussed later, but if you have this indication; you will also hear "**WHOOP WHOOP PULL UP**." It is time to "*GET OUTA THERE*!!!"
CRAM THROTTLES TO THE STOPS, and aggressively PULL NOSE UP TOWARDS 20 degrees.
We will discuss the procedure in greater detail later in the book.

RADIO ALTIMETER INDICATION:
Below 2500, the Radio Altimeter indication is annunciated.
Below 200 feet, the indicator rises to meet the airplane symbol.

APPROACH COURSE:
This course information is set "automatically" and indicates the approach course from the FMC database.

COMMAND BARS:
These provide information from the FMC and indicate the roll and pitch recommended to achieve the flight profile calculated by the FMC. The idea is for the pilot to "fly" the airplane so as to place the indicator square on the intersection of the command bars.

ILS IDENT and DME:
Either from it's database or from pilot insert from the CDU; the FMC tunes the desired ILS frequency. If it CANNOT get a ident, it displays the frequency. If it is able to identify the ILS, it will display the identifier.
If DME available, it will be displayed.

SKYPOINTER and SLIP INDICATOR:
The "TEEPEE" will ALWAYS point directly UP, regardless of the airplanes orientation. The bottom "RECTANGLE" is the slip indicator. Together, they are called the "SAILBOAT."
You use the rudder to keep the rectangle below the sail.

© MIKE RAY 2014
WWW.UTEM.COM

page 41

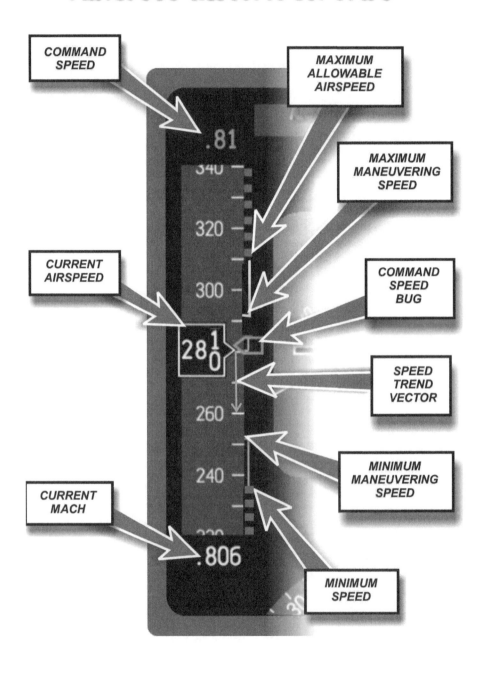

COMMAND SPEED:

This number is the same as that set in the MCP, OR it is the FMC computed airspeed/Mach when the MCP SPD window is blank.

MAXIMUM ALLOWABLE AIRSPEED:

This will display the LOWEST of the following:
Vmo
Landing gear placard speed
Flap placard speed

MAXIMUM MANEUVERING SPEED:

This will display the maneuvering "MARGIN" before buffet begins.

SPEED TREND VECTOR:

This indicates what the airspeed will be in 10 seconds as acceleration or deceleration changes.

COMMAND BUG SPEED:

This number is the same as that set in the MCP, OR it is the FMC computed airspeed/Mach when the MCP SPD window is blank.

MINIMUM MANEUVERING SPEED:

This indicates the maneuver margin to stick shaker or low speed buffet. If there isn't any computed data available or if it is invalid; then this indication is removed.

MINIMUM SPEED:

This indicates the speed at which **STICK SHAKER or LOW SPEED BUFFET** occurs. YIPES!

CURRENT MACH:

This indicates the current COMPUTED MACH. If the computer has no data, the signal is removed. If the MACH number is invalid, then a MACH flag is displayed.

CURRENT AIRSPEED:

This indicates the current AIR DATA computed airspeed. If it is missing, that means that the associated AIR DATA COMPUTER has failed.

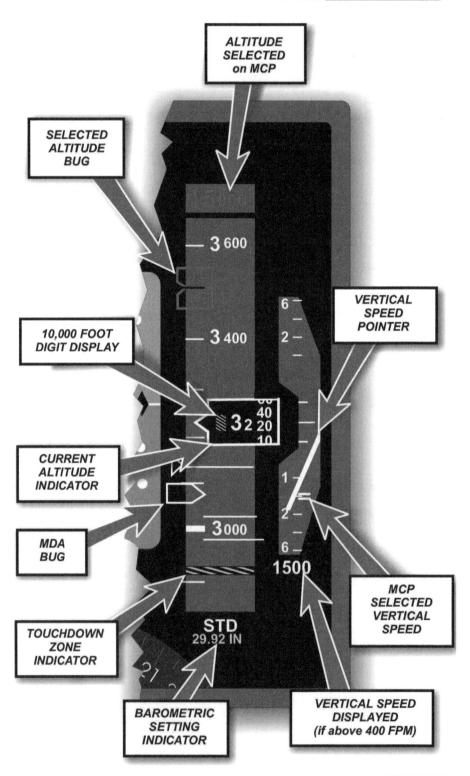

The
ALTITUDE INDICATOR TAPE

ALTITUDE SELECTED ON MCP:
This number is the same as that set in the MCP. The information will appear in a box IF the altitude is within 900 to 300 feet of the selected altitude.

ALTITUDE SELECTED BUG:
This indicator box shows the altitude set in the MCP.
When the selected altitude is offscale, the little box will rest at the top or bottom of the tape.
FYI: The BUG is 100 feet wide.

10,000 FOOT DIGITAL DISPLAY:
When the airplane is below 10,000 feet, the crosshatched box appears.

CURRENT ALTITUDE INDICATOR:
The box indicates the Air Data altitude.
When within 900 to 300 feet of the altitude selected on the MCP, the ALTITUDE BOX will switch to WHITE.
Once within those parametrs, if you should deviate from them by 900 to 300 feet, then the box will turn to AMBER.

TOUCHDOWN ZONE INDICATOR:
An amber rectangle with stripes appears that represents the touchdown zone for the airport selected by the FMC runway.
The upper edge represents the landing altitude.

During cockpit setup (This would be before the runway is entered or the FMC information is available) a "NO TDZ" flag will be displayed to the right and below the altitude tape.

BAROMETRIC SETTING INDICATOR:
This indicates the BAROMETRIC SETTING that is dialed into the EFIS control panel.
This airplane has a great feature that allows the altitude setting (QNH) to be entered before leaving QNE.
If you selected BARO STD on the EFIS, the "STD" is displayed and the preset altimeter setting is displayed in white below the STD.

VERTICAL SPEED DISPLAY:
If vertical speed is greater than 400 FPM, then it will be displayed below the indicator.

MCP selected VERTICAL SPEED:
a "buck tooth" displays the selected Vertcal Speed from the MCP if the V/S mode is engaged.

VERTICAL SPEED POINTER:
The swinging bar indicates the current IRS vertical speed indication.

page 45

747-400 SIMULATOR TECHNIQUES ...

The PFD HEADING / TRACK INDICATIONS

SELECTED HEADING "BUG"
This cursor will select a heading that corresponds to that placed in the **MCP** (Mode Control Panel) with the **HDG SEL** (Heading Select) cursor. This will also correspond to the **HEADING CURSOR** in the **ND** (Navigation Display) also called the **HSI** (Horizontal Situation Indicator.

CURRENT HEADING POINTER
The **CURRENT HEADING POINTER** displays the instantaneous **"HEADING"** of the aircraft and moves when the airplane changes direction, this pointer moves to reflect the change.

MCP SELECTED HEADING
Numeric readout

TRACK LINE
The **"TRACK" indicator** displays the instantaneous **"TRACK LINE"** of the aircraft over the earths surface. The angle between the **CURRENT HEADING** and the **TRACK** represents the **WIND CORRECTION**.

HEADING TRACK REFERENCE
Toggles between **TRUE** and **MAGNETIC** and reflects the mode selected on the **FORWARD INSTRUMENT PANEL** above the center **EICAS** display. Used for operation within the Earth's polar regions.

© MIKE RAY 2014
published by UNIVERSITY of TEMECULA PRESS

The PFD HEADING / TRACK INDICATIONS

CURRENT HEADING POINTER:

This number is the current IRS heading. It represents the direction that the axis of the airplane is pointing.

SELECTED HEADING BUG:

This is the heading selected on the **MCP**. If you have selected a heading outside of the visible limits of the compass rose, the indicator will slew to the side of the instrument that represents the direction of the shortest turn to get to the heading.

This is referred to as the "**BUCK-TEETH**." When you have the airplane operating on autopilot and operating in **HDG SEL**, the airplane will try and turn so that the **HEADING POINTER** will rest on the buckteeth.

Pilot have a gouge for this and say that the "Boat (Triangular aircraft **HEADING** pointer) will dock at Bucktooth Harbor (The Selected **HEADING** on the **MCP**) when **HEADING SELECT** is engaged **ROLL MODE**.

MCP SELECTED HEADING:

This is the numerical value of the **HDG SEL** (Heading) selected on the **MCP**.

TRACK LINE:

This indicates the **TRACK** of the airplane over the ground as indicated by the **FMC**. The difference between the **HEADING** and the **TRACK** is the **WIND DRIFT ANGLE**. We can use this information to our advantage when flying an **NDB (ADF)** or **VOR** approach.

HEADING / TRACK REFERENCE:

Displays whether the system is operating in **MAG** (Magnetic North mode) or **TRU** (True North mode).

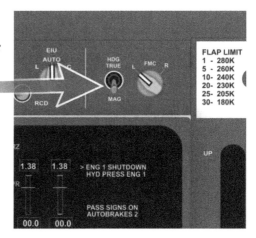

The ND
Navigation Display

There is a lot of information displayed on the ND ... and as long as I worked with the instrument, there always seemed to be something else that I learned. Each situation allows for a different set of symbols and data to show up on the screen. What we will do here is look at just some of the more common indications and identify them.

NAV AID SYMBOL:
When EFIS control panel STA light switch is selected, appropriate navaids are displayed.
If the computer has it selected, it turns GREEN.
When a navaid is manually tuned, the selected course and reciprocal are displayed.

WIND DIRECTION and RELATIVE SPEED ARROW:
The greater the velocity, the longer the arrow. This is really useful when hand flying an approach. Works well during holding, also.
The WIND DIRECTION/SPEED is Magnetic if HDG/TRK is Magnetic;
The WIND DIRECTION/SPEED becomes True if HDG/TRK is True.

GROUND SPEED/TRUE AIRSPEED.
Current groundspeed in knots. Current true airspeed displayed IF above 100 knots.

HEADING BUG.
The "buck teeth" are set using the heading selector of the MCP. If operating in autopilot, and the HEADING SELECTOR is depressed, the airplane will turn so as to put the HEADING POINTER on the HEADING BUG.
The "BOAT" will dock in "BUCK TOOTH HARBOR."

The ND
Navigation Display

MAGNETIC HEADING or TRACK
IN MAP or PLN MODE: Indicates magnetic TRACK. This is called a track-up mode and is the usual operating situation.
in VOR or APP MODE: Indicates magnetic HEADING. This is called a heading mode and is used during approaches.

HEADING POINTER:
Indicates airplane HEADING.

ACTIVE WAYPOINT:
Indicates the ACTIVE WAYPOINT in MAP and PLN mode.
This will be the top waypoint on the LEGS page of the CDU.

ETA TO ACTIVE WAYPOINT:
This is the time that the airplane will arrive at the active waypoint. Same value that is on the LEGS page of the CDU.

DISTANCE TO ACTIVE WAYPOINT:
This is the distance to the active waypoint. Same value that is on the LEGS page of the CDU.

RIGHT ADF POINTER:
There are four indicators that represent the heading to or from a VOR/ADF station. This pointer indicates the bearing from (TAIL) or the bearing to (HEAD) of the tuned ADF station.

VERTICAL DEVIATION INDICATOR:
Displayed during descent ONLY. It shows if the relationship of the airplane to a computed "ideal" descent path. Also known as *"Flight Path Deviation Indicator."* Are we LOW or HIGH on the descent? This little guy tells us.

VOR/ADF IDENTIFIER:
VOR - Displays VOR frequency until station identified. It does this automatically. Then it displays IDENTIFIER and raw data DME.
If only the DME is identified, the identifier is displayed in small font.
ADF - Displays ADF frequency until identified, then displays identifier.

IRS NAVIGATION MODE STATUS:
Displayed when in MAP mode. Displays IRS mode status. If the system transitions to any other status, a green box will highlight the indication for 10 seconds.

FMC POSITION UPDATE STATUS:
Displayed when in MAP mode. Indicates FMC updating status.
DD: DME/DME
VD: VOR/DME
LOC: LOCALIZER

ROUTE LINE:
The MAGENTA LINE indicates the active flight plan as set up in the CDU/FMC.
The DASHED WHITE LINE: It represents the route in the CDU has not been EXECUTED.
The BLUE LINE WITH LONG DASHES represents an "inactive" route.

AIRPLANE POSITION SYMBOL:
The apex of the triangle represents the nose of the airplane. Note that the airplane HEADING will not align with the top of the display when operating in a "track up" mode.

POSITION TREND INDICATOR:
Pilots call this the "SNAKE." Each segment represents the heading of the aircraft in 30 second intervals IF the present "trend" is maintained.

ECU *(EFIS Control Unit)*

Also called the **Electronic Flight Instrument System Panel**. This little unit could more properly be referred to as the **ND** control unit since most of the items on this panel actually control functions on the **ND** (Navigation Display). Here are some comments about the various functions on the unit.

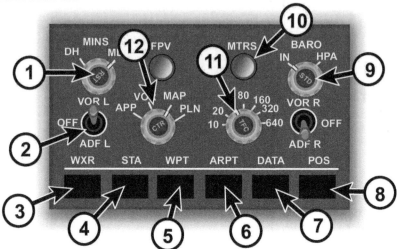

1. **DH/MDA** selector. This knob has three functions
   - **OUTER** part selects either **DH** or **MDA** on the respective **PFD**.
   - **MIDDLE** part selects either **DH** or **MDA** on the **PFD**.
   - **INNER** part pushes to reset the **DH** on the **PFD**.
2. **VOR/ADF** switch. Displays the **VOR** or **ADF** pointer and Frequency, and Identifier on the **ND**. **OFF** removes the displays.
3. **WXR** ... Displays **WEATHER RADAR** data.
4. **STA** ... Displays both high and low **NAV AIDS** in 10 - 40 nm range, and only the high **NAV AIDS** are displayed in the 80 -640 range.
5. **WPT** ... Displays **WAYPOINT**s with **ND** in scales 10 - 40 range.
6. **ARPT** ... Displays **DATA BASE** airports with runways greater than a selectable minimum length.
7. **DATA** ... Displays **ETA** and **FMC** altitude restrictions.
8. **POS** ... Displays **IRS** positions as **WHITE STARS, VOR** radials as solid green lines to the stations tuned on the **CDU NAV RAD** page.
9. **BARO** selector ... Has three knob parts.
   - **OUTER** knob ... Selects **IN** (inches of Mercury) or **HPA** (hectopascals) as the reference on the **PFD**. *IMPORTANT!!!*
   - **MIDDLE** knob ... Selects the barometric setting on the **PFD**.
   - **INNER** knob ... Push to select **STD** (29.92 in HG or 1013 HPA) or selected reference.
10. **MTRS** button ... Meters selector / toggles the display between displaying meters and removing meters.
11. **TFC/ND RANGE** selector.
    - Rotating outer knob selects **RANGE** on the **ND**.
    - Depressing the **INNER BUTTON**, toggles the **TCAS**.
12. **ND MODE** selector.
    - **INNER** knob ... Depressed displays **ROSE** or **FULL** on the **ND**.
    - **APP** ... Displays the selected **ILS** approach.
    - **VOR** ... Displays the selected **VOR** approach.
    - **MAP** ... Displays "track up" display. Includes most information.
    - **PLN** ... **PLAN** mode. Selects a **TRUE NORTH** display. Hooks to the **CDU LEGS** page and allows for stepping through the fixes to evaluate the route of flight.

GET READY

This next section is where we will actually get to climb into the cockpit and start configing the flightdeck for flight. Get ready for the most challenging part of the learning curve and the part that must be completed BEFORE we go fly the jet.

SAILING OUR SILVER SHIP THROUGH SUNSET SKIES
The romance of aviation

Flying high above the vast haze covered earth and seemingly endless sea for hour after endless hour, the whole incredible experience of floating above it all becomes one continuous seeming never-ending experience ... and time becomes compressed and the days become jumbled and mixed together. When I was a young pilot, it just seemed as if all the fantastic flying opportunities would never end, and so I gave only a passing nod to the fact that for us as pilots and aircrew, there is an ultimate conclusion to the flying career. As the fleeting years passed, I was content to being swept up in the joyous experience of escaping from the concerns of our earth borne companions. There was a sense of complacent acceptance since I was never far from the next flight. I simply didn't realize just how quickly my career would be over and that I would then be forever banished from the world of airline aviation. I never truly believed that there would come a time when I would no longer be allowed to sit in my private Captain seat and command this huge flying palace. I just never thought that I would be deprived of the sensuous feeling of fondling the thrust levers, and enjoying the subtle response of the yoke as I gently coaxed it, slightly trembling, towards me.

Ah, but it is true, that these moments of aviation bliss will surely pass ... and here is what I want to encourage you to do. Enjoy the pleasure and the experience. Record the moments as best you can. Take a prodigious quantity of photographs and write an exhaustive journal recounting your adventures as you travel the earth. Don't allow yourself to forget those memories and places where you spent your life.

There is one more thing I would like to tell you about. Since I have retired, I have discovered something called "Flight Simulation", or as simmer veterans call it, "Flight Simming". I was absolutely astonished when I first discovered the depth, fidelity, and accuracy of these inexpensive simulations and immediately saw the incredible potential for training ... and pure enjoyment. I have to confess, that I personally have become a much better pilot now that I have spent time operating these "simulator games". I have come to understand a lot better how the "real" airplane systems operate by being able to create simulated situations and irregularities and then develop solutions to those problems without actually having the potential for creating an in-flight emergency. I heartily recommend you look into adding the Microsoft Flight Simulator X or X-Plane 10 and add-on Boeing 747-400 simulations such as those created by PMDG. A "stand-alone" product like the terrific simulation from Aerowinx 744 PSX. I think you will, like me, become re-addicted to flight in another, more intimate way ... different than before.

THE PRE-FLIGHT SECTION

NOTE TO READERS:
This document will not get involved in a rigorous *"FLIGHT PLANNING"* section that would involve weather, various flight plan formats, fuel considerations, alternates, etc. etc. simply because each airline has it's own special set of rules and protocols ... and besides, it would take about 500 really boring pages to make complete a presentation that would be definitive.

So let's just go right to the airplane and begin the task of trying to make sense out of the long list of items that must be accomplished before we can get this BIG MOTHER airborne.

747-400 SIMULATOR TECHNIQUES ...

AS YOU ARE WALKING ACROSS THE RAMP AND BEFORE WE CLIMB INTO THE JET, CONSIDER ...
EXTRA CREDIT STUFF

- How can you tell from outside the airplane if the **GROUND HANDLING BUS** is powered or if the **APU** is running and electrical power is available. ANSWER: Determine if **BAGGAGE LOADING** or **FUELING** is underway. **NOTE**: Hearing the **APU** exhaust coming out the tail of the jet does not necessarily mean that the **APU** is producing electrical power.

  FYI: This bus is powered from the "**AVAIL**" side of the **APU** or **EXT** power.

- As you approach the jet, look up at the top of the cockpit bulge and ascertain that the **PILOT ESCAPE HATCH IS FLUSH**. It is a good idea for the Captain; because more than likely, in the real world, someone else will be doing the walk-around duties.

- Make certain that the jet is chocked or otherwise hooked to something. Before you mount the steps, get that little chore taken care of. The Check person would be very impressed if you mention that in your oral.

- If during flight planning, you changed the fuel load from the original "suggested" flight plan, it is useful to check with the fueler. If possible (and he speaks your language) and determine if he has received the latest fuel load.

- For your information, once inside the Passenger Loading Bridge, it is customary for the pilots to enter the airplane using the mid-cabin door. Using the First Class door will definitely identify you as a rookie. Bad form!

  Trying to enter through the First Class forward closet is a definite Boo-Boo!

Picture of Mike Ray's private jet.

LET'S GO DIRECTLY TO THE FLIGHT DECK

There are **FOUR THINGS** we must do before we can start the actual airplane setup. Let's begin by looking at these
FIRST 4 MUST-DO ITEMS ....

4

Let's start with the simple stuff first. *There are four (4) things that you should do once you are inside the jet and standing there in the dark wondering what to do:*

Things you gotta do right away:

1 GET SOME LIGHTS ON. In this situation, we will assume that the airplane is STONE COLD and DEAD DARK. Usually, the airplane (and the simulator) will probably have the lights on and be nice and cozy. But if it is not, here is what to do.

2 <u>MAKE CERTAIN YOU ARE ON THE RIGHT JET!</u> Once you can see what you are doing, you will need to determine that you are on the right jet. Let me just tell you that when you are the "*NEW GUY*" you are going to have a lot of curiously unique flight assignments. While the senior guys are flying the cushy passenger flights, you will probably be flying

3 Check and ensure that the "GROUND HANDLING BUS" is active and operating.

4 Check that the BRAKES ARE SET or that the airplane has been properly "CHOCKED" or "RESTRAINED by the push-back tractor".

You didn't tell me that it would be dark in here!

FLIGHT DECK ACCESS LIGHTS

FLIGHT DECK ACCESS LIGHTS

*When you enter the cold dark cockpit, here is where you turn on some lights. This switch is on the **GROUND HANDLING BUS**. Other lights, such as the **THUNDERSTORM LIGHTS**, are on other busses and may not be powered.*

MAKE CERTAIN YOU ARE ON THE RIGHT JET!

And it is my opinion, that as soon as you can get the lights on, the very first thing you should check is the **AIRCRAFT HULL** number. Make **CERTAIN** that you are on the **SAME AIRPLANE** as the **FLIGHT PLAN**.

The AIRCRAFT HULL NUMBER IS ON THE PLACARD UNDER THE LANDING GEAR LEVER.

There are three reasons to check this:
First, to make certain that your flight papers have been prepared for the correct airplane; and
Second, so you won't look like a geek by getting on the wrong airplane. Don't laugh, it happens. **DUH!**
 Third, and this is **MOST IMPORTANT**, this identification **HULL NUMBER** should also be used to match up with the **MAINTENANCE LOGBOOK** and other related airplane documents. It is not good to get airborne with the "wrong" logbook. **<u>FEDERAL REGS say so!!!</u>**

© MIKE RAY 2014
published by UNIVERSITY of TEMECULA PRESS

... and PROCEDURES FOR STUDY and REVIEW ONLY!

THE GROUND HANDLING BUS

At this point in the check-ride, there will be (hopefully) a discussion about The **GROUND HANDLING BUS**. Here are the things you **MUST** know about this bus:

> The **GROUND HANDLING BUS** is powered **_ANYTIME_** and **_ONLY_** when:
> A. **_EXTERNAL POWER_** plugged into #1 receptacle ...or
> B. **_APU GEN #1_** is **_AVAILABLE_**

Why do we care?
...because of some the items on that bus become unpowered if the Ground Handling Bus is not powered:

THINGS ON THIS BUS
- CARGO DOOR OPERATION
- CARGO HANDLING SYSTEMS
- CARGO COMPARTMENT LIGHTS
- FLIGHT DECK ACCESS LIGHTS
- FUELING SYSTEM
- #4 HYD AUX ELECTRIC PUMP

but this is the BIGGIE

The cargo doors and fueling are certainly important, but the **#4 HYD AUX ELECTRIC PUMP** supplies the **AUX BRAKE PRESSURE**. That is **_VERY_** important since it is your **_ONLY_** active brake pressure source until the **ENGINE DRIVEN HYD PUMPS** start producing hydraulic pressure.

The Check-guy wants to make certain that you know the following FACT:

YIPE!

> If the ground Handling Bus is not powered, there is no ACTIVE brake pressure source on the Boeing 747-400 without the engines running.
> (Accumulator MAY NOT BE EFFECTIVE)

There have been a few horror stories about perfectly good 747-400s being parked on some lonely ramp with the parking brakes set properly. About four hours later, the accumulator bleeds off and the big bird takes an unescorted roll down the ramp and rams the first thing it comes to.

> **GOOD CAPTAIN PRACTICE!**
> When you are approaching the parked airplane, **Always** check that the jet is either attached to a tow tractor or chocked satisfactorily.

ARE THE BRAKES SET???

Check and see if the PARKING BRAKE is set.
Even if the wheels were chocked when you arrived at the jet, if a ground person removes them ... you could be standing in the galley sipping your first cup of coffee when the jet starts rolling across the ramp.

Here are some **PARKING BRAKE** things.

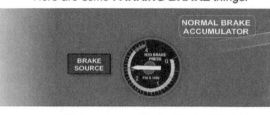

This gauge reflects the "NORMAL BRAKE ACCUMULATOR" pressure.

If the normal and alternate brake systems are not pressurized, then brake pressure is maintained by the brake accumulator.

The brake accumulator is pressurized by **HYDRAULIC SYSTEM 4**.

Sufficient pressure may be stored in the accumulator to **"SET AND HOLD THE PARKING BRAKE; BUT THE ACCUMULATOR IS NOT DESIGNED TO STOP THE AIRPLANE."**

The **BRAKE ACCUMULATOR** provides for Parking Brake **"APPLICATION."**

#4 HYDRAULIC AUX PUMP is powered by the **GROUND HANDLING BUS**, and one of the things it can do is provide pressure to the **BRAKE ACCUMULATOR**.

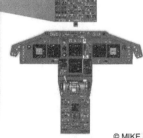

Once the power is on the airplane and the **GROUND HANDLING BUS** is powered, some pilots will:

Turn on the **#4 AUX HYD PUMP** *switch, Monitor the* **BRAKE ACCUMULATOR** *gauge, and when the pressure gets in the green band, they will re-set the brakes.*

page 58

... and **PROCEDURES FOR STUDY and REVIEW** ONLY!

Introducing the ...
FLIGHTDECK SETUP FLOWS

> The whole Checkride, including the Oral exam, simply consists of you trying to convince the wary, cautious, and ever suspicious checkguy that you will be able to go out some dark night and figure out how to make a stone cold Boeing 747-400 work ... and do it safely. That's it! There are no right or wrong answers, there is no particularly exact way they are supposed to question and examine you other than to convince themselves that you will be able to competently operate what is arguably one of the most complex machines ever conceived by mankind. Seems simple enough!

For both **"COGNITIVE EXAMINATION"**
(Pilots refer to it as the **ORAL**) and
"PSYCHO-MOTOR SKILLS DEMONSTRATION"
(Pilots call it the Simulator Checkride or simply the **SIM-RIDE**),

... the Check-guy will be expecting a certain level of proficiency (unknown to you). You will not even know what to expect in the way of "problems". So you will have to make the assumption that you have been given the assignment to deadhead out to some obscure foreign location and fly a jet that has just undergone some really heavy maintenance by some potentially unqualified gremlins. You should assume you are about to enter a cold dark airplane in which the switches and levers may be in any meaningless and improper position.
The Check-Air-Person will be expecting you to know where every switch should be set, and in some cases why.
  That is the reason you bought this book, and that is why I am here. The point of the whole check-ride exercise is to fool the check-person into thinking that you know what you are doing ... or at least look like you know what you are doing. Relax, and realize that you can do this.

Not to worry ... IT'S A PIECE of CAKE!

© MIKE RAY 2014
WWW.UTEM.COM

747-400 SIMULATOR TECHNIQUES ...

WHAT ARE FLOWS ?

The whole concept of "**FLOWS**" is simply the grouping of activities in clusters, then naming them something appropriate to assist the pilot in remembering just "**WHAT AM I 'SPOSE TO DO?**" The flight handbook (while it doesn't use the term "flows") is written in a way that implies that there is a specific "litany" or routine in which the stuff on the flight deck **SHOULD** be done. The pattern that emerges as one reads the "**NORMALS**" section of the Pilot Handbook is what I am going to lay out for you now. What is laid out in these following pages are the steps that we are calling the "**FLOWS**."

When the crew arrives at the jet, here are the flow modules that need to be accomplished to complete the cockpit set-up.

> Once you enter the cockpit:
> **COMPLETE THE FIRST FOUR ITEMS**.

COCKPIT SET-UP FLOW MODULES

INITIAL COCKPIT PREPARATION

*This flow is "GENERALLY" done by the Captain, but the Flight Handbook states that these items **MAY** be assigned to the first officer or the relief pilot.*

EXTERIOR INSPECTION (WALK AROUND)

This flow is "GENERALLY" done by the First Officer or Relief Pilot on international flights, but the Captain MAY do this if she/he desires.

COCKPIT PREPARATION (CAPTAIN'S)

Pilot's obtaining their type rating will be required to complete the Captain's flows from memory. Called the
"S-L-U-G-O-S"

COCKPIT PREPARATION (FIRST OFFICER)

FMS VERIFICATION (FIRST OFFICER)

FMS INITIALIZATION (CAPTAIN)

FINAL COCKPIT PREPARATION

A shared responsibility between all members of the crew.

© MIKE RAY 2014
published by UNIVERSITY of TEMECULA PRESS

... and PROCEDURES FOR STUDY and REVIEW ONLY!

> ### SINCE EVERYTHING MUST BE DONE FROM MEMORY ...
>
> Around the TK University, for years, there has been floating the famous "**SLUGOS**" and "**EGOS**" acronyms; invented by some ancient wiseman to help simple, human pilots in remembering "the famous flows." And who am I to go against convention, so with a little embellishment, and apologies to the author, here are those famous gouges.

The CAPTAIN does the "SLUGOS"

"S" stands for SAFETY SANDWICH.
I am suggesting a sandwich because the safety items are "sandwiched" between two sets of UP-DOWN-UP and UP-DOWN-UP-DOWN flows. Completion of this "S" check also represents the end of the "**INITIAL COCKPIT PREPARATION.**"

"L" IS SORTA THE SHAPE OF THE OVERHEAD PANEL AS YOU COME DOWN.
I have added TWO EXCLAMATION MARKS to emphasis that you has two little added things at the end.

"U" IS JUST LIKE THE "L".
Sorta the shape of the overhead panel as you pass over it.

"G" STANDS FOR GLARE-SHIELD.
Simply a sweep of the hand starting at the light panel on the left side of the glareshield and proceeding to the right end.

"O" REFERENCES THE OXYGEN PANEL.
This is where you start this specific string of items. From left to right, around the "SNAIL", over the "SNAKE", ending up at the "NOISE" overrides.

"S" REFERS TO THE "SPEEDBRAKE."
This is where you begin your circle around the lower console ending up at the the WEATHER RADAR check.

HUH?? WHAZZAT?
I realize that this "SLUGOS" acronym makes no sense to you at this moment, but continue on with the next graphic and you will begin to see that it (hopefully) will begin to form a method that is useful ... maybe.

© MIKE RAY 2014
WWW.UTEM.COM

747-400 *SIMULATOR TECHNIQUES...*

"S" stands for "SAFETY SANDWICH"

This is called the
INITIAL COCKPIT PREPARATION.
We are calling it the *"***S***" check or the "***SAFETY SANDWICH***."*
I call it a sandwich, because the ***SAFETY*** *part of the check is*
"sandwiched" between ***UP-DOWN-UP*** *and* ***UP-DOWN-UP-DOWN***
motion of the flow. ... OK, OK, just use your imagination.

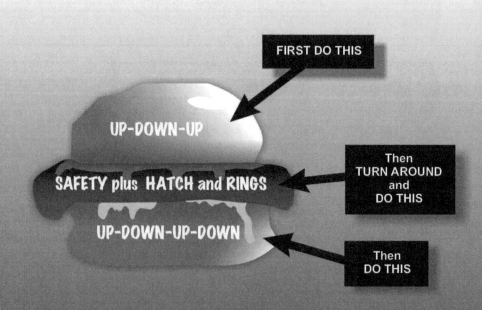

page 62

© MIKE RAY 2014
published by UNIVERSITY of TEMECULA PRESS

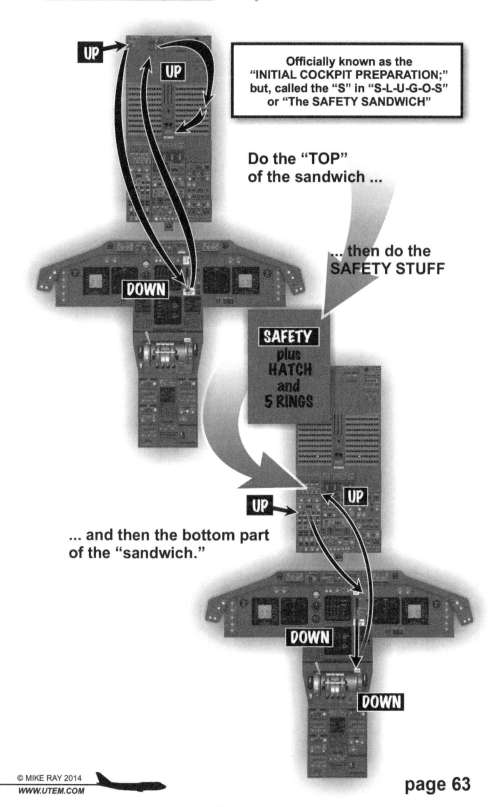

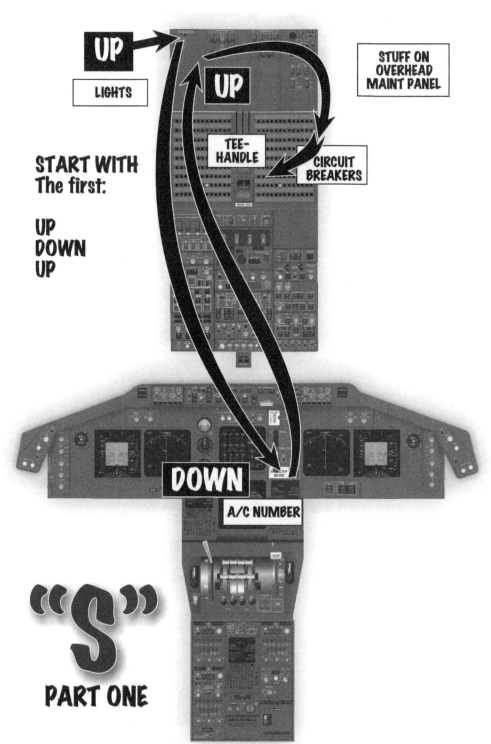

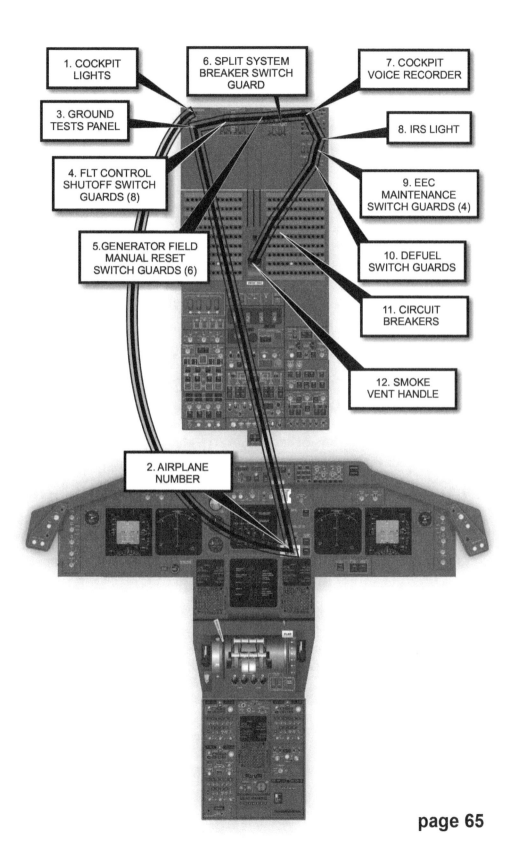

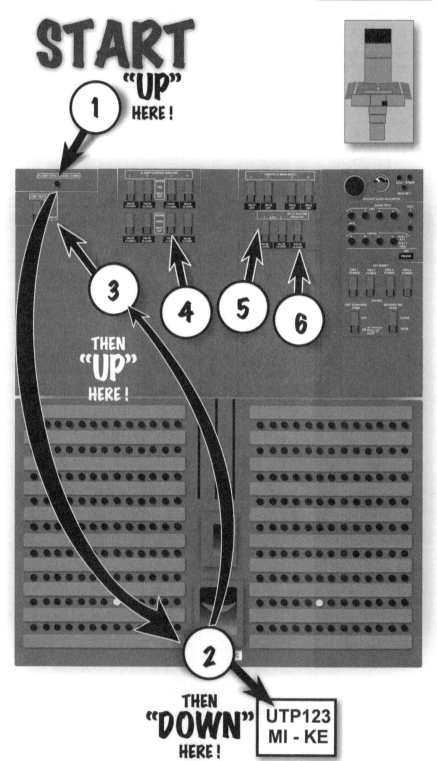

FLIGHT DECK ACCESS LIGHTS

1 When you enter the cold dark cockpit, here is where you turn on some lights. This switch is on the **GROUND HANDLING BUS**. Other lights, such as the **THUNDERSTORM LIGHTS**, are on other busses and may not be powered.

AIRCRAFT HULL NUMBER

2 CHECKED. There are two reasons to check this: First, to make certain that your flight papers have been prepared for the correct airplane; and secondly, so you won't look like a geek by getting on the wrong airplane. Don't laugh, it happens. DUH!

GROUND TEST SWITCH

3 CLOSED. Maintenance function only. There is no need to know anymore about this switch.

4 FLIGHT CONTROL SHUTOFFS

CHECK ALL SWITCHES CLOSED.

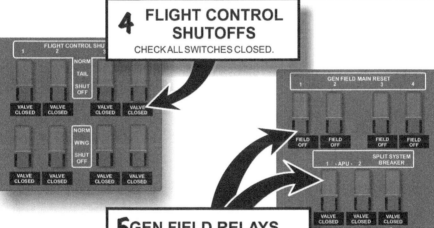

5 GEN FIELD RELAYS

CLOSED. There are 4 for the ENGINES and 2 for the APU for a total of 6 switches.

SPLIT SYSTEM BREAKER

6 CLOSED. This is the breaker that isolates the left and right electrical systems.

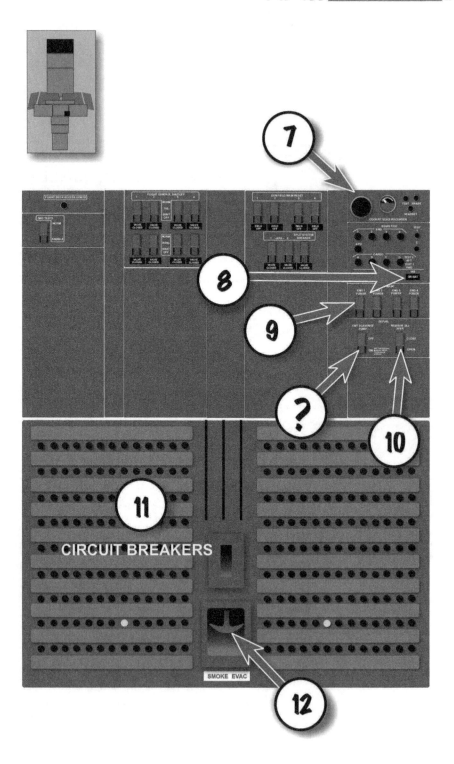

... and PROCEDURES FOR STUDY and REVIEW **ONLY!**

7 COCKPIT VOICE RECORDER

There are at least three different types floating around the system. Each has it's own idiosyncracies. In general, push the button and look for deflection, plug in headset and speak, listen for readback

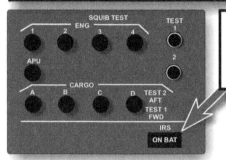

8 IRS "ON BAT" LIGHT

Could be ON. We expect it to be OFF. Maint function only.

9 EEC MAINT SWITCHES

CLOSED. Maintenance function only. There is no need to know any more about these switches.

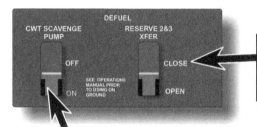

10 RESERVE 2&3 TRSFR SWITCH

CLOSED.

? CWT SCAVENGE PUMPS

Check the guard closed. Some airplanes have electric scavenge pumps and some hydro-mechanical. How do you tell? If this switch is installed, then the hydro-mechanical pumps are installed.

NOTE: *On airplanes equipped with the **HYDRO-MECHANICAL SCAVENGE PUMP** system with **GROUND OPERATION** of the **APU** it is possible to get unwanted fuel transfers. If this switch is installed, turn on the **MAIN TANK #3 AFT BOOST PUMP** after starting the **APU**.*

11 CIRCUIT BREAKERS

Check ALL BREAKERS in or CAPPED ... that includes those on the sidewall by the First Officers seat.

12 MANUAL COCKPIT EXHAUST

Check "TEE" handle stowed.

... and PROCEDURES FOR STUDY and REVIEW ONLY!

13. **CREW EMERGENCY EXIT HATCH**
Since we checked the hatch flush when we approached the jet, all that needs be done is to pull the velcro cover loose in the corner and check the hatch lever is stowed properly.

14. **(5) ESCAPE REELS and "D" RINGS**
They are in a plastic holder directly across from the hatch door at eye level. They can be checked without opening the door. Remember: They are for ONE TIME USE (that is, you need one for each crewmember), and once used, the metal tape has sharp edges.

15. **(6) SMOKE GOGGLES**
CAPTAINS (Located in holder outboard and behind seat)
FIRST OFFICERS (Located in pocket behind seat)
OBSERVERS (4 other goggles located in cabinet in back)

16. **(4) LIFE VESTS** There are 4; one for each position (seat). These are located in the seat back pouch on each seat.
There rear seat has its vest in a pocket to the left of the left armrest.
The test "assumes" that the individual vests are actually ready for use and include *ONLY* a check to see that they are there.

17. **OXYGEN MASKS** Only necessary to check the 2 Observers mask are actually on board.
The Observer seat occupants are responsible for checking:
   Mask, Hose, Regulator for the position.
Captain and First Officers check their own masks.

NOTE: *During the "BEFORE START CHECKLIST" it is expected that the Observers/Crew in the back audibly respond to the appropriate query.*

18. **HALON FIRE EXTINGUISHER**
Pressure should be NORMAL, seal INTACT

19. **CRASH AXE**

20. **FIRST AID KIT**
Seal INTACT

21. **EMERGENCY MEDICAL KIT** *Even if the seal is broken, there are situations where it can be used for dispatch ... see the FOM. Also; there is a second MEDICAL KIT installed; but this kit is required ONLY if the primary kit has been depleted below FAR minimums following a MEDICAL DIVERSION.*

22. **PROTECTIVE BREATHING EQUIPMENT**
Check for *FIRMNESS* and *BLUE DOT*.

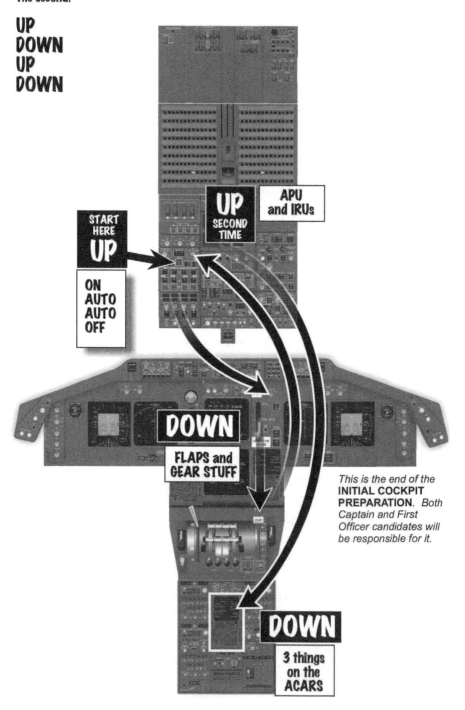

INITIAL COCKPIT PREPARATION continued

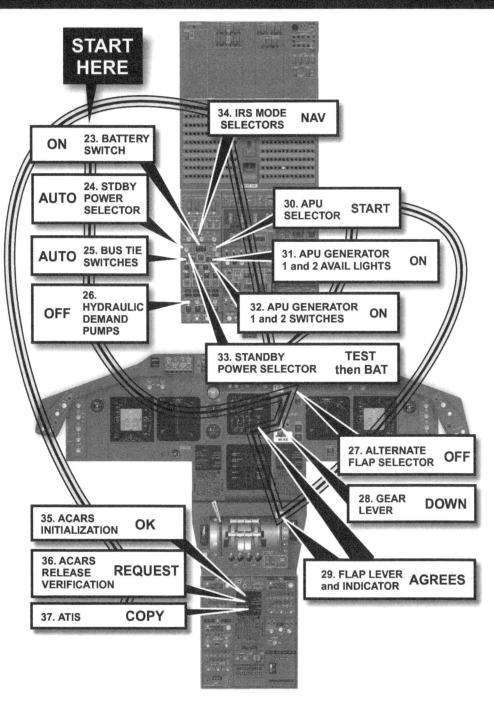

page 73

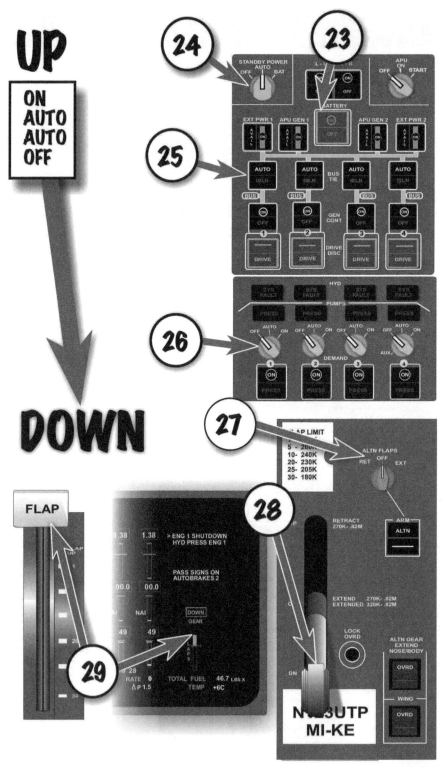

... and PROCEDURES FOR STUDY and REVIEW ONLY!

UP (UPPER PART OF PANEL)

BATTERY SWITCH .... ON

23 NOTE: Selecting the **BAT SWITCH** OFF will cause the **APU** to shut down. On some airplanes, the **APU** will continue to run for 90 seconds *WITHOUT FIRE DETECTION*. However, when using the "**SECURE CHECKLIST**" it is considered SOP to delay shutdown of the battery switch for a full **2 MINUTES** after the APU is shut down. This eliminates the fire detection problem.

STANDBY POWER SELECTOR .... AUTO

24 This time we will ONLY check the switch in the AUTO position.

BUS TIE SWITCHES .... AUTO

25 If the ISLN is indicated in the light bar, pushing the light should re-select the AUTO position. If that does not reset the AUTO position, get maintenance involved.
If one of these remains in ISNL, dispatch is possible but requires coordination with SAM (maintenance).

HYDRAULIC DEMAND PUMPS .... ALL OFF

26 We will visit the Hydraulic Pump switches several times during the set-up. This time ... ALL PUMPS OFF. Some pilots look at the brake press gauge, and if really low, they will turn on the #4 AUX position momentarily to get the parking brake pressure up to a "normal" indication.

DOWN (LOWER PART OF PANEL)

ALTERNATE FLAP SELECTOR .... OFF

27 The mistake that you could make is to have the switch indicating **RET**, a leftover from a previous crews problem.

GEAR LEVER .... DOWN

28 Position of the GEAR LEVER may or may not indicate that the gear are actually down. The check here is to ensure that the gear **_LEVER_** is down ... since this is an EICAS driven machine, an appropriate EICAS indication is the ONLY indication as to where the gear are.

FLAP LEVER ...... AGREES with POS INDICATOR

29 Like the gear, the position of the handle may or may not indicate where the flaps are. It is the EICAS that is the definitive indicator as to flap position. We want to ensure that *BOTH* the HANDLE and the EICAS are showing the same indication.

747-400 SIMULATOR TECHNIQUES ...

30 APU SELECTOR .... START/ON

Starting the APU on this airplane does *NOT* involve any fire test or additional checks. All you gotta do is turn the switch to ON and let it return to run. It is totally automatic and very reliable.

Some APU stuff:

Fuel is obtained from the **#2 MAIN FUEL TANK**.
FUEL PRESSURE is supplied by either:
   #2 AFT BOOST PUMP if AC is available, or
   a dedicated **DC FUEL PUMP** if required.
It is common (and expected) that the #2 AFT BOOST pump light will be OFF on the overhead panel.
The **DC PUMP** is powered by the **APU BATTERY** and starts and stops automatically when required.

The **APU BATTERY** powers:
   STARTER
   AIR INLET DOOR (opens and closes automatically)
   APU CONTROLLER (STBY power is from MAIN BATTERY if needed).
   DC FUEL PUMP
   APU FIRE DETECTION CIRCUIT
BUT ALSO supplies (And here is an ORAL QUESTION)
   <u>BACKUP ELECTRICAL for CAPTAIN'S INSTRUMENTS</u>!

continued on next page ...

> **Some more boring APU stuff:**
> DELAY of 2 minutes after shutdown before shutting off the BATTERY SWITCH, ensures that the Fire Detection system is available during shutdown.
> On some airplanes (N171UA), if the battery switch is shut down, the APU has NO FIRE DETECTION during it's shutdown.

31 APU GEN 1 & 2 AVAIL LIGHTS .... ON

These light bars are the ONLY indication that the electrical power from the APU has the proper VOLTAGE and FREQUENCY. The EICAS MSG (APU RUNNING) only indicates that the APU N1 RPM is greater than 95%.

32 APU 1 & 2 SWITCHES .... ON

Depressing the switches selects the APU power. It has been my experience that it takes about 3 seconds for the switch to react; and even then, additional attempts may be required. The right switch seems to react more easily, and if waiting to shut down the engines, the operation of one generator is sufficient to power the whole airplane.

A QUICK SYSTEM REVIEW

The AUXILIARY POWER UNIT:

A powerful jet engine (PW 901A) in its own right, the APU drives two (2) 90 KVA generators and a massive cooling air blower. The unit sits in the **UNPRESSURIZED** tail cone of the airplane. It is capable of providing electrical power for the entire airplane AND bleed air for such things as air conditioning.

> Here's the BAD NEWS about the APU:
> **CANNOT BE STARTED IN FLIGHT**, but
> If started on the ground, it may operate up to 20,000'.
> In flight, it can supply ONE PACK up to 15,000'.
> **APU SUPPLIED ELECTRICAL POWER
> IS NOT AVAILABLE IN FLIGHT.**

The APU STANDBY BUS:

This bus provides **BACK-UP POWER FOR THE CAPTAIN'S PFD, ND, and FMC**. Normally, these items are powered by the Captain's Transfer Bus, and if all power is lost to that bus; then the **APU STANDBY BUS** is automatically powered by the **APU BATTERY** through the **APU HOT BATTERY BUS** and the **APU STANDBY INVERTER**.

**THE BATTERY SWITCH MUST BE
ON FOR THIS TO OCCUR.**

> If the **AC BUS #1** is also unpowered,
> the **APU STANDBY BUS** is now powered by the **APU BATTERY** and
> **CAN BE EXPECTED TO BE POWERED
> FOR AT LEAST 30 MINUTES.**

STANDBY POWER SYSTEM CHECK

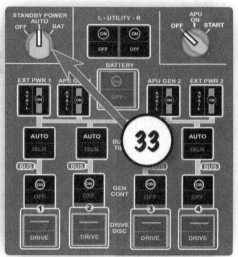

NOTE: Airplane MUST be on the ground with **ALL** busses powered.

ALERT:
If you are interrupted and inadvertently leave the switch in the BAT position, there is a possibility that, if the trickle charger isn't working properly, the battery will discharge about 30 minutes later.

33a STANDBY POWER SELECTORS .... BAT

VERIFY:

 CAPTAIN'S PFD
CAPTAIN'S ND
UPPER EICAS — *REMAIN POWERED*
LEFT CDU

CHECK:
on primary (upper) EICAS the following messages appear:

 BAT DISCH
BAT DISCH MAIN — *MESSAGES APPEAR*

33b STANDBY POWER SELECTORS .... AUTO

VERIFY:

 CAPTAIN'S PFD
CAPTAIN'S ND
UPPER EICAS — *REMAIN POWERED*
LEFT CDU

CHECK:
on primary (upper) EICAS the following messages appear:

 BAT DISCH
BAT DISCH MAIN — *MESSAGES DISAPPEAR*

... and PROCEDURES FOR STUDY and REVIEW ONLY!

IRU also called IRS.

34   IRS MODE SELECTORS (three) .... NAV

All we do here is ... turn the switches to **NAV**.

There is no reason to pause at **"ALIGN,"** *BUT* if you should accidently screw up and go to **"ATT,"** Go all the way back to **"OFF"** and start over. The alignment process is advertised to take **10 MINUTES**. It sometimes seems to take longer. **BE PATIENT.**

SOME IRU FACTS:

1. The *IRS* position can **ONLY** be updated (by pilot entry) **ON THE GROUND** through the **CDU**s.

2. There is **NO** capability (like some other *IRS*s) for the pilot to enter the position on the unit.

3. There is **NO** updating of the *IRS* position, once the units have entered the **NAV MODE**!

4. The **IRS DOES NOT NAVIGATE!** It only provides an **INERTIAL POSITION** input to the **FMC**s. The **FMC**s have navigational computers that use the information.

> Some additional thoughts regarding **ALIGNMENT** process:
>
> The **IRS MUST** go through a **FULL ALIGNMENT** and **RECEIVE INITIAL POSITION** when first turned on.
>
> When an **IRS** mode selector is moved from **OFF** through **ALIGN** to **NAV**:
> > **IRS** conducts 10 second self test (**ON DC light will come on briefly**).
> > **IRS ALIGN MODE** message displayed,
> > **10 MINUTE** alignment process starts.

There is a **30 second FAST ALIGN** feature that can be used on "**THROUGH**" flights with short turn-around times. While parked, you may re-start the initialization by going directly from **NAV** to **ALIGN** and back to **NAV**.

If **ALL THREE "STORED POSITIONS"** or if **ALL THREE IRS**s do not agree, a **REENTER IRS POSITION** message appears.

If **IRS** stored position differs by more than **6 NM** from the entered position, the msg **IRS POS/ORIGIN DISAGREE** is displayed.

It may be necessary to input a correct **INITIAL POSITION** several times to overcome a **POS/ORIGIN DISAGREE** msg.

Here is the **DANGER** in all this.
If you decide to put in another initial position that **DISAGREES WITH THE IRS**; after a coupla attempts it WILL accept your input and input that will agree with yours. Make certain that you are inputting a **CORRECT** position.

DOWN

it's the CDU located "on the console"

ACARS: do 3 things

35 ■ ACARS INITIALIZATION
36 ■ ACARS RELEASE VERIFICATION
37 ■ ATIS

This thing has a bunch of modes, so when you first look at the little screen, it could have some confusing stuff on it. To get it where we can work with it:
FIRST PUSH MENU KEY; then we can continue ...

---

35 ACARS INITIALIZATION .... COMPLETE

It will be necessary to complete the initialization before we go to the **FMCS** to load the flight data. The computer has to have this information **FIRST**.

After selecting **MENU**

SELECT: <ACARS> key **2L**; then

SELECT: INITIALIZATION

Fill in the appropriate information. Be aware that there are 2 pages to the initialization portion.

There's nothing really tricky here, you may not have the fuel on board yet, so skip that and just fill in as much as you can.

The *COMPULSORY FIELDS* are:

FLIGHT NUMBER
DEPARTURE DATE
DEPARTURE AIRPORT
ARRIVAL AIRPORT
FUEL ON BOARD
FUEL/GALLONS BOARDED
CAPTAIN'S NAME/FILE NUMBER

NON-COMPULSORY FIELDS are:

FLIGHT PLAN TIME
ALTERNATE AIRPORTS

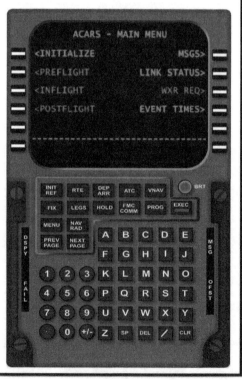

NOTE: There are at least two different types of displays. The **LCD** (Lquid Crystal Display) that displays in **WHITE**. The **CRT** (Cathode Ray Tube) displays are usually in **GREEN**.

36 ACARS RELEASE VERIFICATION.... COMPLETE

SELECT: **REQUEST INDEX**; then

SELECT: **DISPATCH INDEX**; then

SELECT: **RELEASE VERIFY**.

This will send a message to dispatch triggering a series of activities on their part. They will "eventually" send back a message to you on the ACARS. It will take several minutes.

37 ATIS .... OBTAIN

You may obtain the **ATIS** from the **VHF** radio ... **OR** you can obtain it through the **ACARS**:

SELECT: **PREFLIGHT MENU**

SELECT: **ATIS**

ENTER: **AIRPORT ID**
(YYYY international, XXX domestic)

SELECT: **SEND**.

After a few minutes, expect a message via the printer. This (at some airports) will contain the **ATIS**.

TESTING THE RADAR

On some flights, it will be necessary to "test the Radar". There are some considerations before turning on the set. It is OK to use the "**TEST FUNCTION**"; however ...

WARNING !!!
DO NOT OPERATE THE RADAR SET WHEN:
- PERSONNEL WITHIN 15 FEET OF THE RADAR ANTENNA,
- WHILE FUELING or DE-FUELING.

- DO NOT OPERATE THE SET WHEN FACING A WALL, such as sitting at the gate. *The problem is that there will be SIGNAL BOUNCE-BACK and it could burn out the radar sets receiver.*

WARNING!!
DO NOT MOVE THE FLAP or SPOILERS WITHOUT CLEARANCE FROM OUTSIDE OBSERVERS.
As you can see, a truck or service vehicle parked under the wing could have become squished and personnel could be injured or worse.

HF RADIO CAUTION
Even though the HF radio must be checked prior to departure and is also an excellent source for obtaining the most accurate TIME.

- *DO NOT OPERATE THE HF RADIO WHILE BEING FUELED or DE-FUELED.*
- *DO NOT TRANSMIT ON HF IF PERSONNEL WITHIN 6 FEET OF THE VERTICAL STABILIZER*

called the "WALK AROUND" by pilots

EXTERIOR INSPECTION

The exterior inspection must be conducted by a "**QUALIFIED**" crewmember (such as a Captain, lowly Co-pilot, Other qualified First Officer, or Bunkie) prior to **_EVERY DEPARTURE_**, except in the case of a diversion.

While the Captain, of course, bears ultimate responsibility for the thoroughness of the inspection, She/he does not actually have to conduct the inspection. Here is a thought, however, if you are a veteran Captain, and you are paired with a newbie co-pilot ... it might be a good idea to do the walk-around and give the youngster an opportunity to take their time and do their cockpit set-up duties.

One truly significant walkaround item is to check that the static ports and the pitot tubes are **NOT** blocked. It is possible that these tiny openings could be taped over and then the airplane was painted; as a result, it takes an incredible amount of attention to this detail to discern if the camouflaged tape was removed.

ATTENTION:
There have actually been some newly painted commercial airliners that have actually gotten airborne ... and because of this very problem, they crashed.

more CAPTAIN'S SETUP STUFF

After the **ACARS** is complete, the Flight Handbook considers the **INITIAL COCKPIT SET-UP** as being complete. The very next thing listed in the book is the **EXTERIOR INSPECTION**; however, in the "real world" one can expect that the Initial Cockpit Set-up will be completed by the Captain, and the Exterior Inspection to be completed by another crewmember. **Be alert**: *Here is where some confusion may occur in the flows.*

The next thing that the Captain does is the **CAPTAIN FMCS INITIALIZATION AND VERIFICATION**. So, more natural flow is to go right from the three things on the console **CDU** (ACARS) to the **CDU** (FMCS) on the throttle console. The Checkguy will be wanting you to tell him that you have completed the first flow and are ready to do the **NEXT THING**, which is the *EXTERIOR INSPECTION*. In the real world, perhaps you would just flow right up there and go from one thing to the other seamlessly, but on the check-ride, be certain that you are aware that there is a break in the flows.

CAPTAIN'S FMCS SET-UP

Remember:
After the ACARS; DO EXTERIOR INSPECTION, THEN DO THE FMCS

© MIKE RAY 2014
published by UNIVERSITY of TEMECULA PRESS

1 ATC LOG

On **ATC INDEX**, select **<LOG**. Check that all previous messages are deleted. **LS4L**.

If messages exist, use the **ERASE LOG** prompt to remove them.

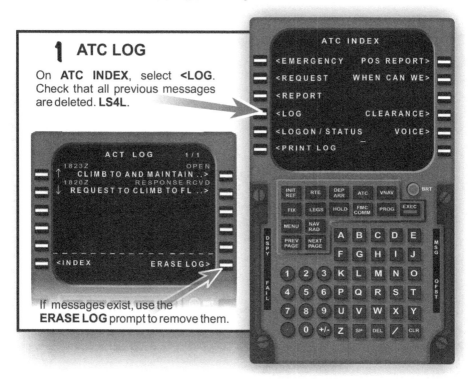

2 FMC COMM

Check that there are **NO UPLINKS** remaining from previous flights.

Select **DELETE** to scratchpad, depress appropriate **LS** button to remove the message.

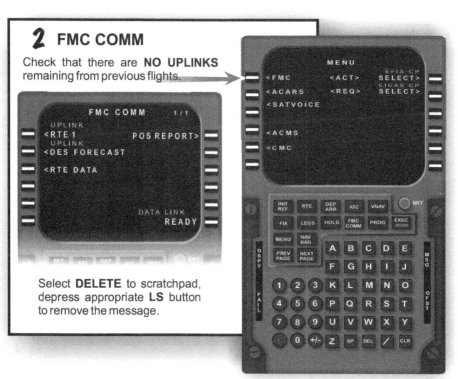

3 ACTIVE NAV DATABASE

To Verify if correct, Select:

IDENT page from the **INIT / REF** page.

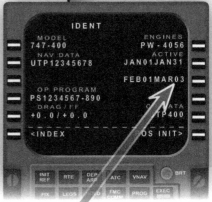

The **VALID NAVIGATION DATE** begins at **0901 UTC** on the first day of the 28 day navigation data cycle. The date and time should be valid at the time of departure.
Changing the **ACTIVE DATABASE** can only be accomplished on the ground.
If the effective **ACTIVE DATABASE** does not match the Captain's clock date, the message **NAV DATA OUT OF DATE** appears in the scratch pad.

TO UPDATE ACTIVE DATABASE DATE:

LS (3R) This will select new date to the scratch pad.
LS (2R) This will change the active date.

4 PERFORMANCE FACTORS

The current BURN FACTORS (**DRAG/FF**) are listed on the FLIGHT PLAN. These should be compared with those on the CDU PAGE "**IDENT**."

To modify the factor if required.
From the **INIT/REF PAGE**:

        select MAINT prompt (6R)
        select PERF FACTORS (2L)
        type ARM into the scratch pad
        Enter at (6R)
        type NEW FACTOR in scratch pad
        Enter at (2L).

Confirm NEW FUEL FACTOR (FF).
        Verify correct factor at (5L).

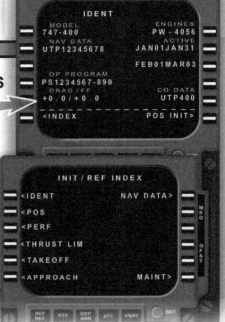

5 complete the POSITION INITIALIZATION

REFERENCE AIRPORT ... ENTER.
Type the four letter identifier for the departure Airport into the **SCRATCH PAD**. Then enter that four letter **ICAO** departure airport identifier by selection the insert button next to **2L** (**LS2L**). For example: **KSFO**

AIRPORT GATE
By entering the airport identifier, this will allow entry of the **GATE CODE** from the **FOM AIRPORTS PAGE** using **LS3L**. It is not essential, but does create a more accurate starting point for the **FMC**.

IRS POSITION (**GPS** preferred).
Select the **IRS POS** by this order of preference:
> GPS (4R)
> GATE (3R)
> REF AIRPORT (2R)
> LAT / LONG (last resort).

UTC (GPS) .... verify.

GPS is the primary time source for the **FMCS**. **UTC (MAN)** indicates that the **FMCS** is getting it's input from the Captain's clock.

If the Captain's clock is more than 12 seconds from GPS time, a scratchpad message will appear:
SET CLOCK TO UTC TIME.

Verify **RNP**
(**REQUIRED NAVIGATION PERFORMANCE**).

Verify **RNP**
(**REQUIRED NAVIGATION PERFORMANCE**).

Values in the small font at **LS3L** depict the normal default mode of operation for **RNP**.
Deleting **MAN** at **LS3L** allows default **RNP** to be engaged.

*If the **RNP** value is different from the default value, a new **RNP** can be **MANUALLY** entered. Values that can be entered range from 0.01 to 99.9.*

VERIFY GPS NAVIGATION HAS INHIBIT DISPLAYED.

On **LS(5R)** of **POS REF 2/3** the word **INHIBIT** should be visible.

NOTES:
*The **FMCS** validates the integrity of the **GPS** signal source. A successful completion of the validation process results in the display of position data at 4R.*

*If the **IRS POS/ORIGIN DISAGREE** message appears in the scratchpad, it is likely that an incorrect position was entered. Re-check and re-enter ... prompt boxes not required for this second entry.*

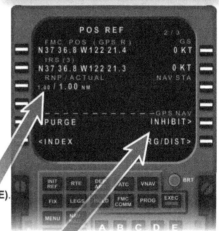

BORING, BUT NICE TO KNOW EXTRA CURRICULAR MATERIAL.

RNP ? HUH! WHAZZAT?

REQUIRED NAVIGATIONAL PERFORMANCE

What is **RNP**? Here's the boring story in a simple form. The Air Traffic Control **(ATC)** system has long wanted some standards on which to develop a more efficient system, called *"Communications Navigation surveillance/Air Traffic Management* **(CNS/ATM)**. In order to do that, they have to have a *"realtime estimate of navigational certainties"* for airplanes operating in that system.

The criteria for *Total System Error* **(TSE)** is:
"The TSE must not exceed the specified RNP value for 95% of the flight time on any part of any single flight."

Example: **RNP-10** means **TSE** less than **10 NM for 95%** of the flight time.

The introduction of **GPS** and **IRU** equipment on the modern Boeing airplanes has provided the needed navigational accuracy and we have that information constantly available. Generally speaking, you should not encounter a problem with exceeding the **TSE** limits.

Be aware that different areas of operation in the Air Traffic Control system demand different **RNP** values and range (currently) from .01 to 20, with .03 or .05 being the most commonly dictated RNP. The value for your flight will be indicated on your individual flight plan.

6 confirm ACARS initialized.

Verify information display is correct.
Use the CDU down on the lower console.

We have done the initialization earlier and now we simply see that the information is correct.

Compulsory information is:

Flight Number
Departure date
Departure airport
Arrival airport
Fuel on board
Fuel/gallons loaded
Captain's name
Captain's file number

Non-compulsory information includes:

Flight plan time
Alternate airports

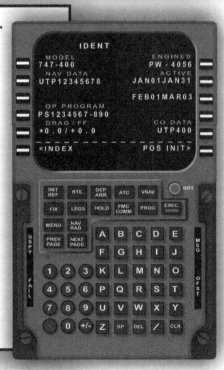

Filling in the route is probably the **MOST IMPORTANT** duty at this point. It must be done completely and scrupulously exact down to each tiny detail. There are several different iterations of the Boeing **FMC** program (PIP, PEG, OLD, NEW, etc.). There are as many techniques as there are pilots for doing this chore, and each airline has outlined their specific protocols in excruciating detail. It will be impossible in a book like this to cover every facet of this operation, so when you read this, just remember that there are other ways of "*LOADING THE ROUTE*". Here is a "sample" **ROUTE UPLOAD**.

1 CAPTAIN enters ROUTE:

There are basically three techniques to enter the route:
1. **DATALINK** using Company Flight Plan computer.
2. **MANUAL** entry from **FPF**

That is, placing the waypints directly from your cleared flight plan.

3. **MANUAL** entry from company database **FMCS** route.

That is a single entry using a "canned" flight plan routing identified by a specific code.

Of course, the route generally dictates the method used. For example, if you are going from Los Angeles to Beijing with a flight plan about 30 feet long, it is nice to use the datalink method. However, if you are going from San Francisco to Los Angeles, a "simple" manual entry would make more sense.

The route of flight is entered by the **CAPTAIN** and **VERIFIED** by the **FIRST OFFICER**.

NOTE:
What is meant by "**CAPTAIN ENTER**" and "**F/O VERIFY**" is this. The intent is for the Captain to enter the flight plan entirely independent of the F/O. That is to say, even though it may make sense for the F/O to "read off" the fixes or "WATCH" the entry ... **this is NOT ALLOWED. THE CAPTAIN MUST DO THIS TASK UNASSISTED**!!! ...and the **FIRST OFFICER MUST VERIFY UNASSISTED**!!! ... any conflicts must then be resolved at this point.

ANOTHER BORING COMMENT

Here we have to make a clear distinction between "*PLANNED*" and "*CLEARED*" as it applies to flight plans. The original flight planning would possibly have taken place hours before and perhaps would have been rendered inaccurate or unavailable at the time of actual flight. So, it is frequently true that actual flight clearances from **ATC** may differ from our flight plan.

The flight plan is under going continual alteration by **ATC** during the entire flight evolution and you can expect possible changes taxiing out or after takeoff or cruise or descent or anytime during the flight. We have to be able to make the appropriate changes in the **CDU/FMC** quickly and accurately **AND** complete the proper cross check to ensure that our flight profile **EXACTLY** matches the one that **ATC** expects.

747-400 *SIMULATOR TECHNIQUES ...*

a TUTORIAL about LOADING the ROUTE:

Without question, when you are just starting out using the Boeing Glass Auto-flight ... it is one of the most intimidating and confusing parts of flying the fabulous 747-400. So we are going to go through a VERY basic loading exercise that will demonstrate some of the problems that pilots have with the FMC Flight Management Computer) and CDU (Computer Display Unit).

DESCRIBING THE BASIC EXAMPLE

The Boeing 747-400 is a "mature" aircraft, and as such has been treated to numerous upgrades and changes to the **FMC PROGRAM OPERATING SYSTEM**. These include PIP, PER, PEGASUS, FANS, POOP, etc.. Rather than address all these variations of the original bare-bones operating modality and attempt to explain all the differences ... all we are going to do here is describe a minimum loading procedure (Departure to Arrival without intervening Route) for a basic, fundamental Boeing 747-400 computer system. The details will be left up to you to integrate in order to accomodate what you will encounter when you are at your airline.

EXAMPLE FLIGHT PLAN

Assume that you are going to fly from **KSFO** to **KLAX**. Here is the proposed flight plan routing we want to enter into the **FMC**.

**KSFO RWY 1L
PORTE3 DEPARTURE . AVE
TRANSITION . SADDE6 .
SMO TRANSITION
KLAX RWY 24R**

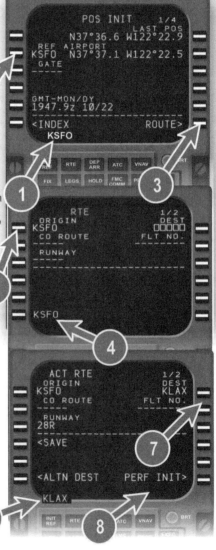

STEP 1: With the **CDU** on the **"POS INIT PAGE"**; using the **CDU KEYPAD**, type **KSFO**. Notice that the letters **KSFO** appear in the **SP (SCRATCH PAD)**.

STEP 2: LINE SELECT the number two button from the top of the right side queue. The notation **LS2R** will be used for this action. This action will cause the contents of the **SP** to be transferred to that position on the **CDU** dispaly. This is the technique that we will be using for this tutorial demonstration.

STEP 3: LS5R (ROUTE>). This will cause the **CDU** to display the **RTE 1** page.

STEP 4: This will cause the **CDU** to display the **RTE 1** page. Type **KSFO** (departure airport) into the **SP** (scratchpad). On some enhanced **FMC** units, the **KSFO** will already be placed in the **SP**.

STEP 5: LS1L: This will place **KSFO** in the **ORIGIN** position on the **CDU** display.

STEP 6: Then type **KLAX** into the **SP** (destination).

STEP 7: LS1R: This will place **KLAX** in the row of boxes under the **DEST** position on the **CDU** display.

STEP 8: The **CDU** will then display the **PERF INIT>** prompt in the lower right hand corner. **STOP!!**
Right here is where I want to make a recommendation. *DO NOT NECESSARILY FOLLOW THE PROMPTS EVERY TIME*!

page 90

© MIKE RAY 2014
published by UNIVERSITY of TEMECULA PRESS

STEP 9: Select the **DEP ARR** key.

STEP 10: LS1L: Line select the button next to the **<DEP** prompt.

STEP 11: This will display the available (line selectable) Runways, database available SIDs, and the Transitions.

Here is my recommendation. Select these items (if known or assigned) and their assigned routing will appear on the **LEGS** page.

This will greatly simplify the loading of the routing as the departure "may" form the first part of the actual airways. In this case, the departure-SID-transition will form the complete routing to the **AVE (AVENAL)** waypoint.
You may have to use the **PREV PAGE - NEXT PAGE** key to "**SCROLL**" through the available choices. For clarification, it is generally suggested that you select the **RUNWAY** first, then the **SID**, and then the **TRANSITION** ... in that order.

Our ultimate goal is to create a continuous stream of waypoints. They will appear on the **LEGS** page. That queue of points will constitute the "**MAGENTA LINE**" when viewed on the **ND**.

STEP 12: The **EXEC (EXECUTE)** light will possibly illuminate (deoends on which software you have). If it does you may depress the key and select the routing. Here is what is important, the routing is not actually a part of the **FMC** memory **UNTIL** you **EXECUTE** the requested routing. Since we haven't completed the whole routing yet, it isn't essential that you EXECUTE this "partial" routing.

STEP 13: If you know what the arrival **STAR** is going to be, you may enter that. It is unlikely in a route of any length that you would have that knowledge, but in this case, we can assume that the routing included in our clearance will be the actual planned route.

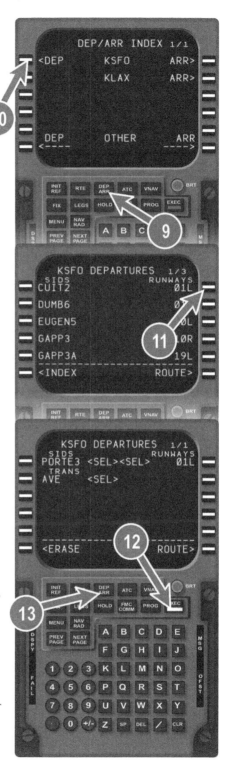

page 91

STEP 13: Select the **DEP ARR** key.

STEP 14: LS2R: Line select the button next to the **ARR>** prompt for **KLAX**.

STEP 15: This will display the available (line selectable) Runways, database available **SID**s, and the Transitions for the arrival airport **KLAX**. Make your selections.

STEP 15: EXECUTE.

STEP 16: Select the **LEGS** key. This will display the **ACT RTE LEGS** page.

NOTE: *If we had waited to* ***EXECUTE*** *the queue ... the title would be* ***MOD RTE*** *in* ***BLUE***. *Since we have already executed our inputs into the* ***FMC***, *the title is now* ***ACT RTE***, *or actual* ***ROUTE*** *(so far).*

STEP 17: Using the **SCROLL** keys, go through the entire **ROUTE** and look for problems such as

> **ROUTE DISCONTINUITY**
> **EXCESSIVE DISTANCE**
> between the waypoints
> **IMPROPER HEADINGS**
> **ANYTHING UNUSUAL**

STEP 18: Note that after **AVE** there is a row of boxes.
After that there is a "**ROUTE DISCONTINUITY**".

In this case we can resolve the discontinuity fairly simply. Here is what to do (in most cases).

STEP19: LINE SELECT a waypoint below "boxes". In this case, we would **LS5L (SADDE)**. This would place **SADDE** into the **SP**.

STEP 20: Line Selecting **4L** (the "boxes") causes the waypoint **SADDE** to be moved up, replacing the boxes, and "closing" up the discontinuity.

Here is the "general" rule about boxes ... they should be filled in or removed! That is NOT always the case as we shall see later.

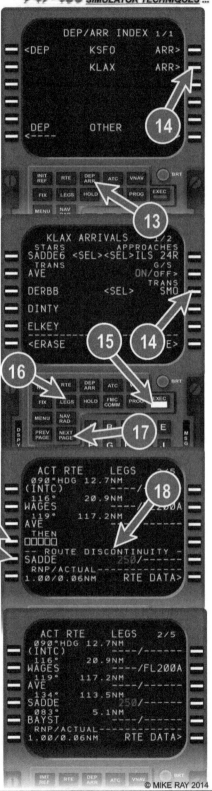

page 92

STEP 21: Select the **NEXT PAGE** key to continue to view the queue of waypoint. You will encounter another **DISCONTINUITY** and row of **BOXES**; only this situation differs slightly from our last discontinuity.

STEP 22: Select a waypoint below the fix below the boxes (**SMO**).

Notice that the waypoint **SMO** is also shown **ABOVE** the boxes. If we were to just place the **SMO** waypoint we have just selected into the boxes, our routing would be to **SMO** then to **JAVSI** and then back to **SMO** before we could continue. This would *NOT* be good. So, in this situation, we would ...

STEP 23: ... select **SMO** and place it on top of the waypoint **SMO** at **LS1L**.

STEP 24: Notice that the page title is "**MOD RTE**" and the **EXECUTE** light is illuminated. This means that the FMC does not yet have your changes in its memory and it will require the **EXECUTE** light to be depressed.

Once we have completed this step and cleared all the "**DISCONTINUITIES**" then we have completed the loading of the **ROUTE**.

STEP 24: Notice that the page title is "**MOD RTE**" and the **EXECUTE** light is illuminated. This means that the FMC does not yet have your changes in its memory and it will require the **EXECUTE** light to be depressed.

Once we have completed this step and cleared all the "**DISCONTINUITIES**" then we have completed the loading of the **ROUTE**.

STEP 25: To proceed to the next step, we selec the **RTE** key. This will open the **ACT RTE** page.

STEP 26: This will also disoplay the prompt for the next page in the loading procedure.can to select the **PERF INIT>** prompt

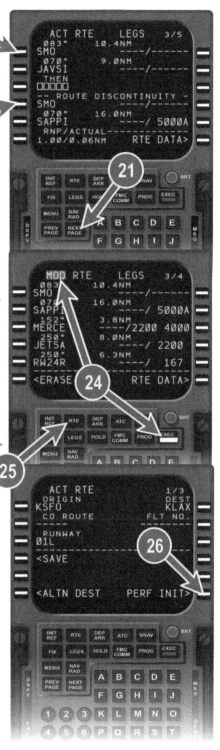

747-400 SIMULATOR TECHNIQUES ...

8 after ROUTE has been entered:

The CAPTAIN does three 3 things:

1. DEPARTURE/ARRIVAL ..... SELECT

Depress the **DEP/ARR** key to enter the:
PLANNED DEPARTURE RUNWAY,
SID (Departure), and
TRANSITION (if applicable).

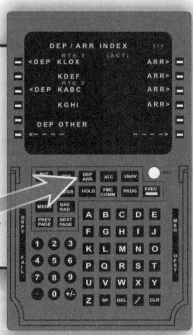

2. ROUTE ..... verify, activate, and execute

Select the RTE key.

At this point, the Captain MUST verify the route of flight is IDENTICAL to the filed route of flight. AFTER verification is complete:

ACTIVATE, and
EXECUTE.

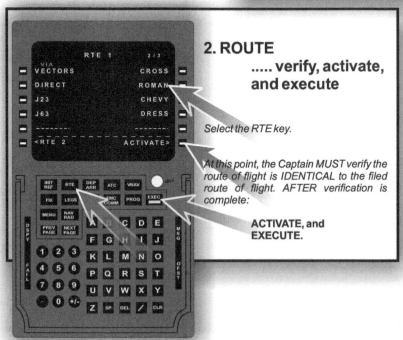

CAUTION:
The route **MUST ALWAYS BE IDENTICAL TO THE "FILED"** route. Flights conducted on **FLEX TRACK ROUTES** also require cross checking the route with the track detail message.

3. WAYPOINT verification:

This is ESSENTIAL for a complete route check. Here are the steps:

1. Select **LEGS** page on the **CDU**.
2. Select **PLN** mode on the **EICAS** Control Panel.
3. Compare the **MAGNETIC COURSE** and **DISTANCE** between each waypoint of the **FMC** and the **AFPAM**.
4. **STEP** through each waypoint at **LS6R** and verify.
5. The leg distances would be within 2nm of the **FPF**.

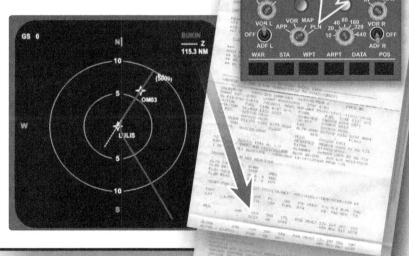

NOTE:
There are requirements to make notations on the **MASTER FLIGHT PLAN** during this evolution, but they are beyond the considerations of this manual. Pilot's going to "INTERNATIONAL" school will be introduced to those techniques.

747-400 SIMULATOR TECHNIQUES ...

FIRST OFFICER'S FMCS verification

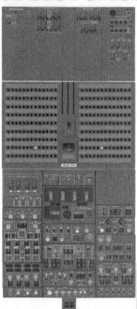

The First Officer has been designated to accomplish the initial inputs to the **ACARS** using the **CDU** on the lower instrument console, As well as the checking of the Captain's entries on the upper First Officer's **CDU**.

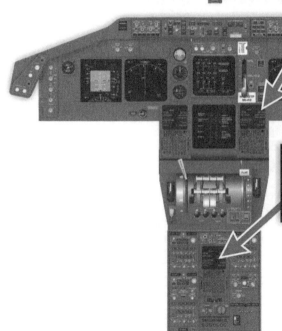

UPPER CDU

LOWER CDU
located on the lower instrument console

... and **PROCEDURES FOR STUDY and REVIEW** ONLY!

1 ATC LOG

Check that all previous messages are deleted.

If messages exist, use the **ERASE LOG** prompt to remove them.

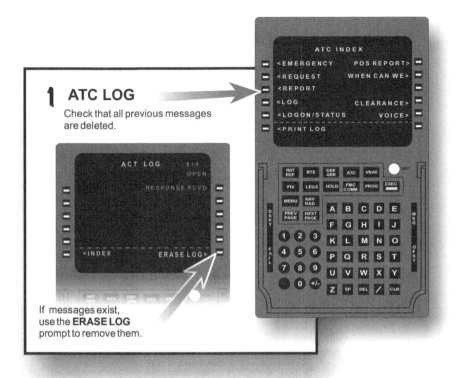

2 FMC COMM

Check that there are **NO UPLINKS** remaining from previous flights.

Select **DELETE** to scratchpad, depress appropriate **LS** button to remove the message.

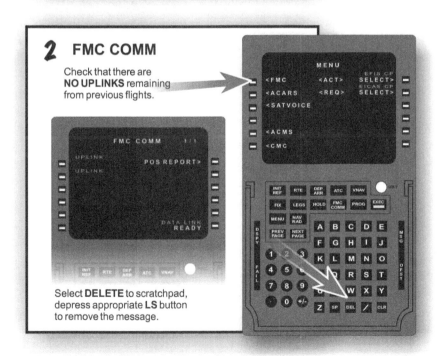

3 ACTIVE NAV DATABASE CHECK

select:
IDENT page from the **INIT / REF** page.

2 THINGS TO CHECK

1. Verify **NAV DATA** page is correct.
This code is called the **NAVIGATION DATABASE IDENTIFIER**

These big airplanes operate on flight segments around the world. Since the amount of **DATABASE** information would be very large if the whole earth were installed at one time, the database of fixes, waypoints, airports and all that is divided into parts. Each part represents a different segment of the database information. Now, I don't know what the protocols are for your airline right now, but these little coded identifiers in the **CDU** represent the database that is installed on the airplane. Here is the problem.
If the jet is going from one database area to another, then the information MUST reflect that new database area. If you should get airborne with the wrong database installed on your airplane, you will have to return and land and get a copy of the correct database hard drive installed before you can continue. This happens occasionally ... and it is really *UGLY!*

2. Verify **ACTIVE DATABASE DATE**.

If the **DATABASE** is "out of date" here is how to:
UPDATE ACTIVE DATABASE DATE:

LS (3R) This will select new date to the scratch pad.
LS (2R) This will change the active date.

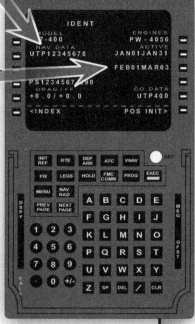
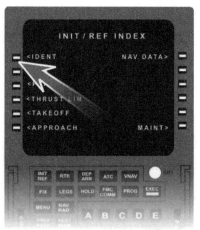

NOTES:
The **VALID NAVIGATION DATE** begins at **0901 UTC** on the first day of the 28 day navigation data cycle. The date and time should be valid at the time of departure.

Changing the **ACTIVE DATABASE** can only be accomplished on the ground.

If the effective **ACTIVE DATABASE** does not match the Captain's clock date, the message **NAV DATA OUT OF DATE** appears in the scratch pad.

4 verify
POSITION INITIALIZATION

REFERENCE AIRPORT ... ENTER.
Enter the four letter ICAO departure airport identifier at (**LS2L**). For example: KLAX

AIRPORT GATE
Enter the airport identifier at **LS3L**, this will allow entry of the **GATE CODE** from the FOM AIRPORTS PAGE.

IRS POSITION (GPS preferred).
Select the IRS POS by this order of preference:
> *GPS (4R)*
> *GATE (3R)*
> *REF AIRPORT (2R)*
> *LAT/LONG (last resort).*

NOTES:
*The **FMCS** validates the integrity of the **GPS** signal source. A successful completion of the validation process results in the display of position data at **4R**.*

*If the **IRS POS/ORIGIN DISAGREE** message appears in the scratchpad, it is likely an incorrect position was entered. Re-check and re-enter ... prompt boxes not required for this second entry.*

UTC (GPS) .... **verify**.

GPS is the primary time source for the FMCS. UTC (MAN) indicates that the FMCS is getting it's input from the Captain's clock.

*If the Captain's clock is more than 12 seconds from GPS time, a scratchpad message will appear: **SET CLOCK TO UTC TIME**.*

Verify **RNP**
(REQUIRED NAVIGATION PERFORMANCE).

Values in the small font at **LS3L** depict the normal default mode of operation for **RNP**. Deleting **MAN** at LS3L allows default RNP to be engaged.

*If the **RNP** value is different from the default value, a new RNP can be MANUALLY entered. Values that can be entered range from 0.01 to 99.9.*

VERIFY GPS NAVIGATION HAS INHIBIT DISPLAYED.

On **LS(5R)** of **POS REF 2/3** the word **INHIBIT** should be visible.

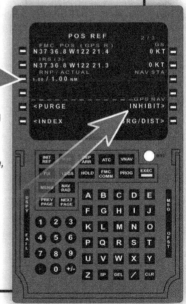

5 verify ROUTE:

On the **ACT RTE** page, verify the **FLIGHT NUMBER**. Notice that the **FLIGHT NUMBER** depicted as part of the **ICAO** filed flight plan is the *REQUIRED* flight number.

> **CAUTION**:
> The improper flight number may inhibit **CONTROLLER PILOT DATALINK COMMUNICATIONS (CPDLC)** or establish **ATC** communications with the wrong airplane.

On the **ACT RTE** page, verify that the route entered by the Captain is correct and agrees with the **FPF, PDC, ATC** clearance, and Charts.

THEN:

On the **ACT RTE LEGS** page, accomplish waypoint verification procedure as follows:

Select **LEGS** page on the **CDU**.

Select **PLN** mode on the **ECU** Control Panel.

Compare the **MAGNETIC COURSE** and **DISTANCE** between each waypoint from the **FMC** and the Flight Plan (**AFPAM**).

Step through each waypoint using **LS6R** to verify accuracy.
The legs distances should be within 2 nm of the **FPF**.

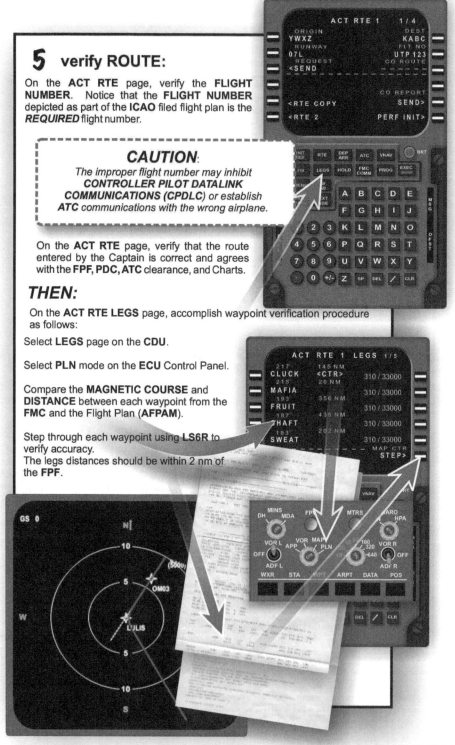

... and PROCEDURES FOR STUDY and REVIEW ONLY!

6 NAVAID inhibits

On the **REF NAV DATA** page,
ensure no unwanted:
    NAVAID INHIBIT,
    VOR ONLY INHIBIT,
    VOR/DME NAV INHIBIT exists.

7 Check VOR/ILS frequencies

On the **NAV RADIO** page,
ensure no unwanted:
    VOR or ILS frequencies are manually tuned.

DISCUSSION:
There are two VOR receivers. They are
automatically tuned by the FMC.

There are three ILS receivers. They are also
automatically tuned by the FMC.

*If you have never seen this system operate, it is
truly spooky. The problem, of course, is that it is
so easy to not check on it and see that it is doing
what it should do because it is truly fantastic.*

*Here is where we check that it is tuning what we
want it too, before we commit to takeoff.*

END OF F/O's FMCS VERIFICATION

CAPTAIN'S COCKPIT PREPARATION

Start at top of left row.

EEC's

danger

check

CALLED THE

"L!!"

CHECK

Here is a more detailed breakdown for the left-brained pilots.

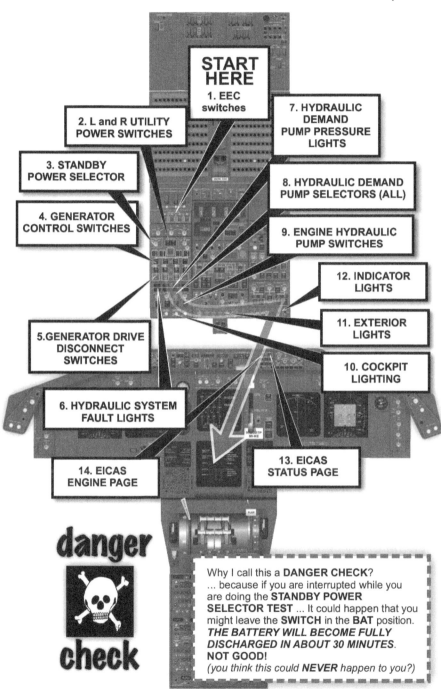

Why I call this a DANGER CHECK?
... because if you are interrupted while you are doing the **STANDBY POWER SELECTOR TEST** ... It could happen that you might leave the **SWITCH** in the **BAT** position. **THE BATTERY WILL BECOME FULLY DISCHARGED IN ABOUT 30 MINUTES**. **NOT GOOD!**
(you think this could NEVER happen to you?)

DANGER check

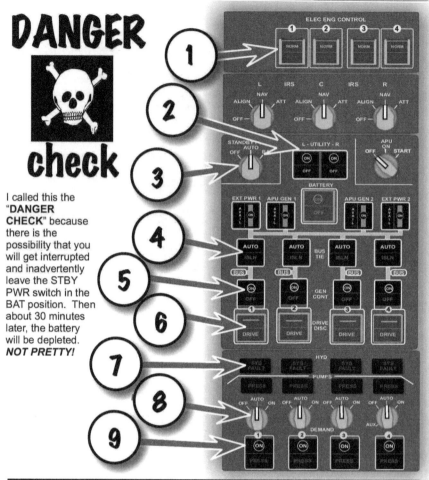

I called this the **"DANGER CHECK"** because there is the possibility that you will get interrupted and inadvertently leave the STBY PWR switch in the BAT position. Then about 30 minutes later, the battery will be depleted. *NOT PRETTY!*

1 EEC SWITCHES .... NORMAL

These switches have two modes:
- PRIMARY uses EPR
- SECONDARY uses N1 RPM

In PRIMARY MODE ... anytime you need MAX THRUST, just push the throttles to the stops. The thrust is continually adjusted automatically.

BUT: In secondary mode, it is possible to **OVERBOOST** the engines. If you manually select secondary mode, retard throttle to avoid overboost.
In secondary mode, continual adjustment of thrust is required during climb.
If one engine is in secondary, it is SOP to place ALL the engines in secondary.

THE BIGGIE!!! AUTO-THROTTLES DISCONNECTS when using secondary mode.

2 L and R UTILITY POWER SWITCHES...... ON

Main ELECTRICAL POWER users are the galleys.

3 STBY POWER SELECTOR: BAT-TEST-AUTO

I call this the danger check because there is the possibility that during this test you can become distracted and *leave the switch in BAT*. This is a NO-NO because the battery will become discharged in 30 minutes and it takes a long time to recharge ... and if you get airborne like that ... **WHEEEEE!**

The concept here is to switch to BAT and observe four things and two EICAS msgs:
- CAPTs PFD remains powered
- CAPTs ND remains powered
- PRIMARY EICAS remains powered
- LEFT CDU remains powered
- EICAS MSG: "BAT DISCH APU"
- EICAS MSG: "BAT DISCH MAIN"

Then switch back to ON and observe these things remain powered and the EICAS msgs go away.

4 GEN CONT switches .... verify ON

There is no reason to expect that these would ever be de-selected. But it is extremely important that they are. Without these guys in the ON position, their respective generators cannot take the bus.

5 GEN DRIVE DISC switches .... GUARDED

Once again, you are just checking to see that these disconnects have not been activated. They will disconnect the generator drive, and there is no way to reconnect them in the air. Be alert.

6 HYD SYS FAULT lights .... ON

System not powered, no pressure: Lights ON.

7 HYD DEMAND PUMP PRESS lights .... ON

Pumps not powered, no pressure: Lights ON.

8 HYD DEMAND PUMP SELECTORS .... ALL OFF

This is the second time we have gone by this panel ... and left these guys unpowered.

9 ENG HYD PUMP switches .... ON

Verify that the PRESS lights are ON

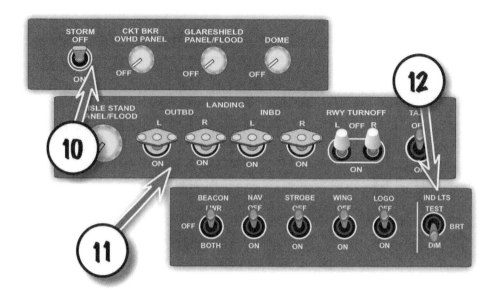

10 COCKPIT LIGHTING .... ADJUST

11 EXTERIOR LIGHTS .... AS REQUIRED

Turn on the WING ILLUMINATION LIGHTS
> *There is a lot of concern about leaving the wing illumination lights on with the passenger loading bridge up to the airplane. Gets hot so ... don't forget to turn them off.*

Turn on the LOGO lights at night only.
NAV LIGHTS are to remain ON at all times

12 INDICATOR LIGHTS .... TEST

You'll have to hold the switch in the test position. Usually you are leaning over the console and the other pilots are doing their thing, then just when you are at full body extension, you have to somehow bend around and check all those lights. wheeeew.

NOTE: *5 UP-5 DOWN rule.* There are 5 lights up and 5 down that do *NOT* come on. They are associated with the fire panels.

... and PROCEDURES FOR STUDY and REVIEW **ONLY!**

DON'T FORGET THESE STEPS !

I have indicated these two steps with the two exclamation marks on the L!!

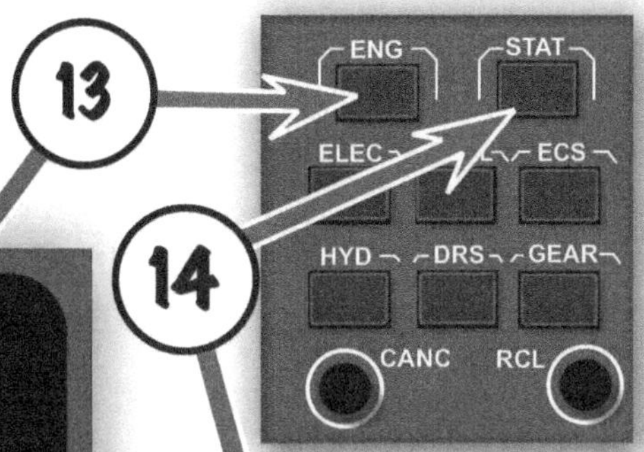

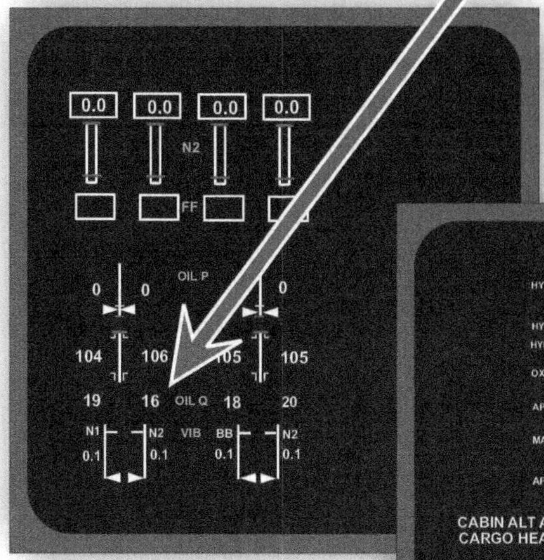

check:
ENG OIL >18 Qts.
ENG INDICATORS NORMAL

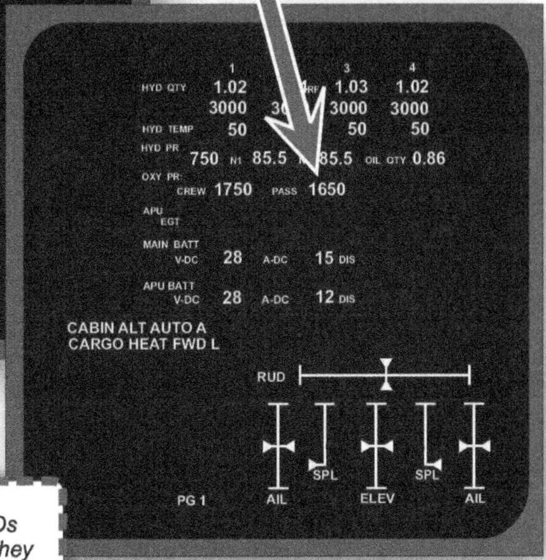

> **NOTE:**
> Some instructors teach that F/Os should check these items when they first get to the cockpit to reduce departure delays; if service is required.

GOOD IDEA!

check:
STATUS MSGS
HYD QUANTITIES
BOTH O2 PRESS > 1650 psi

---

13   EICAS ENG PAGE .... CHECK

On the EICAS DISPLAY SELECTOR PANEL select ENG. Check the LOWER EICAS CRT FOR:
    ENG OIL QUANT ... >18 qts
    ENG INDICATIONS NORMAL

---

14   EICAS STATUS PAGE .... SELECT

On the EICAS DISPLAY SELECTOR PANEL select STAT. Check the LOWER EICAS CRT FOR:
    NOTE ANY STATUS MSG and RESOLVE
    HYD QUANT above RF (>74%)
    BOTH OXYGEN PRESS above 1650 psi

Start at top of center row and make a "U.".

FIRE OVERHEAT TEST SWITCH

Called the "U"

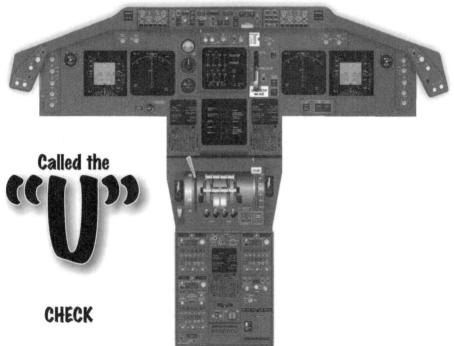

CHECK

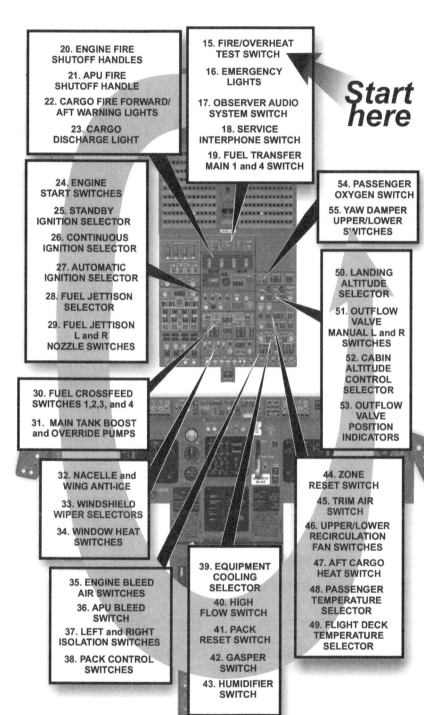

747-400 *SIMULATOR TECHNIQUES ...*

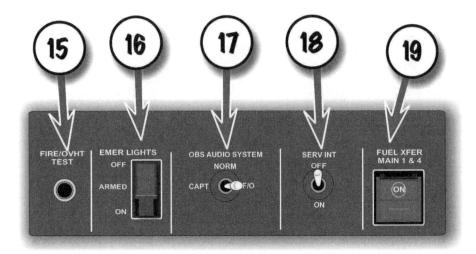

15   FIRE/OVHT TEST switch .... PUSH/HOLD

PUSH and HOLD until the EICAS changes from **"TEST IN PROG"** to **"FIRE TEST PASSED"**
OBSERVE THE FOLLOWING:
    13 lights from the upper and lower fire/ovht panels
    and the Master Warning lights.

16   EMERGENCY LIGHTS switch ...... ARMED

There is a little "paper clip" under the guard and if that is bent or missing, when you close the guard the switch won't arm. So, the technique is to place the toggle switch to arm, and **THEN** close the guard.

17   OBS AUDIO switch .... NORMAL

This IS **NOT** the switch that hooks the radios up to the "entertainment" system. That switch is down on the console. This "OBS" refers to the cockpit observer.

18   SERVICE INTERPHONE switch .... OFF

This has to do with maintenance.

19   FUEL XFER MAIN 1 and 4 switch .... OFF

These switches are **NOT** part of the NORMAL fuel management system. New guys sometimes confuse this switch with the FUEL CROSSFEEDs selection which is located on the Fuel panel. These are not part of the pilots **NORMAL** cockpit operation.

20 ENG FIRE SHUTOFF handles .... IN

There are actually three things to do:
- HANDLES all the way in
- LIGHTS in handles OUT
- A and B BOTTLE low press (discharge) lights OFF

21 APU FIRE SHUTOFF handle .... IN

Light in the handle should be OFF and
the APU bottle discharge light (low pressure) should be out

CARGO FIRE FWD/AFT WARNING LIGHT ...... OFF

22 The Checkguy LOVES this system. What are some questions he could ask? Here is a quick system review.

There are four bottles of extinguisher (A,B,C,and D). Each is armed with its own squib and has its separate discharge port. Pushing the **CARGO FIRE ARMED** switch selects either the fwd or aft cargo pit and arms all four bottles. Pushing the **CARGO FIRE DISCH** switch discharges bottles A and B into the selected cargo area.
The C and D bottles discharge *AUTOMATICALLY 30 minutes* later OR on landing.
The bottles provide about *2 hours and 30 minutes* suppression time.

23 CARGO DISCHARGE light .... OFF

The EICAS msg:
BTL LOW CGO A and B; and/or
BTL LOW CGO C and D msg will be illuminated when the bottle pressures are LOW.

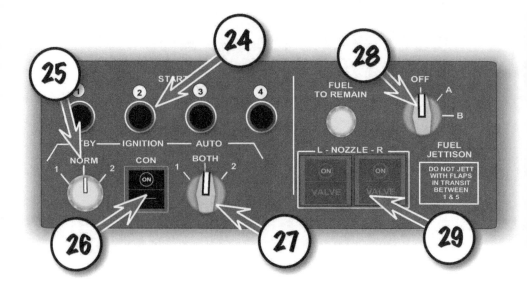

24   ENG START SWITCHES .... IN

Verify that the **START VALVE LIGHTS** are not illuminated.

The start light illuminates whenever the START VALVE is open.
If this light is illuminated, then we must assume that the start valve is stuck open and contact maintenance.

25   STBY IGNITION SELECTOR .... NORM

If this switch is in NORM and the entire system fails, then the MAIN STBY BUS will *CONTINUOUSLY* provide power to IGNITER 1 when the FUEL CONTROL SWITCH is in RUN.
What that said was this ... in order for the automatic backup system to operate, this switch must be in NORM.

26   CONTINUOUS IGNITION SELECTOR ..... OFF

This causes the igniters selected with the AUTO IGNITER switch to operate continuously, whenever the FUEL CONTROL SWITCHES are ON.
This is used when penetrating weather, when you are using engine anti-ice, when encountering turbulence, or when ever you feel like it.

27 AUTO IGNITION SWITCH .... 1 or 2

SELECT IGNITER 1 for ODD FLIGHTS
SELECT IGNITER 2 for EVEN FLIGHTS.
The concept here is to alternate use of the igniters so as to equalize wear cycles.

What these guys do is just set up the automatic system. The ignition system isn't activated unless:

FUEL CONTROL SWITCH ... RUN
ENGINE IN START CYCLE (less than 50% N2)
TE FLAPS DOWN (or more correctly, not up)
NACELLE ANTI-ICE ... ON

Here's an ORAL QUESTION: *When do the engine igniters operate?*

28 FUEL JETTISON SELECTOR .... OFF

When A or B is selected; then the EICAS FUEL TO REMAIN display is displayed, and the JETTISON TIME is displayed on the FUEL SYNOPTIC.

29 FUEL JETTISON NOZZLE selectors .... OFF

If this switch is pushed, the associated nozzle valve opens.

CAUTION
The FUEL JETTISON VALVES can be opened with these switches even though the FUEL JETTISON selector is OFF.

With the selector switch in either A or B and this switch is pushed, then a whole lot of other stuff happens. *You will not be required to recite them from memory, but here is the list FYI.*

6 OVERRIDE/JETTISON pumps are activated
    in tanks 2 and 3 and CENTER TANK.
4 Jettison valves open.
2 STAB TRANSFER/JETTISON pumps.
4 STAB TRANSFER/JETTISON valves open.
ARMS 1 and 4 MAIN TRANSFER VALVES
    (these guys open when tank 2 or 3 reach 20,000#)
Opens associated NOZZLE valve.

747-400 SIMULATOR TECHNIQUES ...

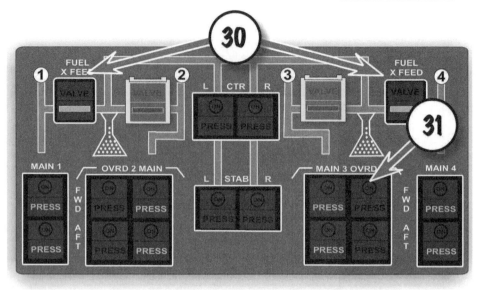

Before the engine start with the **FUEL PUMP SWITCHES ALL OFF** ... the **LOW PRESSURE** lights are illuminated only on the **MAIN TANK PUMP** switches, and, the **LOW PRESSURE** lights are extinguished on the **OVERRIDE, CENTER TANK**, and **STABILIZER TANK PUMP** switches.

the 120K switches

30 FUEL X-FEED SWITCHES 1 and 4 (NOT 2 and 3)

IF the fuel load is less than 120,000 pounds:
        1 and 4 X-FEEDs should be ..... CLOSED
IF the fuel load is greater than 120,000 pounds:
        1 and 4 X-FEEDs should be .... OPEN

Re: **2 and 3 X-FEEDS**. I cannot think of a time when you will have to **CLOSE** them. It is **NOT** part of the normal fuel panel operations to **EVER** raise the guards and **CLOSE** those switches. If, of course, the plastic guards are open and the X-feeds are selected, the **#2 and #3 X-FEED** valves should be "**OPENED**" and the guards closed.

PERSONAL NOTE: On an operational line trip, I would certainly like to know why those X-feeds were closed and involve maintenance in reviewing the situation. However, for a checkride situation, mention it to the CHECK-PERSON if they are closed.

31 MAIN TANK BOOST PUMPS and OVERRIDE PUMP SWITCHES (ALL) ... OFF

NOTE: If the APU is running, on airplanes with **HYDRO-MECHANICAL SCAVENGE PUMPS**, the **#3 AFT FUEL BOOST PUMP** should be **ON**. The reason is that with the APU running, unwanted fuel transfer could occur. This will (probably) NOT be a required item on the checkride.

... and PROCEDURES FOR STUDY and REVIEW ONLY!

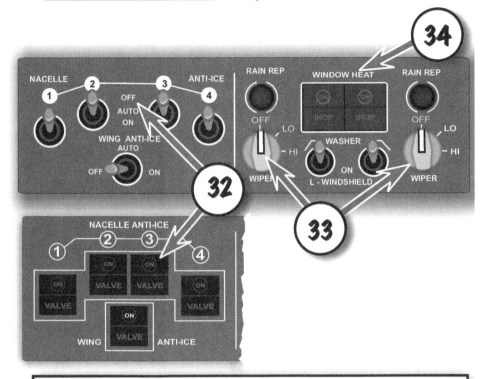

32   NACELLE and WING ANTI-ICE switches and VALVE LIGHTS .... OFF

ENG NACELLE ANTI-ICE operates **ONLY** when the engine is running and causes continuous ignition to operate.
WING ANTI-ICE does **NOT** operate on the ground and is worthless whenever flaps are extended.

NOTE: There are a couple of different panel layouts, including these two

33   WINDSHIELD WIPER SEL .... OFF

34   WINDOW HEAT switches .... ON

ONLY the forward windows (CAPT and F/O 1L and 1R) get anti-ice and anti-fog protection through these switches. The side windows (2L,2R,3L,3R) get anti-fog only and they are on anytime the airplane is powered.

If the INOP light comes ON, cycle the switch OFF for 10 seconds to reset.

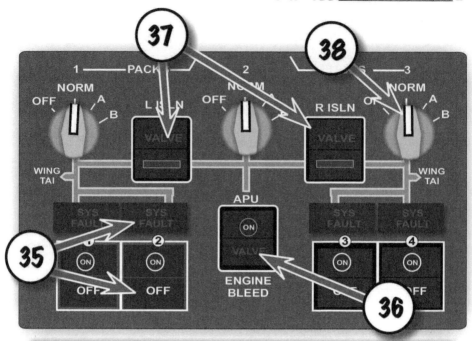

35 — ENG BLEED AIR switches .... ON

Verify that the **BLEED OFF** lights are **ON**, and **SYS FAULT** lights are **OFF**.

36 — APU BLEED switch .... ON

Verify that the VALVE light is .... OFF (not illuminated)

37 — LEFT and RIGHT ISOLATION switches ..... OPEN

Verify that the VALVE BARS are in view and that the VALVE lights are OFF.

38 — PACK CONTROL SELECTORS .... NORMAL

There are TWO CONTROLLERS for each pack. By selecting NORMAL, you have set the system up so that if one fails, the other takes over automatically. It automatically selects alternate controllers on alternate flights.
If we select A or B, then it becomes the primary controller and the automatic backup system selects the other one in the event of failure.
Going to OFF momentarily resets the protection logic in the event of a problem.

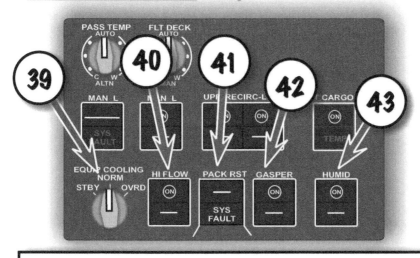

39   EQUIP COOLING SELECTOR .... NORMAL

In the NORMAL position, prior to engine start, based on OAT, the cooling air is either exhausted overboard or into the forward cargo compartment. With a "SINGLE INTERNAL FAULT," equipment cooling valves close to allow internal circulation of the air.

40   HIGH FLOW SWITCH .... OFF

OFF allows PACK AIR to be controlled automatically.

41   PACK RESET SWITCH .... OFF

Verify SYS FAULT light is OFF.

42   GASPER SWITCH .... ON

43   HUMIDIFIER SWITCH .... ON

Not available on N105UA and N106UA.
Water comes from the potable water system. The system automatically comes on at top of climb and shuts down 2 hours prior to top of descent.

Here are a couple of things to be aware of:
(1) On some airplanes, shortly after takeoff, a strong odor can be detected near door 1L. It is usually associated with the humidifier and may be reported by the Flight Attendants as an "electrical" smell.
(2) Sometimes it makes a loud vibration or humming noise.

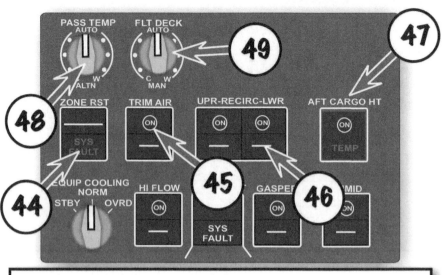

44   ZONE RESET SWITCH .... OFF
Verify SYS FAULT light is OFF

45   TRIM AIR SWITCH .... ON

46   UPPER and LWR RECIRC FANS .... ON
There are 4 RECIRC FANS (2 overhead and 2 under floor). In cruise, when the packs go to normal flow, the LWR RECIRCS come on to supplement normal flow.

47   AFT CARGO HEAT switch .... OFF
Aft cargo compartment temperature is controlled using BLEED AIR. OFF position shuts off air to the compartment. In ON position, the temperature is controlled automatically using bleed air.

48   PASSENGER TEMP SEL .... AUTO
Put the indicator straight up for new guys. You old timers, you know what you want to do.

49   FLIGHT DECK TEMP SEL .... AUTO

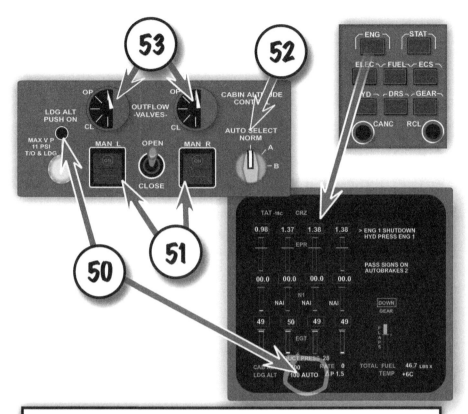

50 LANDING ALTITUDE SELECTOR .... AUTO
Select **ENG DISPLAY** on the **LOWER EICAS** and
If necessary "**TOGGLE**" the "**LDG ALT PUSH ON**" button (if installed) and
VERIFY that the word **AUTO** is annunciated in the lower part of the **EICAS**.

51 OUTFLOW VALVE MAN L and R .... OFF

52 CABIN ALTITUDE CONTROL SEL .... NORM
Automatically selects A or B controller on alternate flights. Also automatically selects the secondary controller in the event of malfunction of the primary controller.

53 OUTFLOW VALVE INDICATORS .... OPEN
The needles should be pointing **UP**.

The **OUTFLOW VALVE** controllers read the landing airport from the airport entered in the **CDU** as destination, and calculates landing altitude from that. If you should have to divert ... remember to place the diversion airport in the **CDU**.

747-400 SIMULATOR TECHNIQUES ...

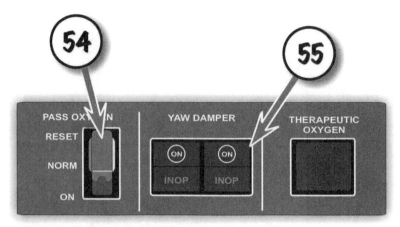

NOTE: Some airplanes have a "**THERAPEUTIC OXYGEN**" selector installed.

---

54 **PAX O2 SWITCH** .... **NORMAL**
Verify that the switch is:
 GUARDED and
 SAFETIED

QUICKIE OXYGEN SYSTEM REVIEW

The system **AUTOMATICALLY** activates if the **CABIN ALTITUDE** exceeds **14,000 Feet**. It can be activated **MANUALLY** when the cockpit **PASS OXYGEN** switch is selected to the **ON** position.

The masks are located in Passenger Service Units (**PSU**) the seatback or overhead panel within reach of the passengers;as well as in the lavatories, Flight attendant stations, and the crew rest areas. Now here is something that you should know:
On "SOME" airplanes the oxygen flow begins immediately when the masks are deployed; but on others, the mask must physically be pulled by the user so that a "trigger" is activated.

To **STOP** oxygen flow and reset the system: Hold the **PASSENGER OXYGEN** switch to the reset position momentarily. This will electrically **CLOSE** all **FLOW CONTROL UNITS** if the **CABIN ALTITUDE** is below **12,000 Feet**. There is NO manual reset capability for the oxygen system.

If below 14,000 Feet, the oxygen will flow at a reduced rate commensurate with the cabin altitude.

The pressure in the oxygen bottles is indicated on the secondary **EICAS** panel.

---

55 **YAW DAMPER UP/LWR switches** .... **ON**
The INOP lights will be illuminated until IRS alignment is complete. The first IRS to align will illuminate both YAW DAMPER "ON" lights and extinguish the "INOP" lights.

... and PROCEDURES FOR STUDY and REVIEW **ONLY!**

SIDE BAR *discussion*

Since we have mentioned setting up a **DIVERT** airport as our **DEST** (destination) in the **CDU** in order to set up the **LANDING** altitude for the **PRESSURIZATION** system, I thought this might be a good place to more clearly describe that process. Since once the **FMC** recognizes the new landing altitude, it will automatically position the outflow valves to conform to cabin altitude rate limits and differential pressure limits so as to achieve proper landing cabin altitude during the climb-out and descent.

HOW TO ...
SET UP THE CDU/FMC FOR A DIVERT

Anytime you are required to **CHANGE DESTINATION** ... such as taxi-out during a diversion, selecting an alternate, or if an emergency requires landing at some other place than your destination ... here is how to tell the **FMC** (Flight Management Computer) where you are going. This will allow a whole lot of data from the onboard database to be processed and available for your use. Altitude of the airport, runways available, approach profiles, distance, fuel burn, routing, and all that. Simple to do once you have seen how it can be done. Let's take a look.

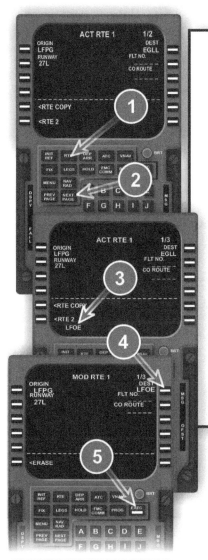

5 STEPS to DESIGNATE THE DIVERT AIRPORT AS DESTINATION

STEP 1: Select the **RTE** page key on the **CDU**.

STEP 2: Select the **NEXT PAGE** key on the **CDU**.

STEP 3: TYPE the letters for the divert airport into the **SCRATCH PAD (SP)**. For our example, we will type **LFOE**.

STEP 4: **LINE SELECT** CDU KEY 1 RIGHT (this is written in shorthand like this: **LS1R**). This will replace the previous destination airport (**EGLL**) with our new destination **LOFE** in the **CDU** display.

STEP 5: *IMPORTANT!!!* Depress the **EXECUTE** key. The **EXECUTE** light should be illuminated, and until you select it, even though the new destination is shown on the screen, it is NOT placed in the computer until you have "**EXECUTED**" it. Whenever the execute light is illuminated, it means that the screen information has not been inserted and that you will have to **EXECUTE** it before it becomes a part of the computer program.

You can now select the **DEP ARR** page and select from the available runway/approach routing options.

© MIKE RAY 2014
WWW.UTEM.COM

747-400 SIMULATOR TECHNIQUES ...

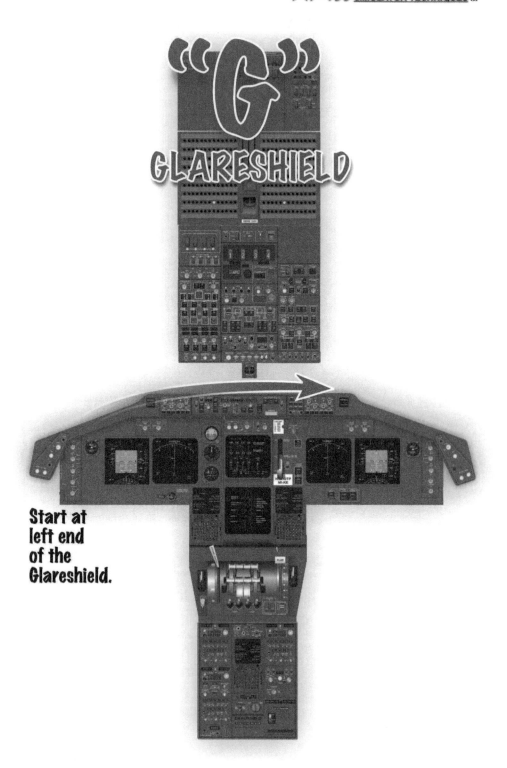

GLARESHIELD

Start at left end of the Glareshield.

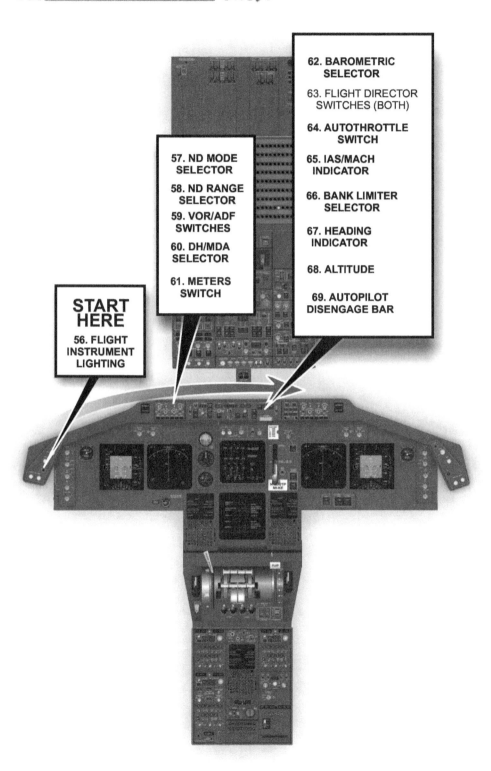

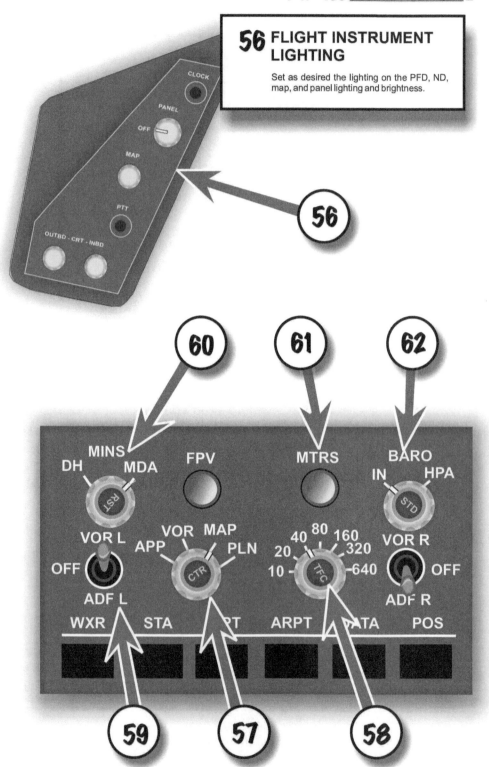

57 ND MODE SELECTOR .... MAP

The ND controls are located on the **EFIS CONTROL PANEL**. The ND MODE Selector has the following features:
    When inner knob pushed, **FULL COMPASS ROSE** displayed.
      Pushing it again, restores **EXPANDED MODE**.
    OUTER knob used to select MODE.
    APP when selected sets up the displays for the approach;
      including ILS freq, identifier, course, and DME.
      Displays EXPANDED COMPASS MODE.
    APP with CTR pushed ... restores COMPASS ROSE display.
    VOR when selected sets up the displays for the VOR approach;
      including VOR freq, identifier, course, DME, and
      TO/FROM indication. Displays EXPANDED COMPASS MODE.
    VOR with CTR pushed ... restores COMPASS ROSE display.
    MAP Selects TRACK UP display, EXPANDED compass mode,
      FMC map information, airplane position and heading,
      active waypoints and vertical path deviation at T/D.
    This is the mode you will be operating in MOST of the time.
    MAP with CTR pushed ... restores COMPASS ROSE display.
    PLN selects a static TRUE NORTH up display.
      Good for route displays only.

58 ND RANGE SELECTOR .... AS DESIRED

Either 10, 20 or 40 nm. This gives the best after takeoff information and allows display of the airport symbol.

59 VOR/ADF SWITCHES .... AS DESIRED

Allows the VOR/ADF information to display on the ND.

60 DH/MDA SELECTOR .... AS REQUIRED

OUTER knob - *Selects either DH or MDA on PFD.*
MIDDLE knob - *Selects the ALTITUDE for the PFD display.*
INNER knob - *PUSH to reset the DH ALERT on the PFDS.*

61 MTRS SWITCH .... AS REQUIRED

PUSH - *to display METERS on the PFD ALTITUDE display.*
Second PUSH - Removes meters from the display.

62 BARO SELECTOR .... AS REQUIRED

OUTER knob - *Selects IN or HPA in the PFD.*
MIDDLE knob - *Selects the numerical barometric setting.*
INNER knob - *PUSH and STD is annunciated on the PFD. This allows the anticipated barometric values to be preset while the unit displays 29.92 Push again, and the preset value is displayed..*

747-400 SIMULATOR TECHNIQUES ...

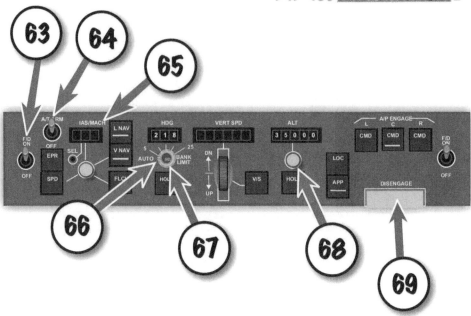

63 F/D SWITCHES .... BOTH ON

ON - allows diplay of the FLIGHT DIRECTOR COMMAND BARS under NORMAL conditions.
If they are already ON, cycle them OFF and then back ON.
ON THE GROUND - Turning one switch ON and the other switch OFF, activates **TO/GA** roll and pitch modes.
IN FLIGHT - The first switch ON, if no autopilot engaged, default modes are:
  HEADING HOLD, and
  VERTICAL SPEED, or
If bank angle greater than 5 degrees:
  ATTITUDE, and
  VERTICAL SPEED.

64 AUTOTHROTTLE SWITCH .... OFF

A/T ARM - *Arms the autothrottles for engaged if appropriate modes are operating.*
  NOTE: *The autothrottle disconnects if:*
    more than one engine is inoperative,
    a dual FMC failure, or
    an EEC is in the ALT MODE.
OFF - *Disarms autothrottle ... preventing engagement.*
It also prevents ENGINE TRIM EQUALIZATION.

65 IAS/MACH INDICATOR .... SET

Set V2 if known, do not make rapid inputs when setting V2 for takeoff.

It displays **200 KTS** when first powered on the ground. The selected speed shows up on the SPEED TAPE of the PFD.

66 BANK LIMIT SELECTOR .... AUTO

AUTO - Limits the bank angles for the AFDS in HEADING SELECT MODE between 15-25 degrees, depending on the TAS, flap position, and V2.
5 - 10 - 15 - 20 - 25 - Limits the bank angle regardless of AIRSPEED.

NOTES:
ATC expects that while in the terminal areas and below 18,000 feet that all turns will be made at 25 degrees. Checkpilots may FROWN on using the AUTO mode when working below 18,000 feet for that reason.

Bank angles at altitude are critical and can increase the airplane's susceptibility to "STALL." For that reason, large bank angles at altitude with a heavy airplane are to be avoided.

67 HDG INDICATOR .... SET

This knob has TWO functions.
FIRST: It can be rotated to set the desired HEADING. This will be displayed on the PFD, and on the ND as the "BUCK TEETH."
SECOND: When PUSHED in, the HEADING SELECT MODE is engaged and if the autopilot is engaged, the airplane will turn towards the selected heading.

> **CAUTION:**
> TO/GA roll mode should NOT BE USED AFTER TAKEOFF if the departure clearance specifies "MAINTAIN RUNWAY HEADING."

68 ALTITUDE .... SET

Set in the first restriction, if known. This would be on the SID, or assigned by ATC, or part of the clearance.

69 AUTOPILOT DISENGAGE BAR ...... ARM (UP)

Check that the bar is engaged in the UP position.
Would you believe it ... flights have departed and, being unable to engage the autopilots, were subjected to lots of stress ... until they realized that this "bar" was out of position.

"O" OXYGEN

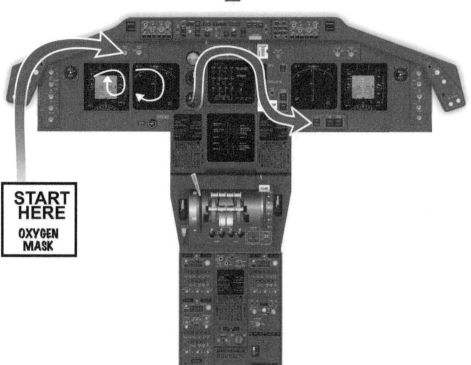

START HERE — OXYGEN MASK

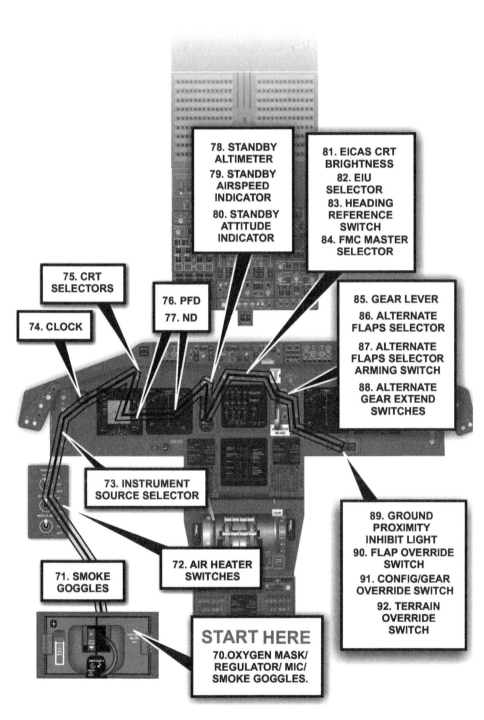

747-400 SIMULATOR TECHNIQUES ...

THE INCREDIBLY COMPLICATED OXYGEN MASK SET-UP

70 CHECK .... OXYGEN MASK, REGULATOR, and MICROPHONE.

One of the MOST complicated procedures on the Set-up is the Oxygen mask checkout. In brief form, here is what to do:

> Select FLT interphone
> Select SPKR on
> Adjust FLT and SPKR volume

STEP 1: PUSH and hold DOWN, RESET/TEST lever. and while holding it down ...

STEP 2: Observe **YELLOW CROSS** flicker.

STEP 3: Push and Hold **EMER/TEST** selector and observe **YELLOW CROSS** appear steadily;

STEP 4: While still holding the **RESET/TEST** lever Push the **PTT (push to test)** switch and
verify the sound of **O2** flowing across the mask mic.

STEP 5: ... then Release the **EMER/TEST** selector
Release **RESET/TEST** button:
Observe **YELLOW CROSS FLOW IND** is **BLANK**
and **O2 Flow** stops.

STEP 6: Verify that the **N/100% lever** is

Now, put it all back together:

> MASK O2 @ 100%
> MASK HOLDER doors closed
> MASK/BOOM switch to BOOM

WARNING
Removing the mask from the container to check for O2 flow and Mask Mic operation is neither desired nor required. Once the mask is out of the container, it requires considerable effort to get it all back together. It is a lot like trying to cram a 6 foot Boa Constrictor into foot long lunch box!

71 CHECK .... SMOKE GOGGLES

Check the SMOKE GOGGLES for clarity, glazing, proper fit, and serviceability. If they are wrapped in plastic, remove them and stow them for quick accessibility.
When you re-stow the goggles, make certain that they are within arms reach.

72 FOOT, SHOULDER, and WNDSHLD HEAT SWITCHES .... AS DESIRED

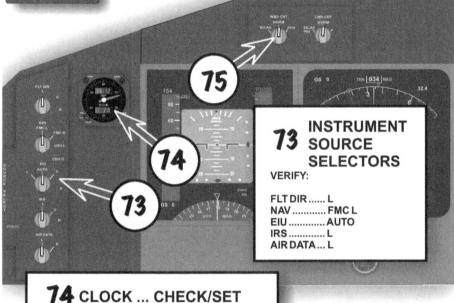

73 INSTRUMENT SOURCE SELECTORS

VERIFY:

FLT DIR L
NAV .... FMC L
EIU .... AUTO
IRS L
AIR DATA ... L

74 CLOCK ... CHECK/SET

OBSERVE ... Hrs and MINs correct
PUSH DATE SWITCH, observe correct
PUSH again, observe second hand at zero.
Position ET sel to RESET,
Release to HLD, observe ET/CHR display blanks

75 CRT Selectors .... NORM

INBD CRT selector and LWR CRT selectors should be set to Norm.

IRS ALIGNMENT MUST BE COMPLETE TO CONTINUE ➡

747-400 SIMULATOR TECHNIQUES ...

> **IRS ALIGNMENT MUST BE COMPLETE TO CONTINUE**

76 PFD .... CHECK

Check PFD is displayed on the CRT and there are no flags. Then:

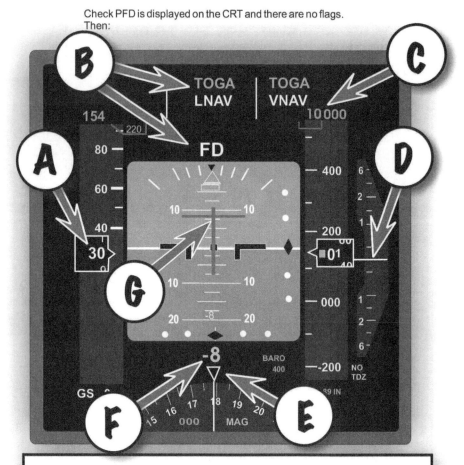

A: AIRSPEED IND .... 30 KTS

B: FD and TO/GA .... Annunciated

C: ALTITUDE IND .... Alt set in MCP

D: VERTICAL SPEED .... ZERO with NO FLAGS

E: HEADING .... Agrees with MCP

F: RADIO ALTITUDE .... -8

G: ATT and COMD BARS ..... level and 8 deg UP

... and PROCEDURES FOR STUDY and REVIEW ONLY!

!-------------------------------!
: **IRS ALIGNMENT MUST**   :
: **BE COMPLETE TO CONTINUE**   :
!-------------------------------!

77 ND .... CHECK

Check ND is displayed on the appropriate CRT and there are no flags. Then:

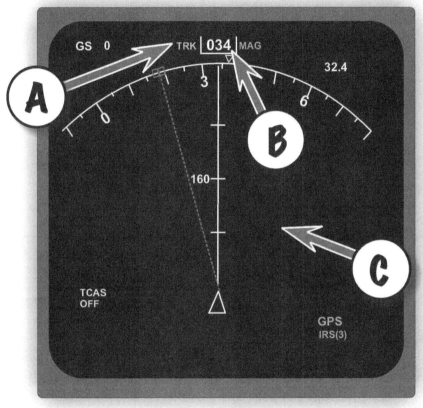

NOTE:
PLAN MODE will work even if IRS' are NOT aligned.

A: **PROPER HEADING AND TRACK**

B: **CHECK HEADING agrees with STBY COMPASS**

C: Verify **NO FLAGS** showing.

page 133

the STANDBY INSTRUMENTS !

These three instruments can be expected to operate for 30 minutes when everything else fails.

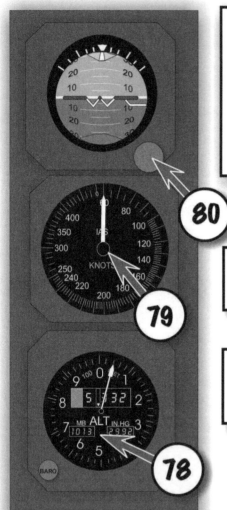

80 STBY ATT IND ... CHECK
Observe NO FLAGS and LEVEL ATTITUDE.
Cage if necessary.
NOTE:
If unable to cage, or attitude continually precesses after caging, suspect battery discharged and get maintenance involved.

79 STBY IAS ... CHECK
Observe "ZERO" indication.

78 STBY ALT.... CHECK
Set current altimeter setting and observe "proper" altitude displayed.

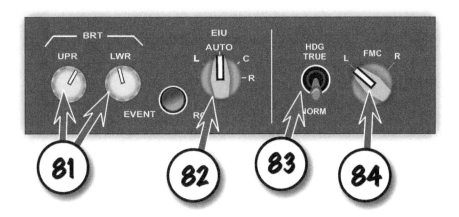

81 EICAS CRT BRIGHTNESS .... as desired
The interior knob on the right CRT controls the WX display portion of the display.

82 EIU Selector .... AUTO
Selecting the AUTO position prepares the EIU to automatically shift to the other symbol generator should the original symbol generator fail. Without this in auto, the CRTs on the Captains side would go blank if the symbol generator failed. If, however, only one CRT fails, this is an indication of the failure of the CRT and there is no way to fix that in the air. Call SAMC and the other guy flies.

83 HEADING REF switch .... NORM
The TRUE setting is used only two times that I can think of:
    (1) Vectors in NCA (Northern Canadian Airspace) are given in true, and
    (2) Flying an approach in Greenland requires the use of 'true" headings.

84 FMC MASTER selector L
BIG DEAL! This selects the Left FMC as master.

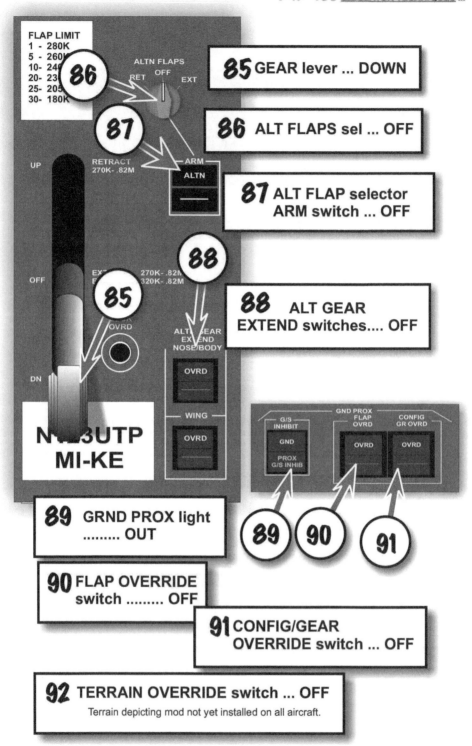

MAXIMUM FLAP SPEEDS

Here are some of the LIMIT AIRSPEEDS for this airplane.

MAXIMUM FLAP EXTENDED SPEEDS (V_{FE})

The V_{FE} are the certified design **MAXIMUM** speeds for extending the flaps. The speeds are usually found on the **PLACARD** just above the **GEAR LEVER**.

FLAPS	1	5	10	20	25	30
V_{FE}	280	260	240	230	205	180
V_{FE} - 10	270	250	230	220	200	170

However, extending the flaps at those speeds has proven to result is an excessive wear on the flap linkages and related fragile mechanisms. Also the **FLAP OVERSPEED** warning can be triggered and that requires an extensive inspection after landing. As a result, most airlines prefer for the pilots to us the V_{FE} - **10** Knots airspeed restriction to minimize those issues. So it is intended that the pilots commit these V_{FE} - **10** Knot figures to memory ... and they are the values that will probably be a part of the **ORAL EXAM**.

> **MAXIMUM FLAP EXTENDED ALTITUDE**
> **20,000 Feet**

MINIMUM CMS SPEED

> **30 REF + 80 KNOTS**

NOTE: If the **GROSS WEIGHT** of the airplane **EXCEEDS 750,000** Pounds
... **LIMIT BANK ANGLE** to **15** Degrees until established at **30 REF + 100 KTS**.
The **AFDS** (Autopilot Flight Director System) will automatically limit the bank for you until reaching **30 REF + 100 Knots**. This means that the **AFDS** Flight director commands can be followed.

MAXIMUM GEAR SPEEDS

MAXIMUM GEAR SPEEDS (V_{LO} and V_{LE})

The V_{LO} are the certified design **MAXIMUM** speeds for extending the **LANDING GEAR**. Notice the limiting airspeed for flying around with the gear extended (V_{LE}) is significantly different from the operating speeds once the gear has been extended; however, the **MACH** number (**.82**) is the same..

V_{LO} - **MAXIMUM RETRACTION AIRSPEED**	270 / .82
V_{LO} - **MAXIMUM EXTENSION AIRSPEED**	270 / .82
V_{LE} - **MAXIMUM EXTENDED AIRSPEED**	320 / .82

"S" SPEED BRAKE

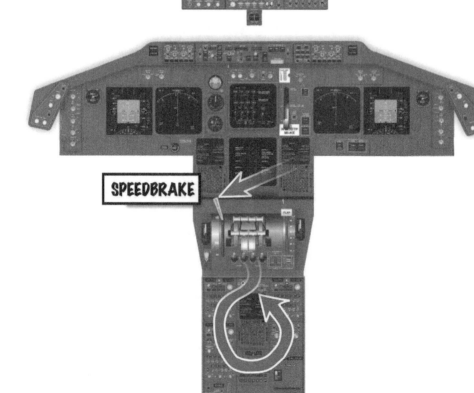

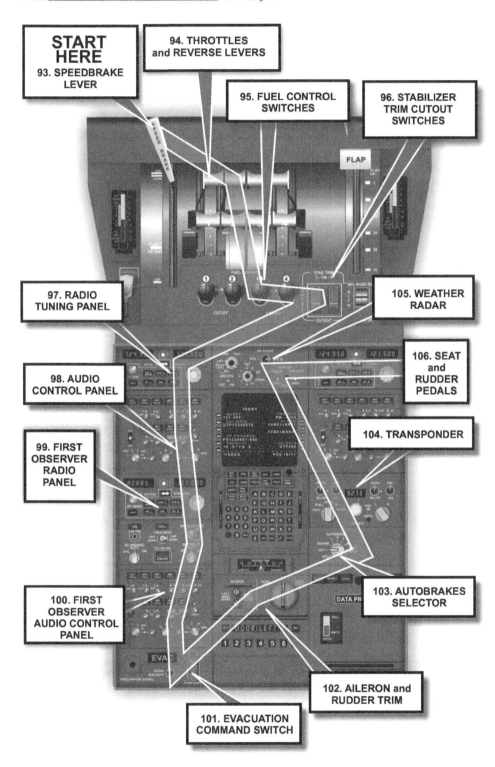

93 SPEED BRAKE LEVER .... DETENT

The lever MUST be full forward. Some pilots will bump the lever with the palm of their hand to ensure that it is seated.
If it is not, it will trigger the **TAKEOFF WARNING** horn when the throttles are advanced for takeoff.

94 THROTTLES and REVERSE LEVERS .... CLOSED/DOWN

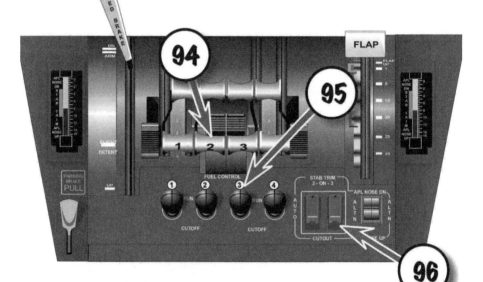

95 FUEL CONTROL SWITCHES .... CUTOFF

When the **FUEL CONTROL SWITCH** is in cutoff:
   FUEL VALVE and **SPAR VALVE** close,
   IGNITION discontinues,
   HYDRAULIC DEMAND PUMP operates (if selected)

96 STAB TRIM CUTOUT switches .... AUTO

Verify switch guards closed. On some of the airplanes, the guards are fitted with a safety wire. It is likely that, in the simulator, the wire will be broken.

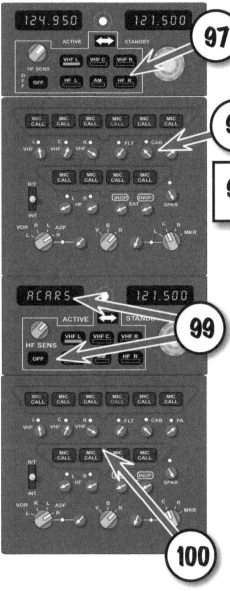

97 CAPT RADIO TUNING PANEL ... SET

*Observe the **OFF** light is **NOT** illuminated.*
Set the radio panel as desired.

98 AUDIO CONTROL PANEL ...... SET

99 FIRST OBSERVER RADIO TUNING PANEL ...... SET

*Observe **OFF** light is **NOT** illuminated.*
*Normally, this radio will be set up with **ACARS** in the active frequency window since ACARS will nearly always be used.*

100 FIRST OBSERVER AUDIO CONTROL PANEL .... SET

*Verify **FLT MIC** is selected.*

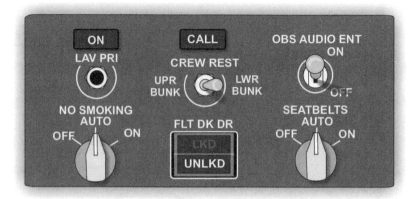

NO SMOKING SIGN selector .... ON

Usually, **ALL** Flights are designated as **NON-SMOKING** Flights.

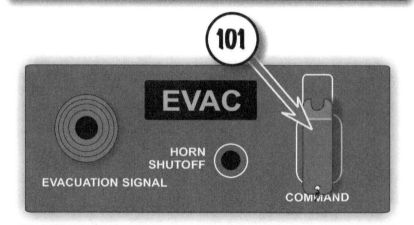

101 EVAC COMMAND SWITCH .... OFF

Verify switch guard closed.

SYSTEM STUFF:
*Any station on the airplane with an alarm control panel can initiate an alarm. Once initiated, every alarm station on the airplane will sound. Each individual station can extinguish the alarm **AT THAT STATION ONLY!**
What that means to us as pilots is this. If we should initiate the alarm from the cockpit, we need only to let it sound only momentarily to confirm operation. We may then shut off the alarm using the switch and feel confident that the alarm will continue to sound throughout the airplane.*

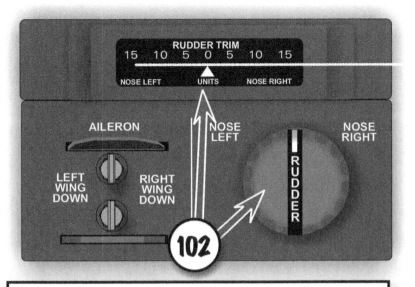

102   AILERON and RUDDER TRIM .... ZERO

If the RUDDER TRIM indicator indices are not visible or are not working; reset the circuit breakers. If you look at the rudder trim left over from the last leg and see a coupla degrees; suspect that the trim will be needed and take out only about one half of the indicated trim.

To check the AILERON trim, use the yoke indices and operate the trim switch until the yoke indice matches the zero mark.

DISCUSSION:
The rudder trim may be operated even when the AUTOPILOT is operating, and in the case of an engine out operation, it is essential that you continue to trim, even though you have the autopilot operating. The technique, in that case, is to "trim the sailboat" on the PFD. It is also desirable to use the foot pedals to "fly the sailboat" even with the autopilot engaged. If the "sailboat" is kept aligned, the yoke will be more or less level. The autopilot will not trim the rudder. This becomes critical during engine out turns using only the autopilot, you will have to "push" ot trim he rudder..

RUDDER BIAS: *However, during Auto-coupled approaches using multiple autopilots below 1500 feet, the rudder is engaged and operated by the autopilot. It is the **ONLY** regime where the autopilot operates the rudder.*

BEWARE - CAUTION

SERIOUS PROBLEM! *During a CAT II or CAT III approach with multiple auto-pilots, on engine out go-around the autopilot will trim the rudder so as to keep the airplane in trim.* **BUT!!!!!** *The autopilot will "kick off" rudder pressure that it had been holding and allow the rudder to center when:*
  -Above 400 feet and another roll mode selected
  -when ALT capture annunciated
 -when autopilot shut off.

YIPE!

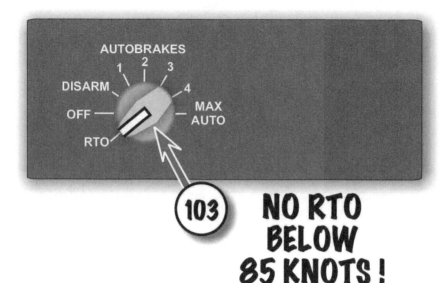

NO RTO BELOW 85 KNOTS!

YIPE!

103 AUTOBRAKE SELECTOR .... RTO

The RTO setting automatically applies **MAXIMUM BRAKING** when:
- **THROTTLES** moved to **IDLE**
- above 85 **KNOTS**
- **AFTER** take-off roll initiated.

WARNING: Since the **NORMAL BRAKES** are supplied from the **#4 HYDRAULIC SYSTEM**, and the **ALTERNATE BRAKES** system has all the capabilities of the normal systems except **AUTO-BRAKES**: It is **IMPORTANT** when you bring the hydraulic systems on line, that the **#4 HYDRAULIC PUMP** is allowed to pressurize the system before the **#1** system is brought on line. A "shuttle valve" arrangement may allow the system to be pressurized by the **#1** system, and therefore you would not have **RTO** and wouldn't know it. **NOT GOOD!**

... and PROCEDURES FOR STUDY and REVIEW ONLY!

104 TRANSPONDER TEST AND SET-UP

A TCAS system test can only performed on the ground.
IRSs must be in the NAV mode.

Holding the switch in the test position will allow FAILED systems to be listed on the ND.

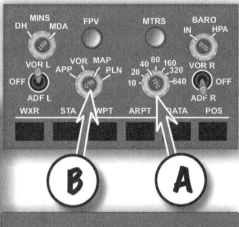

To set up for the **TCAS SYSTEM TEST:**

A ND RANGE SEL-10 NM

B ND MODE SEL-MAP

Observe the **OFF** light is *NOT* illuminated.
Set the radio panel as desired.

C TCAS/ATC switch ..... TA

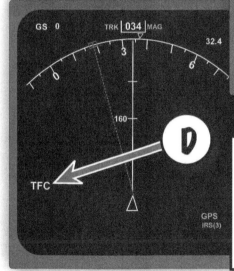

ND .... verify TFC displayed

If TFC is not displayed, push the TFC button which is skillfully disguised as the center of the ND range selector knob.

NOTE: *The traffic display will be interrupted for several seconds after the TCAS/ATC selector is once again moved to TA or TA/RA.*

After the TEST is completed

E **SET TRANSPONDER to NORMAL/ABOVE**

Consider using the ABOVE setting for departure. This accomplishes two things: it declutters the display, but more importantly, it allows to look ahead for traffic as we climb into it.

DISCUSSION:
If the **TCAS/XPDR** selector is held in "**TEST**" or positioned to "**TRANSPONDER, TA, or TA/RA**" after the test is started, the transponder function will be interrupted, but the test sequence will continue normally. **TCAS** is inhibited any time the transponder is in the "**TEST, STANDBY, or TRANSPONDER MODE.**" The traffic display on the **ND** is inhibited for up to 12 seconds after the selector is moved to **TA** or **TA/RA**.

© MIKE RAY 2014
WWW.UTEM.COM

105 TEST THE WEATHER RADAR

While it is strictly not required on ALL routes; some routes specifically require a weather radar test; and, of course, if there is potential weather a test is a great idea.

A. MODE selector to MAP, VOR, or APP.

B. RANGE SELECTOR as desired.

Observe that the selected range mid-distance appears near the center of the ND track line.

NOTE:
Maximum range for weather radar operation is 320 nm.

C. TILT KNOB FULL UP.

D. FUNCTION SELECTOR TEST position.

Using the "TEST" position provides a more complete evaluation of the RADAR than NORM position.

E. WX RADAR switch ON (EFIS panel)

F. OBSERVE BOTH NDs

Observe the green, yellow, magenta, and red test pattern, and that the word **WXR TEST** should be annunciated on the CRTs on both sides of the cockpit. AGAIN. *Having both EFIS selectors in WXR tests the "OTHER" radar head.*

G. SYS TRANSFER SYS L or SYS R

H. WXR switch OFF

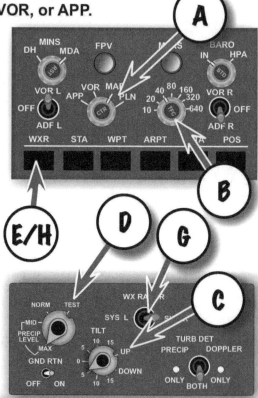
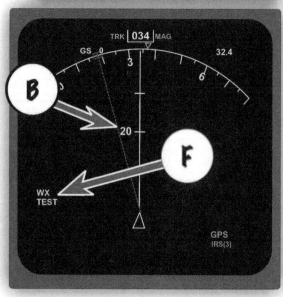

Some comments about the ...
WEATHER RADAR

The primary function of the Weather Radar is to serve as an aid to **AVOID SEVERE WEATHER, NOT AS PENTRATION TOOL**.

There is simply no way that I can describe how to operate the **RADAR** in this book. There are available manuals with hundreds of pages on the subject. But, let me cut right to the heart of the matter and simplify it greatly for you: The most important item is

"TILT MANAGEMENT".

Here are some simple guidelines:
NUMBER 1. Call in sick and stay home on stormy days ... just kidding. However, I will say that more than once I have gone into holding well clear of potential weather and "waited out the storm" rather than attempt a penetration, particularly at low altitude, in order to reach my destination. Specifically, I would rely on other airplane PIREPs, contact the Dispatcher for NOTAMS, as well as the ATC recommendations before I would attempt to continue into areas of known weather. **AVOID SEVERE WEATHER !**

Even at cruise altitude, always be aware that **SEVERE TURBULENCE** can exist well above the actual top of a well developed convective condition (cloud) ... even though the airplane may actually be flying in clear air. Here is where you can get into a corner if you elect to climb to an altitude that is higher than your recommended maximum. You may encounter vertical gusts that will exceed your stall margin and you could experience an uncontrolled altitude deviation (referred to as an "air-pocket by my Grandmother) or worse yet an upset (YIPE!). This is a very sensitive issue, since the airplane has such tremendous power, it can actually "out-fly" the wing.

If there is a "**RADAR SHADOW**" ... AVOID ... AVOID ... AVOID flying into or above that area. You can assume that if you can paint the weather behind a well defined cell that it will be at least one level more hazardous than depicted. This is due to "**ATTENUATION**", (which is the reduction in intensity of the return due to the opacity of the cell) which is an indication of really strong activity.

DURING TAKE-OFF ... if there are **TSMS** or convective activity along EITHER or BOTH (1) your planned path of flight and/or (2) your **EOSID** (Engine Out Escape Routing), in order to properly use the **RADAR** to evaluate the situation you may have to taxi onto the runway and hold in position so as to align the radar towards the activity. This would be a common check-pilot evaluation item ... and certainly a good idea in real life.
...and just as an obvious recommendation; during climbout more "DOWN TILT" is required; and likewise during decent, more "UP TILT" may be useful. **DUH!**

Probably the most information is gained by using the 40 mile scale, although changing the scales may be useful in rapidly developing weather scenarios. Be aware that each time you change scales, you will have to re-orient the tilt.

© MIKE RAY 2014
WWW.UTEM.COM

THE LOWLY FIRST OFFICER'S REALLY SIMPLE COCKPIT SET-UP

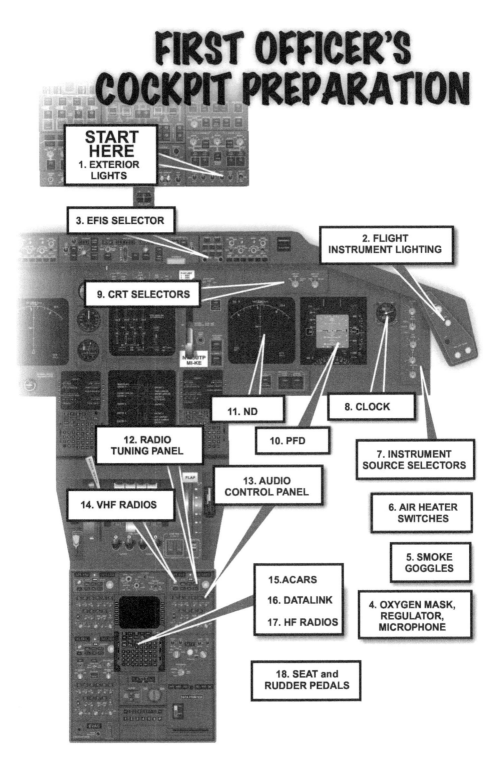

747-400 SIMULATOR TECHNIQUES ...

AFTER RETURNING FROM THE EXTERIOR INSPECTION ...

1. EXTERIOR LIGHTS

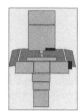

1 **EXTERIOR LIGHTS** .... as required

Here are the "**LIGHT RULES.**"

 1. **NAV** lights *MUST* remain **ON** at all times.

2. **BEACON (ANTI-COLLISION)** lights *MUST be ON* when:
   ENGINES are RUNNING or
   AIRCRAFT is MOVING.

3. **BEACON (ANTI-COLLISION)** lights are *NOT* to be checked during the walk-around.
   Maintenance has the responsibility of checking them.

4. **LOGO** lights are usually turned on **ONLY** during after dark.

2. FLIGHT INSTRUMENT LIGHTING

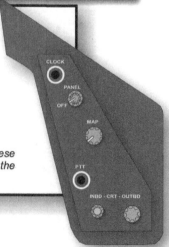

2 **FLIGHT INSTR. LIGHTING** .... SET

You set the:
   PFD (referred to as OUTBD CRT)
   ND (referred to as INBD CRT)
   map
   Panel lights
as desired.

I found that I would continually be changing these during a flight because the ambient lighting in the cockpit is constantly changing.

... and PROCEDURES FOR STUDY and REVIEW ONLY!

3. EFIS CONTROL PANEL

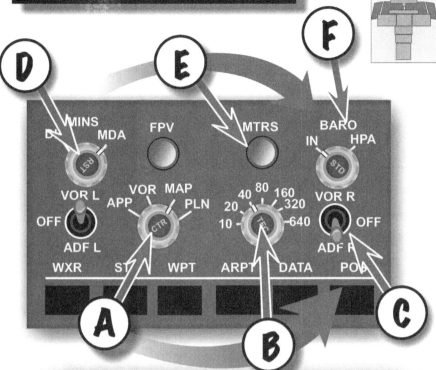

3 EFIS CONTROL PANEL .... SET

There are 6 things to set-up on this panel.

A. **ND MODE** selector .... **MAP**
B. **ND RANGE** selector .... as required.
   *Using 10 mile scale gives
   the best information for single engine tracking.*
C. **VOR/ADF** switches .... as required.
D. **DH / MDA** selector .... as required.
E. **METERS** switch .... as required.
   *In some countries, meters **IS** the unit of measurement.
   You MUST be aware of that information.*
F. **BAROMETRIC** selector .... as required.
   *Domestically, IN HG (inches of mercury) is the standard.
   At other places in the world it varies from country to country,
   and may be hPa (hecto-pascals).*
   The difference could be disastrous.

You **MUST** be alert. So, here is something very important. you should also:

CHECK THE OTHER GUYS
EFIS CONTROL PANEL.

747-400 SIMULATOR TECHNIQUES ...

THE INCREDIBLY COMPLICATED OXYGEN MASK SET-UP

3 CHECK .... OXYGEN MASK, REGULATOR, and MICROPHONE.

One of the MOST complicated procedures on the Set-up is the Oxygen mask checkout. In brief form, here is what to do:

Select FLT interphone
Select SPKR on
Adjust FLT and SPKR volume

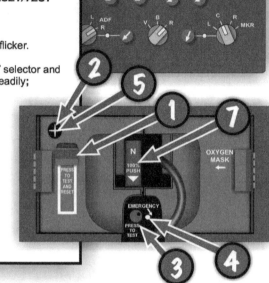

STEP 1: PUSH and hold DOWN, RESET/TEST lever.
and while holding it down ...

STEP 2: Observe **YELLOW CROSS** flicker.

STEP 3: Push and Hold **EMER/TEST** selector and observe **YELLOW CROSS** appear steadily;

STEP 4: While still holding the **RESET/TEST** lever
Push the **PTT (push to test)** switch
and
verify the sound of **O2** flowing across the mask mic.

STEP 5: ... then Release the **EMER/TEST** selector
Release **RESET/TEST** button:
Observe **YELLOW CROSS FLOW IND** is **BLANK**
and **O2 Flow** stops.

Now, put it all back together:

MASK O2 @ 100%
MASK HOLDER doors closed
MASK/BOOM switch to BOOM

WARNING

Removing the mask from the container to check for O2 flow and Mask Mic operation is neither desired nor required. Once the mask is out of the container, it requires considerable effort to get it all back together. It is a lot like trying to cram a 6 foot Boa Constrictor into foot long lunch box!

4 CHECK .... SMOKE GOGGLES

Check the SMOKE GOGGLES for clarity, glazing, proper fit, and serviceability. If they are wrapped in plastic, remove them and stow them for quick accessibility.
When you re-stow the goggles, make certain that they are within arms reach.

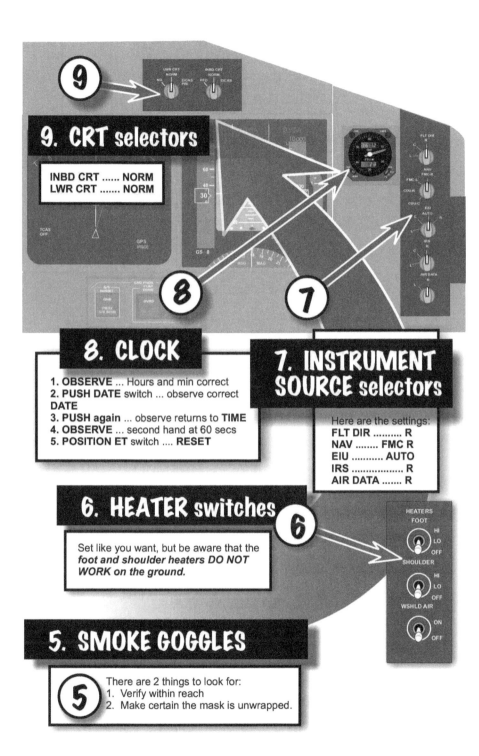

747-400 SIMULATOR TECHNIQUES ...

> **IRS ALIGNMENT MUST BE COMPLETE TO CONTINUE**

10 PFD .... CHECK

Check PFD is displayed on the CRT and there are no flags.
Then:

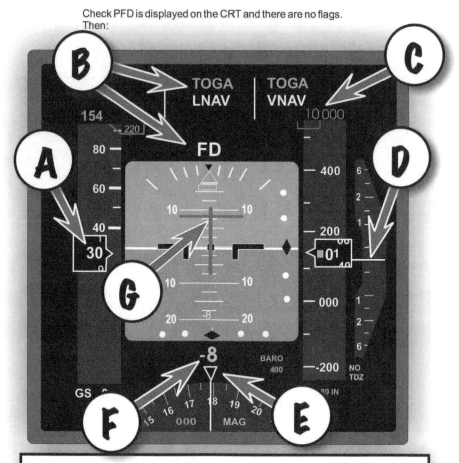

A: AIRSPEED IND .... 30 KTS

B: FD and TO/GA .... Annunciated

C: ALTITUDE IND .... Alt set in MCP

D: VERTICAL SPEED .... ZERO with NO FLAGS

E: HEADING .... Agrees with MCP

F: RADIO ALTITUDE .... -8

G: ATT and COMD BARS ...... level and 8 deg UP

> **IRS ALIGNMENT MUST BE COMPLETE TO CONTINUE**

11   ND .... CHECK

Check ND is displayed on the appropriate CRT and there are no flags. Then:

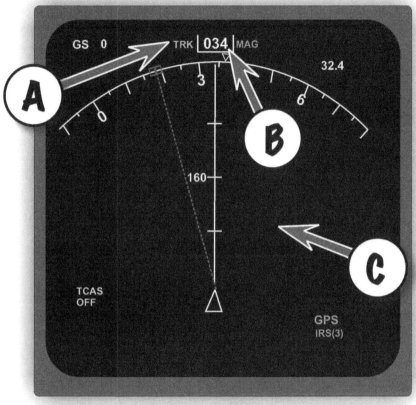

NOTE:
PLAN MODE will work even if IRS' are NOT aligned.

A: **PROPER HEADING AND TRACK**
B: **CHECK HEADING agrees with STBY COMPASS**
C: Verify NO FLAGS showing.

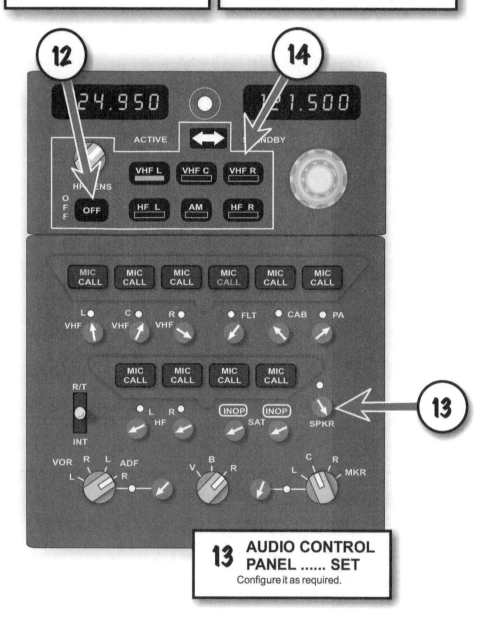

ACARS: Do two things.

15 SATELLITE VOICE CALL setup .... DO IT!

Select **MENU** on **CENTER (CONSOLE) CDU**.

SELECT: <SAT> or **<SATVOICE>**; then

LINE SELECT: DIRECTORY (LS2L for VOICE 1)

Select the first Attached **FACILITY** with **SATCOM** capability.

Repeat the procedure for **VOICE 2** for **DISPATCH** in use.

16 DATALINK ... CHECK

Select **LINK MANAGER** on the center **CDU**.

Verify **AVAIL** status is in view for **VHF** and **SATCOM**

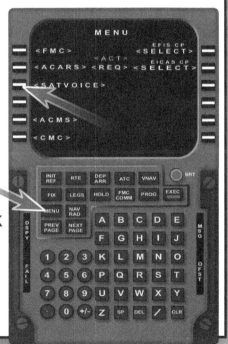

17 LEFT and RIGHT HF RADIOS .... CHECK

If required, (see criteria outlined in the company literature):

TUNE HF radios,
Select appropriate frequencies (normally on the 10-7 page) in the **STANDB**Y frequency indicator and position frequency transfer switch to the **ACTIVE** frequency. Momentarily press a **PTT** switch.

If these checks are required ... they **MUST** be accomplished **PRIOR** to departure.

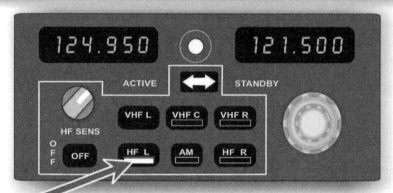

Momentarily press a **PTT** switch.

A piercing tone will be emitted while the **HF** transmitter is tuning. The tone *IS NOT* transmitted on the frequency; it is an internal broadcast and only you can hear it. The tone last for a few seconds and then should be replaced with a "**FRYING EGG**" sound once the HF frequency is tuned and useable.

18 SEAT and RUDDER pedals .... adjust

END of the First Officers Cockpit Set-up.

... and PROCEDURES FOR STUDY and REVIEW ONLY!

FINAL COCKPIT PREP

A lotta people will show up at these last "busy" moments. They all have to be treated individually ... and it is your job to coordinate all this flood of information and personnel and still make the departure time as best as you can. **FAA** Check personnel, **CSR** (Customer Service representatives) with special instructions and concerns, Cockpit riders, and the list goes on and on. Each one MUST be dealt with before we close the door and begin the actual push-back procedure.

Somewhere along in this part of the preparation for flight, the cockpit will be visited by the **"A" FLIGHT ATTENDANT** or **CHIEF PURSER**. They have a rather complex social structure that few, if any, pilots understand. However, it was always my experience that no matter how busy I was ... it was a great idea to stop what I was doing and turn and give them my full attention ... *IMMEDIATELY*! There are several pieces of information that I want from them, and they want some stuff from me also. There has to be a moment in the flow of activities for this exchange to take place ... and while I personally prefer to talk with the 'A' or Chief Purser as soon as I get on board ... frequently they are occupied with their set-up duties and do not welcome an intrusion by the Captain at that point in the "loading" or embarkation..

The **FUELER** will come to the cockpit and present you with the Fueling Form. You MUST properly confirm that the appropriate amount and type of fuel has been boarded. The procedures are complex and they are iterated somewhere else. I don't know what your SOPs are for that process, but the check-airman will want you to tell him what they are ... so a word to the wise.

A **MAINTENANCE PERSON** will deliver the **LOGBOOK** to the cockpit. STOP what you are doing and give them your attention!!! While He/She has probably already confirmed that the logbook is the proper one for the aircraft ... COMPARE the **AIRCRAFT NUMBER** with that from the **LOGBOOK**. You MUST HAVE the correct logbook with all maintenance items signed off on board **BEFORE** you can initiate **PUSHBACK**. Generally, they will brief you about any unusual maintenance situations.

747-400 SIMULATOR TECHNIQUES ...

A discussion about charts in general:

The "Standard" by which most others are judged (and certainly the biggest supplier) are those charts developed by the **JEPPESON COMPANY**. As a result, I have selected those for inclusion in my examples and remarks. The "jeps" as they are called are similar enough to all the others as to make my comments and observations about them valid for virtually all charts.
Each airline, however, has the responsibility to select their own charts or chart manufacturing company.

10-7

The pages have identifying numbers at the top of each different category of page. Here are some of the representative pages:

the 10-7 chart

Lists the available operations facilities, frequencies, and services available plus a treasure trove of other useful information. It may actually be several pages long. Generally used during the pre-departure phase.

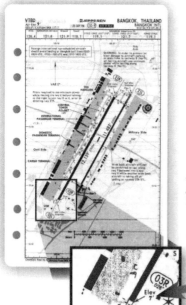

the 10-9 chart

Generally, a map of the airport. It contains taxi routes and general information regarding hazards and other considerations.
Also re-iterates frequencies, 4 letter identifier, lat/long, etc.

The approach end depiction contains some information that we need before departure:

ELEVATION at a specific point -this is essential for checking the altimeter.

MAGNETIC HEADING of the runway.

the 11-1 chart

The FIRST of the "APPROACH" plates contains most of the information needed to fly the MOST desirable approach. By that I mean, generally speaking, the approach plates are grouped in the order which represents the "best" flyability, all things considered. You will find that CAT-III approaches sometimes occur at the end of the approach plates even though they may represent the most precise landing minimum.

At smaller airports, the information that we represented as being on the 10-9 page is actually placed on the back of the 11-1 page.

the 10-2 and 10-3 charts

The 10-2 charts (**STAR**) ARRIVAL and the 10-3 (**SID**) STANDARD INSTRUMENT DEPARTURE charts. These present the specific details such as frequencies, radials, distances, altitude associated with the TITLES. That is, if you are assigned the "ALBOS 2A DEPARTURE," then ATC will expect that you will fly so as to conform to the specific decsription of that departure profile **WITHOUT** further clearance information.

Your clearance might be "... via the ALBOS 2A departure ...." and that would completely define all the details of your expected flight path after take-off.

page 161

About 20 minutes prior to pushback ...
or as indicated on the airport information sheet ...

GET CLEARANCE

There are basically two ordinary ways to get the clearance.
> Using the VHF radio, and
> Using the center CDU and PDC
>> (Pre-departure clearance)

Here is a short primer on how to operate the radio head.

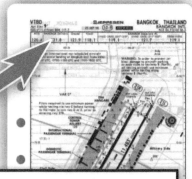

1: DETERMINE desired frequency (In this case we will use the 10-9 page to obtain CLEARANCE DELIVERY frequency)

2. SELECT RADIO SWITCH for the radio to be tuned.

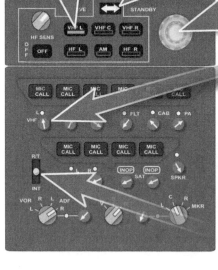

4. DEPRESS the FREQUENCY TRANSFER SWITCH to make the selected frequency active.
NOTE: The frequency that was the "old" active frequency, now becomes the standby.

3. DIAL IN FREQUENCY using the STANDBY FREQUENCY SELECTOR KNOB.

5. ADJUST associated receiver volume control knob.

6. PUSH appropriate button to select the appropriate transmitter.
NOTE: Only one switch can be selected at a time.

7. **HAVE OTHER PILOT MONITOR** the clearance.

8. Make tranmission and request clearance. Use either:
> **YOKE** switch,
> **REMOTE** pendant, or
> **DEPRESS** transmit switch on
>> **AUDIO CONTROL PANEL**.

9. **WRITE IT DOWN**. Be prepared for a rapid transmission from the controller. Sometimes the "Aviation English" will be difficult to interpret. Make certain that you understand and obtain a complete clearance.

GETTING THE CLEARANCE using PDC

Another (**BETTER**) way to get the clearance is using the **ACARS CDU**. Those airports where this service is available will be indicated on the 10-9 or 11-1 airport pages by the notation "**PDC**" in the legend of the airport diagram page.

The initial call-up frequency at a station where a **PDC** has been received will be either:
on **ATIS**, or
REMARKS on the **AIRPORT PLATE**, or
by **NOTAM**.

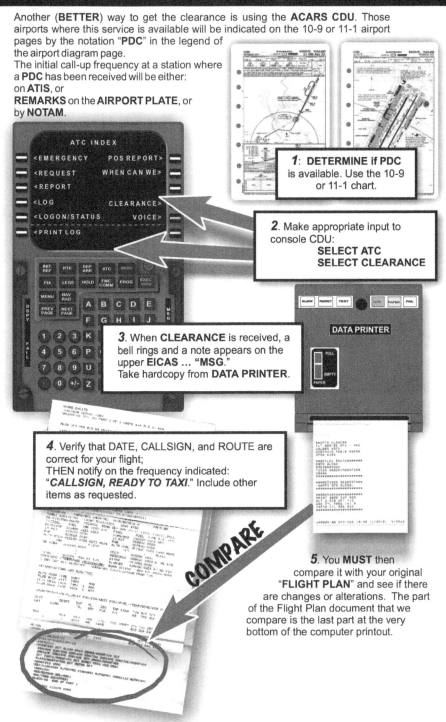

1: **DETERMINE if PDC** is available. Use the 10-9 or 11-1 chart.

2. Make appropriate input to console CDU:
SELECT ATC
SELECT CLEARANCE

3. When **CLEARANCE** is received, a bell rings and a note appears on the upper **EICAS** ... "**MSG**."
Take hardcopy from **DATA PRINTER**.

4. Verify that DATE, CALLSIGN, and ROUTE are correct for your flight;
THEN notify on the frequency indicated:
"**CALLSIGN, READY TO TAXI.**" Include other items as requested.

5. You **MUST** then compare it with your original "**FLIGHT PLAN**" and see if there are changes or alterations. The part of the Flight Plan document that we compare is the last part at the very bottom of the computer printout.

SET INITIAL CLEARED ALTITUDE

Either the Captain or the F/O will set the initially assigned altitude into the **MCP**.
CROSS CHECK with the other crew member(s).
THIS IS VERY IMPORTANT!!

This is the printout from the ACARS Printer on the console.

NOTE: *You will **NEVER** depart anywhere ... ever ... without an assigned altitude. If you are not certain what the assigned target altitude is; stop and find out.*

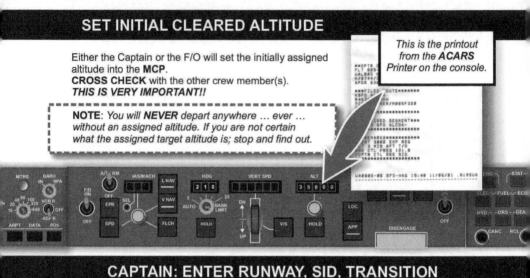

CAPTAIN: ENTER RUNWAY, SID, TRANSITION

The Captain is expected to make these entries ... and the First Officer is to VERIFY that the entries made by the Captain are correct.

NOTE:
ONLY SELECT THE RUNWAY ON THE DEP/ARR PAGE !!!
NEVER enter the runway on ROUTE PAGE 1.

HOW TO DO IT:

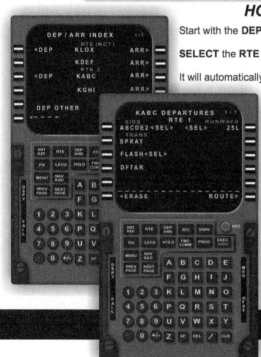

Start with the **DEP/ARR** key.

SELECT the **RTE 1 DEP AIRPORT**

It will automatically scroll to the **DEPARTURE PAGES** for that **AIRPORT**.

Then simply select the **RUNWAY**, **SID**, and the **TRANSITION** (if required).

The ORDER of the entries is ***IMPORTANT***: RUNWAY, SID, and then TRANSITION.

NOTE: *IF there is a runway change after the initial execution; You MUST re-enter the SID and TRANSITION also even though they may remain unchanged.*

F/O VERIFY

page 164

F/O: WAYPOINT VERIFICATION

Any new waypoints resulting from changes or additions to the flight plan route **MUST** be checked by the First Officer using "**WAYPOINT VERIFICATION PROCEDURES**."
ADDITIONALS 6-57

1. SELECT LEGS PAGE ON THE CDU
2. SELECT PLN MODE ON THE EFIS CONTROL PANEL
3. COMPARE THE MAGNETIC COURSE AND DISTANCE BETWEEN EACH WAYPOINT FROM THE FMC AND THE AFPAM.
4. STEP THROUGH EACH WAYPOINT AT LS6R TO VERIFY ACCURACY.
5. LEG DISTANCE SHOULD BE WITHIN 2NM OF THE FPF.

F/O: CDU NAVIGATION RADIOS CHECK

The Boeing guys decided that the pilots needed help, and so they have installed "**SELF TUNING RADIOS**" on this airplane. Sometimes, however, the "automatic magic tuner" will tune the wrong radio. So, we have to check and see that the proper radios are tuned. If they are not, then we have the option to manually tune the radios. So you might ask,

How do we "**MANUALLY TUNE**" the radios?

Select the **NAV/RAD** key and when the "**NAV RADIO**" page is displayed, fill in the appropriate blanks by typing the numerical value for the frequency into the scratch pad and then line selecting it to the appropriate window.

NOTE:
The **ADF** radios do **NOT** have an autotune capability and must **ALWAYS** be manually tuned, if needed.

F/O SET FMCS PERFORMANCE

To access this page:
LS **PERF INIT** (if displayed) or push **INIT REF** key.

COST INDEX:

Type "100" to the Scratch Pad
Line Select 5L
(*A COST INDEX of 100 will give us VNAV climb and descent speeds compatible with ATC and Flight Manual data*).

RESERVES:

If "NO ALT" required; I suggest you enter MINIMUM PLANNED LANDING FUEL (MPLF) 19.0#.
Just a comment here, your call, but it is OK to put in the MINIMUM DESIRED LANDING FUEL (MDLF) 11.1# .

If "ALT" required enter burn to most distant alternate PLUS MDLF.

At certain airports which require a designated minimum arrival fuel, enter that value.

If ALT is deleted or added enroute; UPDATE the RESERVES fuel line.

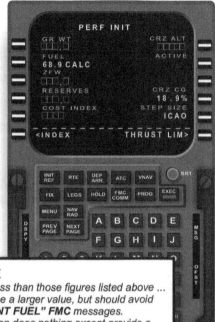

NOTE
RESERVES fuel entry **SHOULD NOT** be less than those figures listed above ... however, the Captain may elect to choose a larger value, but should avoid values that result in **"INSUFFICIENT FUEL" FMC** messages.
Also of interest ... the **RESERVE** selection does nothing except provide a warning message on the **CDU** when the fuel onboard reaches that figure.

ZFW (Zero Fuel Weight):

Get the **ZFW** from the **PLANNED TAKEOFF DATA MESSAGE** supplied from the flight planning materials. The **GROSS WEIGHT** will be automatically entered.

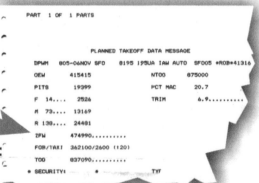

CRZ ALT (Cruise Altitude):

Get the CRZ ALT from the Flight Plan.

STEP SIZE :

As desired.

CRZ CG :

DO NOT CONFUSE THIS WITH TO/CG (Take-off CG).
DO NOT PLACE TO/CG HERE!!! ***NEVER !!!***

... and PROCEDURES FOR STUDY and REVIEW ONLY!

F/O SET THRUST LIMIT DATA ENTRY

To access this page:
LS **THRUST LIM>** (key 6R)

If "**REDUCED THRUST**" is desired, a normal situation:

Type in the desired **ASSUMED TEMPERATURE** into the scratch pad and line select to **1L (SEL** temp line).

EXAMPLE
For a "C" degree entry, just type in the numeric value, such as 48. If you wish to insert fahrenheit, the entry must be followed by an "F" to distinguish it from centigrade.

If "MAXIMUM RATED TAKEOFF THRUST" is desired, AFTER a reduced temperature has been set in the FMC:

 PUSH the DELETE key and put "DELETE" in the scratch pad. Line selct 1L.   then

 Line select CLB> (Line select 2R). This will ensure that CLB is armed.

 Either CLB 1 or CLB 2 may have been automatically selected by the FMC depending on the weight and assumed temperature. We want the most fuel efficient climb thrust setting, so...

Line select 2R.

CAPT check THRUST LIMIT DATA

On the **THRUST LIM** page, the Captain is to verify the thrust limit entries, including the displayed EPR, and compare them with the *Planned Takeoff Data Message*.

CAPT- F/O initialize TAKEOFF DATA

To access this page: **LINE SELECT TAKEOFF>** (LS6R)

> **NOTE**
> Takeoffs made with flaps 10 will provide better acceleration, a higher climb rate, and earlier flap retraction; thereby improving overall fuel economy and cost of operation.

The FIRST OFFICER will make the following entries:

▶ **FLAP SETTING ACCELERATION HEIGHT**

▶ **E/O ACCELERATION HEIGHT**

▶ **THRUST REDUCTION**
example: 5 for flaps 5 OR AGL altitude.

▶ **V SPEEDS**

NOTE: The "V speeds" will probably be displayed in "SMALL" font size. They should be made "**BOLD**" font size. To do this, simply depress the key next to the item.

▶ **TAKE-OFF CENTER OF GRAVITY**
Here is where the T/O CG (Take-Off Center of Gravity) figure is selected by the pilot.

▶ **TRIM SETTING**

This number should agree with the information that perhaps came with the flight papers. If there is a disagreement, further investigation (call to load planning) should result in the conclusion that the CDU figure is correct. If not, then suspect that there is some improper input made to the CDU load information.

▶ **POSITION SHIFT (if required)**

The runway position indicated on the default setting is the **RUNWAY LANDING THRESHOLD**. Unless you have **GPS** positioning, if you depart from any other position, you must enter a **POS SHIFT** to ensure that the airplane symbol on the **ND** updates to the actual aircraft position or *WHEN TOGA IS PRESSED, A MAP SHIFT WILL BE INDICATED*.

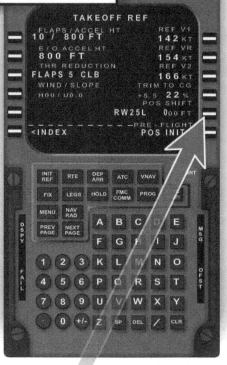

> **GOOD CAPTAIN HABIT**
>
> As a Pilot ... *We must ALWAYS be suspicious of important information and ALWAYS seek to resolve ambiguity whenever possible.*

CAPT - F/O VOR/ADF POINTERS

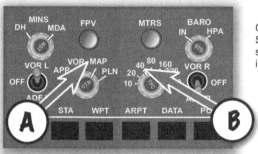

On the **EFIS** panel;
Set the **ADF** and **VOR** switches to display the desired information on the **ND**.

CAPT - F/O check ND (NAV DISPLAY)

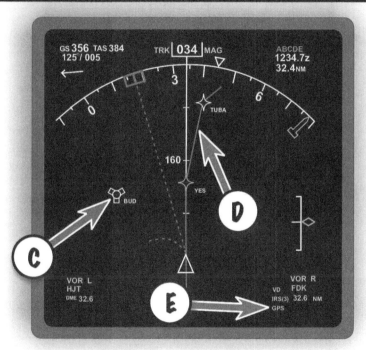

VERIFY:

A — the indicator is in the **MAP** mode,

B — scale set to *APPROPRIATE* value (**20** or **40** nm usually good)

C — location of the tuned **VOR/DME** stations,

D — active route as a solid magenta line between waypoints,

E — **GPS** in **SMALL FONT** above the **IRS (3)**,

The map display, map scale, and LEGS page sequence should be consistent with the departure procedure.

FUEL STUFF
FUEL "WEATHERING"

You may never have heard of the potentially hazardous situation called "**FUEL WEATHERING**". Here is a simplified description of what it is ... and this description applies to the fuel in all airplanes, not just the 747-400. When an airplane is fueled, the turbulence induced by the process mixes or "traps air" in the mixture. This is generally not a concern at lower altitudes (such as during take-off); however, due to vibration and the air pressure reduction during the climb, the air bubbles tend to come out of solution. These air-bubbles reduce the ability of the engines to initiate "suction feed" of the fuel (if required), especially at high thrust settings and at altitudes above 20,000 Feet. Although I could not find any reference specifically about the 747-400, on some airplanes actual engine shutdown or uncommanded power reductions have been attributed to this problem.

> It is generally accepted, however, that by the time the airplane reaches cruising altitude, most of the trapped air has been released and the air-bubbles no longer present a problem, even should suction feed become necessary.

comments regarding
COLD FUEL

THE BASIC RULE:
"... inflight fuel tank temperature **MUST BE** maintained at least 3 degrees Celcius above the freezing point of the fuel being used."

If a portion of the flight is planned through areas with outside air temperatures **BELOW -65 DEGREES C**, the potential exists for reaching the "operational fuel temperature" limit.

Generally speaking, the **MINIMUM FUEL TEMPERATURE for TAKE-OFF is -43 Degrees C** or higher temperatures if allowed by fuel being used.

While operating in areas where the **OUTSIDE AIR TEMPERATURE** is -65 degrees C or less and anticipating forecast exposure to this situation for an hour or more ... consider taking these actions:

FUEL TYPE	MIN. TEMP	MAX. TEMP
JET A	-37° C	+54° C
JP 5	-43° C	
JET A-1	-44° C	
JET B	-44° C	+43° C
JP-4	-45° C	

- **CHANGE ALTITUDE** to warmer air.
  Just a note here, warmer air may not necessarily be at a lower altitude.
- **INCREASE SPEED.**
  Expect fuel temp increase of about 1/2 degree C for each .02 Mach increase.
- **RE-ROUTE**.

Here is the bottom line: "There have been **NO KNOWN CIVIL JET AVIATION ENGINE SHUTDOWNS** due to operations below the "fuel freeze specification limit"."

... and PROCEDURES FOR STUDY and REVIEW ONLY!

IMPORTANT !

POUNDS v. KILOGRAMS

It will <u>ALWAYS</u> be necessary to be vigilant when flying around the world ... because the **FUEL LOADING** is measured by (at least) two different internationally sanctioned measurement standards. The American **POUND (#)** or the European **KILOGRAM (Kg)**. Here is an important "*RULE OF THUMB*".

**1 Kg (roughly) = 2 #   or
1 # = ½ Kg.**

Here is the problem: If you "**REQUESTED**" **KILOGRAMS** and if the fuel load was put on the airplane using **POUNDS** ... you would have only about HALF the fuel that you wanted (and thought you had on board). ***THINK ABOUT THIS!!!!***
<u>You could take-off with only about *1/2 THE AMOUNT OF FUEL* that you thought you had!!!</u>

A 767 got airborne with a faulty fuel indicator system that showed that fuel in #s instead of Kgs and the problem went un-noticed. They wound up running out of fuel and were forced to make a "**DEAD STICK**" landing on empty tanks! It was a miraculous story that ended successfully. *<u>PAY ATTENTION!</u>*

FUEL LOADING notes:
- Fuel in the **CTR** (Center) tank is <u>**ONLY**</u> allowed when *ALL MAIN AND RESERVE TANKS ARE FULL*, except when **CENTER TANK FUEL** is payload or structural check.
- **RESERVE TANKS 2 and 3 MUST BE FULL** when the actual taxi weight is greater than **740.000 pounds**.
- **ALL WING TANKS** (MAIN 1,2,3,4 and RESERVE 2,3) **MUST BE FULL** for operations with **STABILIZER TANK FUEL**.

NOTE:
<u>*If engines are running, ALL PASSENGERS MUST BE DEPLANED prior to fueling.*</u>

However: Fueling may be accomplished with passengers on board, being boarded, or deplaned under the following conditions:
- *ALL ENGINES are shut down*.
- A loading bridge (jetway) or passenger stand is in position at the airplane.
- The minimum number of *Flight Attendants MUST BE ON BOARD*!

FURTHER ... If a loading bridge or adequate passenger stand in NOT available, fueling may still be accomplished if:
- *ALL ENGINES are shut down*
- All doors are **ARMED**.
- Flight Attendants are stationed at their assigned exits.
- Communications are maintained between **FUELER and COCKPIT** using the **MAINTENANCE INTERPHONE**.
- Flight Attendants are notified when fueling begins and ends.

F/O BOARDED FUEL QUANTITY check

These procedures are discussed in greater detail in the FOM Maintenance/Fueling/Loading 7.20.6.

NOTE:
This check assumes that the crew has received an FSF (Fuel Service Form). If there is no FSF, then the boarded fuel check is not applicable.

1 **VERIFY PRESERVICE FUEL FROM THE FSF BY COMPARING IT WITH THE ARRIVAL FUEL.**
PRESERVICE fuel may be less than the arrival fuel due to APU ground use, maintenance taxi, engine runs, etc.

2 **CONVERT THE FSF FUEL ADDED FROM GALLONS TO POUNDS BY MULTIPLYING THE TTL (total fuel from trucks) BY THE FUEL DENSITY.**

3 **THE CALCULATED FUEL ON BOARD IS THE SUM OF STEPS 1 AND 2.**

4 **COMPARE THE CALCULATED FUEL ON BOARD FROM STEP 3 WITH THE TTL REQUESTED FUEL** (CLEARED fuel plus TAXI fuel).

5 **CALCULATED FUEL ON BOARD SHOULD EQUAL TTL REQUESTED FUEL PLUS OR MINUS THE ALLOWED TOLERANCE** (gallons converted to pounds).

TECHNIQUE
Check the fuel load at the time the fueler delivers the FSF to the cockpit. Discuss any discrepancies at that time.

NOTE
If the calculated fuel on board is out of tolerance with the TTL requested fuel, then the FSF should be checked for errors and for indications of over servicing or under services. Check pre-service amounts, fuel density, math, TTL FROM TRUCKS, and for possible missing maintenance FSFs. If the discrepancy cannot be determined, the fueler should be notified and the fuel tanks dripsticked to confirm fuel on board.

AWARENESS NOTE
There is always the possibility that somewhere in your travels to some place on the earth unknown to you, they just might use Kgs (Kilograms) instead of Lbs (Pounds) to measure the fuel load.
IT HAS HAPPENED!!!
And if the crew is not vigilant, they will depart without enough fuel.

... and **PROCEDURES FOR STUDY and REVIEW** *ONLY!*

FUEL NOTES:

MAXIMUM FUEL IMBALANCE

Max. Diff between **TOTAL FUEL** in **MAINS 2** and **3** and **MAINS 1** and **2** after tank to engine criteria met.	**6000**
Max. Diff between **TOTAL FUEL** in **MAINS 1** and **4**	**3000**
Max. Diff between **TOTAL FUEL** in **MAINS 2** and **3**	**6000**

CARRYING EXCESS FUEL: Here is the sample ... During a typical 6000 nm flight, 55 Lbs of additional fuel burn required for each 100 Lbs of excess fuel. So, if you put on 10,000 Lbs of extra fuel, you would burn 5500 pounds and arrive with 4500 Lbs of fuel left as extra.

> **55% OF EXTRA FUEL WILL BE BURNED TRANSPORTING THAT EXTRA FUEL**.

RUNNING ENGINES AT THE GATE and **DURING TAXI**: Each engine burns approximately 24 Lbs of fuel per minute after start and during taxi. Consider shutting down engines for ramp delays of taxi. Here are the parameters to consider:

> **AIRPLANE GROSS WEIGHT.**
> - **THREE ENGINE** taxi OK if weight below **650,000 Lbs.**
> - **TWO ENGINE** taxi OK if weight below **600,000 Lbs.**

... and anything else that comes up including but not limited to TAXIWAY GRADIENT, CONDITION, COMPOSITION, RAMP and CONGESTION issues, ENGINE ANTI-ICE REQUIREMENTS, PNEUMATIC, HYDRAULIC, ELECTRICAL system demands, ENGINE WARM-UP REQUIREMENTS, and so forth.

> As a general rule:
> "If an engine can be shut down for **7 (SEVEN) MINUTES**, the fuel savings will exceed the cost of STARTER and STARTER VALVE wear."

RUNNING APU INSTEAD OF ENGINES DURING PROLONGED DELAY: Running **ONE ENGINE AT IDLE** and then using it to crossbleed start the remaining engines is **LESS EXPENSIVE** than shutting down **ALL ENGINES**, starting the **APU** and using the **APU** to start engines ...
UNLESS THE DELAY IS MORE THAN 10 MINUTES!

> **RULE OF THUMB**:
> **5 MINUTES of TAXI FUEL = 1 MINUTE of IN-FLIGHT CONSUMPTION.**

FLAPS 10 TAKE-OFF:
Takeoffs made with flaps 10 degrees provide better acceleratioon, higher climb rate, and earlier flap retraction. This allows for better fuel consumption.

some OBVIOUS FUEL SAVING TIPS from the Flight Handbook:
- Slower speeds in turns decrease the turn radius and miles flown.

- Close adherence to FLAP RETRACTION schedule saves fuel.

- When a "GATE HOLD" or "DEPARTURE FLOW" delay is in progress, if the gate is available, remain at the gate under GROUND POWER.

WHEN FUELING IS COMPLETE ...
CAPT sets up FUEL PANEL

The FABULOUS "NO-BRAINER" FUEL SET-UP RULE:

Some incredibly smart person (maybe it was a pilot) figured out a simple way to remember how to set up the fuel panel. These simple two situations cover the whole range of fuel loads from totally full to minimum fuel. That rule goes like this.

Some pilots call it the

120 K RULE.

IF there is **MORE THAN 120,000 #** of fuel on board, then:

Turn on **ALL BOOST PUMPS, OVERRIDE PUMPS, CENTER WING PUMPS**, and **STABILIZER PUMPS** *that contain fuel*.

And, **OPEN #1 and #4 CROSSFEEDS**.

BUT

IF there is **LESS THAN 120,000 #** of fuel on board, then:

Turn on **ALL MAIN TANK PUMPS**, and **CLOSE #1 and #4 CROSSFEEDS**

NAME IDENTIFICATION OF PUMPS.

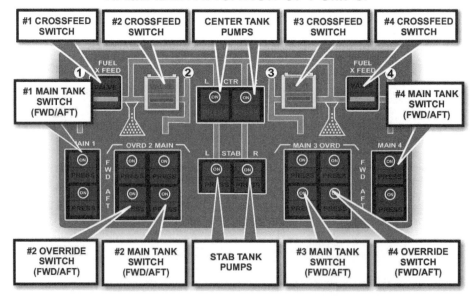

... and PROCEDURES FOR STUDY and REVIEW ONLY!

HOW THE FUEL SYSTEM WORKS:

Before the engine start with the **FUEL PUMP SWITCHES ALL OFF** ... the **LOW PRESSURE** lights are illuminated only on the **MAIN TANK PUMP** switches,
and, the **LOW PRESSURE** lights are extinguished on the **OVERRIDE, CENTER TANK**, and **STABILIZER TANK PUMP** switches.

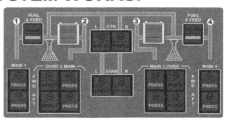

In our example, we will start with ... **FULL FUEL TANKS:**

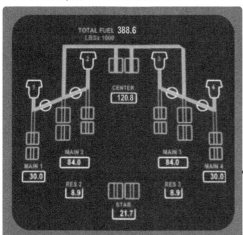

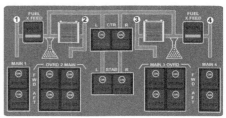

Full fuel load of 388,600 #. All tanks are full. The setup is "**ALL PUMPS ON, #1 and #4 X-FEEDS OPEN**."

FYI: Approximately 35 seconds after the Flaps have reached their selected Take-off position on the ground, the #2 and #3 X-FEED valves automatically close, momentarily flashing yellow while in transit.
Then after flaps are retracted, #2 and #3 X-FEED valves will re-open and we will have the selected situation shown above. I interpret this to mean that all take-offs are made with engine #2 and #3 TANK to ENGINE with or without fuel in the Center Tank.

@ 80,000# in the CENTER TANK:

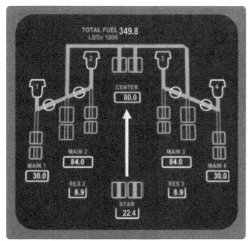

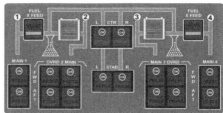

When the **CENTER TANK** gets to **80,000 #**, then fuel is automatically pumped from the **STAB TANK**
to the to the **CENTER TANK** and
from there into the **SYSTEM**.

At this point **NO MAIN TANK** fuel has been used.

WHEN THE TWO STAB TANKS GET EMPTY:

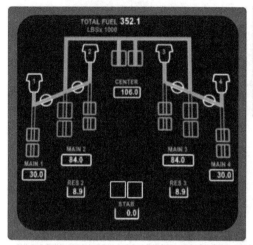

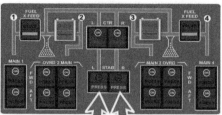

When the **STAB TANK** gets **EMPTY**, the outline on the synoptic turns yellow, the **PRESS** lights illuminate on the **FUEL PANEL**, and **ADVISORY MESSAGES** display on the **UPPER EICAS**.

PILOT ACTION is REQUIRED!

When **BOTH FUEL PMP STAB** messages are displayed and **FUEL QUANTITY INDICATIONS** show that the **STABILIZER TANK** is **EMPTY**, then Pilot action is to
SHUT OFF THE TWO STAB TANK FUEL PUMPS.

WHEN THE CENTER TANK GETS TO 2000#:

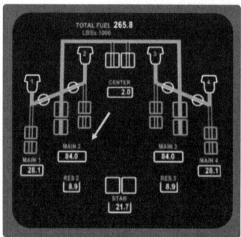

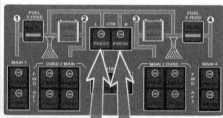

When the **CENTER TANK** gets to **2000#**, the **SYNOPTIC** for those tanks turns yellow, an arrow appears indicating that the remainder of the fuel is draining into the **#2 MAIN TANK**. An **EICAS** message displays, and the **LOW PRESS LIGHTS** appear on the fuel panel.

When **EITHER FUEL OVRD CTR** message is displayed and **CENTER TANK** fuel quantity **LESS THAN 2000#** ... PILOT ACTION REQUIRED:
TURN OFF THE CENTER TANK FUEL PUMP SWITCHES.

WHEN MAIN TANK 2 OR 3 LESS THAN 40,000#

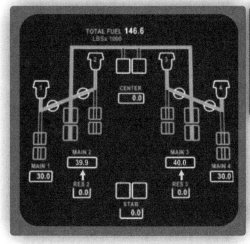

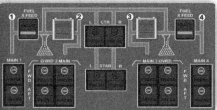

When the fuel in **MAIN TANK 2 or 3** decreases to **less than 40,000#**, fuel transfers automatically from the **#2 and #3 RESERVE TANKS**.

No pilot action is required.

WHEN TOTAL FUEL LESS THAN 120,000#

When the **TOTAL FUEL** drops to **120,000# or below**, the **OVERRIDE PUMPS** on the **SYNOPTIC** turn yellow, the **EICAS** displays A**DVISORY MESSAGES, LOW PRESS** is illuminated in the **OVERRIDE PUMP SWITCHES**.

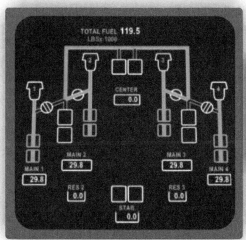

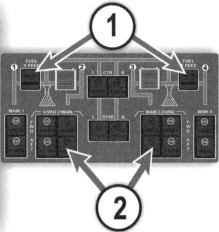

When **TOTAL FUEL LOAD LESS THAN 120,000#** ... **PILOT ACTION REQUIRED:**
1: CLOSE #1 and #4 FUEL X-FEED switches, and
2: SHUT OFF OVERRIDE PUMP SWITCHES.

747-400 SIMULATOR TECHNIQUES ...

OVERWEIGHT LANDING

During the checkride ... specifically when you are cleared for a "**MAX ATOG**" (At Takeoff Gross Weight); you can expect that you are about to encounter:
- **HIGH SPEED ABORT**
- **V1 CUT**
- **V2 CUT.**

Since you will be required to demonstrate "*Engine Failure Of the Most Critical Engine at the Most Critical Phase of flight while at Maximum Take-off Gross Weight*", it is a wise Captain that realizes when cleared onto the runway at gross weight, that "unexpectedly" he will be confronted with this situation. He should be mentally prepared for this event ... and that is a really good mindset to have in the "real" world also. In addition to all the challenges associated with the resolution of the immediate problem ... there eventually will be concerns about landing the airplane safely.

Since the airplane will have been at MAX ATOG, the weight of the will be extremely heavy and that must be addressed. Consideration must be given to

REDUCING THE WEIGHT OF THE AIRPLANE!

One of the potential solutions will be to **JETTISON (or dump) FUEL**. I would point out that even though we will be expected to address the issues associated with over-stressing the airframe and maintaining operation of the airplane within specifically defined limitations; even though ...

> **THE AIRPLANE IS CERTIFIED FOR LANDING AT MAXIMUM GROSS WEIGHT!**

Let me tell you what I just said. Generally speaking, if the engine failure does **NOT** demand an immediate return to the field ... you will be expected to "dump fuel" to reduce the weight of the airplane. If, however, in your opinion, you **MUST LAND IMMEDIATELY**, then you will be expected to **DECLARE AN EMERGENCY** and land the airplane. The airplane is perfectly capable of routinely making an **OVER-WEIGHT** landing without structural failure. However, here is the "structural" limitation for operating the airplane with a **NON-EMERGENCY** situation!

MAXIMUM "STRUCTURAL" LANDING WEIGHT: 630,000 Lbs

NOTE: *This is the **MAX STRUCTURAL LANDING** weight over at a specific major airline. However, each airline may have designated their own maximum landing weight.*

If you are required to make an **OVER-WEIGHT** landing, then there are some guidelines in your manual that suggest that if you are landing overweight ... then **REFERENCE AIRSPEED** be adjusted. That chart is called "**LANDING REFERENCE SPEEDS**". Here is the relevant parts of that chart that you might look at and be familiar enough with it to think about increasing your airspeed. I doubt whether you will have the time to reference the chart during the check-ride evolution.

LNDG FLAP REF SPEED KIAS	GROSS WEIGHT-over structural limit (1000 Pounds)						
	640	680	720	760	800	840	880
25 REF	162	167	172	177	182	187	192
30 REF	155	160	165	170	175	180	184

Let's take a moment to look at this chart. Since we are specifically considering an "IMMEDIATE" return to the airport for landing, then we should be familiar anough with the numbers at the "HIGH" or right hand side of the chart to be able to recall and use the information. Here is what I mean by that, obviously you can't memorize the whole chart, and it is highly unlikely that you would be able to even locate it in your Flight Handbook. Since we will "probably" be using a 30 degree flap (and this is the most conservative figure anyway) ... I suggest that we look at one number from this chart. ***Here is that number: 184 Kts;*** it is the adjusted **FLAP 30 REFERENCE SPEED** for landing a **STRUCTURALLY OVER-WEIGHT 747-400.**

... and PROCEDURES FOR STUDY and REVIEW ONLY!

STRUCTURAL versus PERFORMANCE
LIMITATIONS

Before we consider landing overweight ... we have to think about other operational parameters than just the "structural" capability of the airplane. It is all well and good that the jet won't be damaged ... but what about the ability to stay on the runway. I can't imagine trying to land an 840,000 pound 747-400 with a Vref speed of 184 Kts on a short runway ... and getting it to stop within the confines of the airport, let alone stay on the concrete. So let's talk some more about this situation.

Right after take-off, the airplane will be at **MAX ATOG** and **REALLY HEAVY** ... so it is unlikely that you would be taking off on a runway that does not have the performance capability to support an "overweight" landing scenario. So, during the check-ride scenario, returning to the airport of departure would probably be a realistic and viable option. And I would add that this is the likely situation when operating in the real world also.

However, after hours of droning along in cruise, if we had to divert to some dinky airport in some remote location, the airplane would have (more than likely) burned down enough fuel so that a structurally overweight landing would not have to be a consideration. However, we would then be concerned with the "performance capability" of that airport to accommodate a 747-400. Assuming everything still worked on the jet and that it wasn't damaged to the point of requiring an IMMEDIATE landing, we would have to determine if landing at a specific destination would be the best and most safe solution.

We have "**PERFORMANCE LIMIT**" charts that supply "general" operational guidelines for determining "**ALLOWABLE LANDING WEIGHT**" selecting runways predicated on:
- Airport altitude,
- Outside Air Temperature (degrees C)
- Gross Weight of the airplane.

There are "**RUNWAY LIMIT CHARTS**" that are predicated on:
- Field Altitude
- Runway Conditions,
- "FAR" Landing Field Length of Runway
- Headwind/Tailwind.

Another chart which is probably more useful is the "**DEMONSTRATED LANDING DISTANCE**" chart. This chart is intended to be used as a guide to evaluate landing options. Its values are predicated on:
- Airplane Gross Weight
- Pressure Altitude
- Runway Condition

Here is what it says: While a lighter 747-400 (GW about 440,000 Pounds) would require only 4100 Feet of runway (Wheeewww!) ... a 747-400 at Maximum Structural weight (630,000 Pounds) "could" land in as little as 5700 Feet.

Let me interpret that for you:

> **The 747-400 "could" land at an airport with a runway length of 6000 Feet.**

In reality, during line operations, seldom do the PERFORMANCE LIMIT WEIGHTS create a problem. Looking at the charts it seems to me that:

> **IF AIRPORT BELOW 4000 Feet ELEVATION, and TEMPERATURE LESS THAN about 90° F (30° C), then**
> You are OK to land right up to
> **850,000 Pounds Gross Weight.**

Your flight planning materials should already include potential diversion airports along your route of flight. Generally speaking, whereever you find yourself in the world, there will be company preferred "**DESIGNATED DIVERT AIRPORTS**" listed in the company literature ... or suitable information can be obtained by contacting the company **DISPATCHER**. Always available is diversion information from the controlling **ATC** facility. This discussion is in no way intended to be definitive ... and it is just a teaser to get you thinking about the options.

747-400 SIMULATOR TECHNIQUES ...

LIGHTEN UP!
FUEL JETTISON

Here are some thoughts about the JETTISONING FUEL.
(Pilots refer to it as "FUEL DUMPING")

KNOW WHERE THE PROCEDURE IS LOCATED. Although the clearly defined steps to accomplish the "**FUEL DUMP**" are somewhere "in the book"; it is hidden in some obscure section. It is **NOT** a part of the **QRC!!!**. If you start yelling for the **QRC**, that only tips off the check-guy that you don't know your **EMEGENCY PROCEDURES**. In the Handbook and literature that I have consulted, the "**FUEL JETTISON**" (or **FUEL DUMP**) procedure is located in the "**NON-EICAS IRREGULARS**" section ... and in your company literature it will be listed in a similarly labeled section. You **NEED** to know where the procedure is located in your companies literature! Wildly and aimlessly thumbing through the mass of available flight information pages, anxiously looking for the directions on what to do, could be check-ride suicide!

Once you have decided that the inflight problem you have experienced is under control and you determine that the airplane can continue flight long enough to dump fuel down to landing weight .... get **ATC** involved. Request vectors to the designated fule dump area. These "FUEL JETTISON areas are specifically designated areas and altitudes where the various Federal agencies have agreed that fuel dumping can take place. This has to do with Environmental and population density considerations and are generally not known to the pilots.

Of course, if you are declaring an **EMERGENCY** (such as the loss of two engines while at **MAX GROSS**) you can do just about anything you need or want to do, including dumping fuel wherever you are. You must decide early on what course of action is necessary to get the airplane back on the ground successfully.

While a "**FUEL JETTISON**" problem seems fairly straight-forward and simple; there are some parts of the exercise that need some explaining. Since the checkride fuel dump scenario usually results from the failure of an engine at take-off with Max Take-off Gross weight (MAX TOGW or MAX ATOG or whatever your airline calls it); it is most likely going to result in attempting to return to the airport of departure for landing. This should initiate two thoughts:
 - concern for "LANDING OVERWEIGHT" and
 - cause your mind to consider "FUEL JETTISON".

I mention this process this way, because in the heat of the "**ENGINE FAILURE**" exercise, some pilot candidates overlook this requirement. Oh sure, the airplane is certified to land overweight and if you make a normal landing nothing will break. However, unless you have a "get-back-on-the-ground-ASAP" situation and declare an "**EMERGENCY**" ... you will be expected to "**DUMP FUEL**" and be at or below "**MAX LANDING WEIGHT**" when you make your landing. It is probably a good idea, anyway, to contact ATC and request vectors for fuel jettison. This way, you can establish a **HOLDING PATTERN** and while you are dumping fuel, sort out a plan for resolving your irregular or emergency event ... and complete your preparation for the Engine out Landing and/or Missed Approach.

However, I would be quick to point out that there are also the **ENROUTE SITUATIONS** that may require additional thinking on your part. Here are some thoughts about fuel jettisoning in these cases. As a general statement, most likely events that occur after some time has gone by since take-off result in enough fuel burn to have reduced the weight of the airplane significantly. This means that dumping fuel in order to achieve reductions in weight so as to comply with published limitations may not be needed. However, if the situation requires a non-normal approach and landing situation with potential for a non-normal result (such as a gear failure); it is my view that it may be useful to dump down for avoid fuel problems should the integrity of the tanks be compromised.

DO NOT jettison the fuel too early in the evolution. It would be a mistake to "run out of gas" on the way to your divert field because you dumped it along the way too soon. It is my recommendation that you carry the fuel to a point closer to the divert field so that you will have enough to establish holding pattern and assess the situation. This assessment would include weather as well as airport length, availability of airplane handling, and unknown conditions, etc.

© MIKE RAY 2014
published by UNIVERSITY of TEMECULA PRESS

... and PROCEDURES FOR STUDY and REVIEW ONLY!

> It is almost a certainty that the CHECK-RIDE will require a "**FUEL JETTISON**" demonstration. The (most likely) place where this will occur is during a **MAXIMUM ATOG TAKE-OFF / ENGINE FAILURE** situation. Stay alert!

How MUCH gas ya gonna dump?

Here is one way to calculate how much fuel to dump:
- The **ZERO FUEL WEIGHT** is the weight of the airplane without any **FUEL**.
- The **GROSS WEIGHT** at any given time is the weight of the airplane and the **FUEL ON BOARD**
- The **MAX LANDING WEIGHT** is: 630,000 Lbs

We need to know what is the "**ESTIMATED GROSS WEIGHT AT LANDING**. Here are some thoughts on that:
- Estimate the **TIME** it will take to get to the **DIVERT AIRPORT**.
- Take the **FUEL BURN RATE**. You can estimate that by looking the FF (fuel flow indicators).
- Multiply the **BURN RATE** times the **TIME TO DIVERT FIELD**. That will equal **FUEL** that will be burned.
- Subtract that from the **PRESENT AMOUNT of FUEL**.
- ADD this to the **ZFW** (Zero Fuel Weight) ... this will give you the estimated **GROSS WEIGHT** of the time of the airplane arrival at the divert airport. If that is greater than 630,000 Lbs; then you may want to dump fuel. If **NOT**, then no fuel dump may be required.
- **SUBTRACT 630,000** from the **GROSS WEIGHT** figure and that will be the estimated **AMOUNT OF FUEL TO DUMP**.

Divide the amount of fuel to dump by **4650** and get the number of minutes required to get the airplane down to landing weight.

With this information, we know when to start dumping in order to arrive at out **DIVERT FIELD** with the weight of the airplane below the **630,000#** limit?

What this means is that as we get within about 20 minutes of landing we should be dumping fuel, otherwise we are going to be landing "overweight".
Of course, if we are *JUST AFTER TAKE-OFF AND RETURNING TO LAND*, then we would not have to include time to our divert field and the calculation becomes much simpler. In fact, since we don't have to concern ourselves with enroute burn calulations, the airplane **FUEL JETTISON** system will do all that for us automatically.

Here is how to setup and accomplish a "**FUEL JETTISON**".

SAMPLE PROBLEM

FUEL DUMP CALCULATION

CURRENT GW = 849,600
− FOB   382,800

ZFW   466,800

TIME TO LANDING
   4:30

F/F = 30,000 PPH
FUEL BURN TO DEST
4.5 × 30 = 135,000 #

849,600
−135,000
―――――
714,600
−630,000
―――――
 84.6 # to dump.

The FUEL JETTISON RATE is 4650 # per MIN

84,600 ÷ 4650
= 18.19 minutes

© MIKE RAY 2014
WWW.UTEM.COM

page 181

HOW TO ... DUMP FUEL

747-400 SIMULATOR TECHNIQUES ...

STEP 1: Calculate "**FUEL TO REMAIN**" after dump complete.
This is kinda tricky ... you don't want to place the amount you want to dump;
but rather the amount you want to "remain" or be left in the tanks
AFTER YOU DUMP FUEL!!!

The goal is to get the airplane **GROSS WEIGHT** to **630,000 Lbs**
(Maximum Structural Landing Weight limit) or less
by the time you land the airplane.

Example: **630,000** Lbs minus **ZFW** = *Fuel remaining after dump*.

STEP 2: Select **JETTISON CONTROL** selector .... A or B.
Either one will activate the system.
If **FUEL JETT A** or **B** message is displayed on the **EICAS**, select the other system.

STEP 3: FUEL TO REMAIN .... Set using the "**FUEL TO REMAIN**" selector knob
and confirm the amount selected is displayed on the **EICAS ENG** page.

STEP 4: Open the plastic covers and
select **BOTH JETTISON NOZZLE VALVES** switches .... **ON**.

STEP 5: OVERRIDE PUMP 2 and 3 switches .... Verify **ON**.
It is possible that **FUEL OVRD 2** or **FUEL OVRD 3 EICAS**
message may be displayed. Ignore the message.

STEP 6: Select the **FUEL SYNOPTIC**
from the **EICAS** selector.
The **JETTISON TIME** in minutes will be
displayed on the **FUEL SYNOPTIC**.

page 182

© MIKE RAY 2014
published by UNIVERSITY of TEMECULA PRESS

... and PROCEDURES FOR STUDY and REVIEW ONLY!

The **FUEL JETTISON** system will automatically dump fuel down to the selected level. Jettison TIME is initially estimated using pre-programmed dump rates. 90 seconds after the jettison begins, the system starts updating the time predicated on actual fuel quantity changes.

Jettison is automatically terminated when TOTAL FUEL QUANTITY decreases to the FUEL TO REMAIN quantity. The JETTISON system will automatically deactivate all the all the respective override, transfer, and jettison pumps.

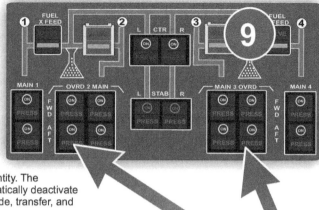

NOTE:
During fuel dumping with an **ENGINE SHUT DOWN**, the **OUTBOARD** may get slightly out of balance and a message **FUEL IMBAL 1-4** will be displayed. ***DO NOT DELAY LANDING TO BALANCE FUEL***

The jettison control system controls **FUEL BALANCING** in **MAIN TANKS 2** and **3** as the fuel is dumped.

If an **ENGINE IS SHUTDOWN**, the **OUTBOARD MAIN TANKS** may become imbalanced. This is due to the **MAIN TANKS 1** and **4** supplying the engines.

NOTE:
DO NOT EXTEND or RETRACT FLAPS
between positions 1 and 5
during fuel jettisoning!

When the FUEL JETTISON is completed:

STEP 7: FUEL NOZZLE VALVES .... BOTH OFF.

STEP 8: FUEL JETTISON control selector ... OFF

STEP 9: Setup FUEL PANEL for the new fuel load.

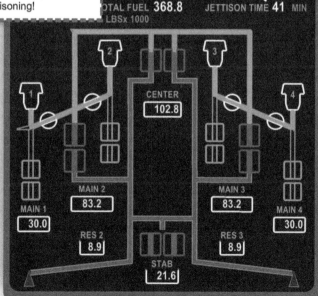

BRIEFING THE FLIGHT

At this point in the evolution of our "trip," we need to brief. It is suggested that there be two separate venues of briefing information. It is a good idea to break up the information so that the PF (Pilot Flying) does one list and the PNF (Pilot-Not-Flying) does the other. It is the Captains responsibility to ensure that all applicable items are completed prior to take-off.

CAPT or F/O BRIEF TAKEOFF DATA

- NOSE NUMBER / LOGBOOK / MRD
- RELEASE NUMBER
- ZFW / TOGW
- TRIM
- BLEED CONFIGURATION
- ALTIMETER SETTING
- TEMPERATURE
- FLAP SETTING
- EPR / N1
- RUNWAY LIMIT / PERF LIMIT
- WIND CORRECTION
- AIRSPEED CORRECTIONS
- T-PROCEDURES (DPWM)

CAPT or F/O BRIEF PRE-DEPARTURE BRIEF

- CREW DUTIES (Cockpit/cabin)
- FLIGHT PLAN CHANGES
- NOTAMS / POSBDs
- WINDSHEAR / RUNWAY CONDITION
- REJECTED TAKEOFF
- ENGINE FAILURE PROFILE
- T-PROCEDURE
- TAKEOFF PROFILE
- CLEARANCE / SID
- TERRAIN / OBSTACLES
- TRANSITION ALTITUDES
- TRAFFIC WATCH
- OMC BRIEFING

A few comments regarding the importance of these briefs to the simulator check-ride. The instructor will want particularly to hear you articulate your plan for a potential engine failure on take-off, and the same, of course, would be true in the real world.

> *For example, if a Thunderstorm is sitting on your potential "escape" routing (T page), but doesn't actually affect your planned departure routing. It would be "good practice" to delay your departure until both paths are acceptable.*

ABOUT 5 MINUTES BEFORE "P" TIME

"P" time is right on your flight plan and represents the planned pushback time. If there is going to be a delay in the pushback, you should notify CLEARANCE DELIVERY, GROUND CONTROL, and/or COMPANY so that they can make the necessary adjustment to their plans.

At this point in the flows, it is time to prepare for pushback. There is not a specific place where this evolution should start, and the Captain must use her/his judgement. Captain does five things:

> **SEAT BELT SIGN ON**
> **FLIGHT DECK DOOR LOCK**
> **HYDRAULIC DEMAND PUMPS**
> **HYDRAULIC PRESSURE INDICATOR**
> **PARKING BRAKE**

When the First Officer sees the Captain start these 5 items, then she/he does the three items:

> **SEND COMPANY REPORT**
> **LOAD WINDS**
> **ATC LOGON**

Here is a place to make some notes regarding last minute stuff that you think would be useful additions to my suggested "***BRIEF***" items.

F/O SEND COMPANY REPORT

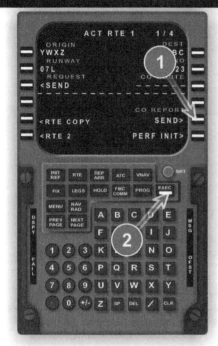

After the route has been verified by the Captain and determined that it is identical to the filed route;
ACTIVATE ROUTE,
EXECUTE, then

STEP 1: Line select "SEND>" (key 6R), *when illuminated*
STEP 2: PUSH EXEC key.

This will send the active route of flight to Dispatch.

In the flow of the set-up, at this point that should have been completed and now we are ready to tell **DISPATCH** that we are ready. We do this by sending the "**ACTIVE ROUTE OF FLIGHT**" to the dispatcher.

F/O LOAD WINDS

Once the route of flight is activated, Request the **WINDS**.
STEP 3: push **LOAD> (6R)**

When the winds are loaded, the **EXECUTE LIGHT** illuminates.

STEP 4: Push **EXECUTE** and the data is loaded into the **FMCS** (onboard computer).

If there are a lot of pages, sometimes you will get a message **WIND DATA UPLINK READY**. You have to continue to load those pages until all the pages are loaded.

> **CAUTION**
> ***DO NOT MAKE MORE THAN ONE WIND REQUEST!***
> *Here's the reason, Multiple wind requests may inhibit other datalink communications. Normally, the uplink will take one to three minutes.*

F/O perform ATC LOGON

NOTE:

The **ATC LOGON** needs to be established between 15 and 45 minutes prior to entering into the area providing **ATC** datalink services. This is not necessarily "at the departure" gate. You may have to **LOGON** to different **ATC** datalink sites as your trip progresses. This is difficult to describe; probably the best way to understand this is to fly a trip and then it becomes obvious what is intended. **NOT brain surgery!**

To access this **ATC INDEX** page: select **ATC** and you will get this page:

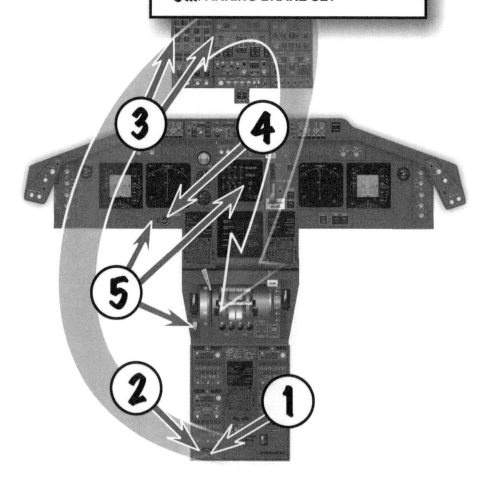

SEATBELT SELECTOR ... ON

FLIGHT DECK DOOR LOCK
PUSH the switch/light, and **OBSERVE** the **LKD** light **ON**

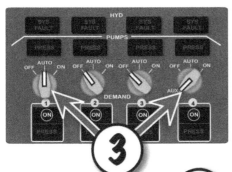

HYDRAULIC DEMAND PUMP SELECTORS
#4 selector ..... AUX
#1 selector ..... AUTO

HYDRAULIC PRESSURE INDICATOR
OBSERVE *3000* PSI

Indicates **HYD BRAKE PRESS** available from the **NORMAL BRAKE SYSTEM**; however, when the normal system isn't available, it will indicate **ACCUMULATOR** pressure.
NOTE 1: *The accumulator can **ONLY** be pressurized by the **#4 HYD** System.*
NOTE 2: *The accumulator can **SET AND HOLD** the **PARKING BRAKE.***
NOTE 3: *The accumulator is **NOT** designed to **STOP THE A/C**.*

PARKING BRAKE

While holding pressure on brake/rudder pedals, lift up parking brake lever. **VERIFY** lever remains locked in **UP** position.

EICAS msg
PARK BRAKE SET

BRAKE SOURCE LIGHT ... OFF

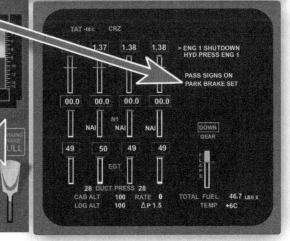

BEFORE START CHECKLIST

CHALLENGE (F)	RESPONSE
Departure briefing	Complete (C)
FMCs, radios	Programmed, set, verified (C, F)
IRSs	NAV, aligned (C)
Hydraulic demand pumps	No. 1 auto, No. 4 aux (C)
Fuel panel	____ Pumps on, ____ crossfeeds opened (C)
Fuel quantity	____ Pounds, cleared with ____ Pounds (C)
Oil quantity	Normal (C)
Oxygen check	Complete (C, F)
Altimeters	_____ , Set (C, F) (In/hPa)
Parking brake	Set, pressure normal (C)
Fuel control switches	Cutoff (C)
Autobrakes	RTO (C)
Cabin signs	On (C)

BEFORE PUSHBACK CHECKLIST

CHALLENGE (F)	RESPONSE
Doors	Closed (C)
Cabin preparation	Complete (C)

BEFORE TAKEOFF CHECKLIST

CHALLENGE (F)	RESPONSE
Flaps	____ Planned, ____ indicated, detent (C)
Control check	Complete (C, F)
LNAV, VNAV (as required)	Armed (C)
Nacelle anti-ice	On/off (C)

———————— MANIFEST CHANGES ————————

Trim	____ , ____ , ____ , Set (C)
Weight, speeds	Checked (F), set (C, F)
FMCs, radios	Programmed, set for departure (C, F)
Thrust	____ Reduced/max EPR, set (C)
MCP	V_2 ____ , heading ____ , altitude ____ , set (C)

———————— FINAL ITEMS ————————

Cabin notification	Complete (F)
Transponder	TA/RA (F)
Autothrottle	Armed (F)
EICAS	Recalled, cancelled (F)

AFTER TAKEOFF CHECKLIST
(To be checked *ALOUD* by the pilot not flying)

Landing gear lever	Off
Flaps	Up

APPROACH DESCENT CHECKLIST
(To be checked *ALOUD* by the pilot not flying)

Approach briefing	Complete
FMCs, radios	Programmed, set for landing
EICAS	Recalled, cancelled
Airspeed .... _____ Flaps, _____	, set (Ref)
Autobrakes .... Level _____	/Off
———————— *TRANSITION LEVEL* ————————	
Altimeters .... _____	, Set (In/hPa)

FINAL DESCENT CHECKLIST
(To be checked *ALOUD* by the pilot not flying)

Cabin notification	Complete
Landing gear	Down, green light
Speed brakes	Armed
Flaps .... _____ Planned, _____	Indicated

PARKING CHECKLIST

CHALLENGE (F)	RESPONSE
Parking brake	Set, pressure normal (C)
Stabilizer trim	6 Units (C)
Fuel control switches	Cutoff, fuel flow zero (C)
Transponder	Standby (C)
Radar	Off (C)
Emergency exit lights	Off (C)
Aft cargo heat	Off (C)
Fuel pumps	Off (C)
Anti-ice	Off (C)
Window heat	Off (C)
IRSs	Off (C)
Hydraulic demand pumps	Off (C)

747-400 **SIMULATOR TECHNIQUES ...**

CHECKLIST MANIA !

Just a couple of (maybe obvious) notes regarding the reading and completion of checklists. It is considered a no-no to memorize checklists. In general most of the checklists are **"CHALLENGE - RESPONSE"**, that is to say, one of the pilots reads the item and the other pilot responds. There are other checklists that require one pilot, referred to as the **PNF** (Pilot Not Flying) or **PM** (Pilot Monitoring), to read the item aloud, then the other pilot, usually the **PF** (Pilot Flying, will check to see that the "challenge" item was accomplished. The concept of the checklist is to ensure that the items, when they are read aloud, are completed.

> If not completed, the checklist should be stopped until the item has been completed. Delays, subsequent changes to items, or interruptions to the flow of the checklist necessitates re-starting the checklist from the beginning. **BIG CHECKRIDE ITEM!**

Specifically the **_After Take-Off Checklist_**, while it should not be memorized ... the **PNF** should be so familiar with the items on the checklist that they can be accomplished with a minimum of time and effort. It is especially true during the rush of the check-ride that **_ALL_** applicable the checklists must be accomplished quicky, completely, and with complete accuracy.

It is also important, in the cases where there is an irregularity or emergency that requires several checklists, that the pilot flying (**PF**) or the captain be familiar enough with the procedure and the airplane systems to be able to request the checklists by name in the proper order.

For example, during the **ENGINE FIRE** procedure ... there are several other procedures needed to be addressed in order to complete the whole process. These would include the **QRC** procedures for **FIRE ENG ()/ SEVERE DAMAGE/SEPARATION** then the procedures for **FUEL JETTISON** and **THREE ENGINE APPROACH and LANDING**, Then the **EICAS** messages need to be addressed in turn, and then the systems have to be reconfigured and the **DIVERT** routing (if needed) has to be input into the **CDU/FMC**, and then ... the list just goes on and on.

EACH CHECKLIST MUST BE COMPLETED EXPEDITIOUSLY AND ACCURATELY !!!

CAPT - F/O complete BEFORE START CHECKLIST

BEFORE START CHECKLIST

CHALLENGE (F)	RESPONSE
Departure Briefing	Complete (C)
FMCs, radios	Programmed, set, verified (C,F)
IRSs	NAV, aligned (C)
Hydraulic demand pumps	No. 1 auto, No. 4 aux (C)
Fuel panel ____	Pumps on, ____ crossfeeds opened (C)
Fuel quantity ____	Pounds, cleared with ____ pounds (C)
Oil quantity	Normal (C)
Oxygen check	Complete (C,F)
Altimeters	____,Set (C,F) (In/hPa)
Parking brake	Set, pressure normal (C)
Fuel control switches	Cutoff (C)
Autobrakes	RTO (C)
Cabin signs	On (C)

GETTING READY FOR ...
PUSHBACK

This will be either HECTIC or relaxed ... and a lot of what is going on is outside the framework of the flight crew and the tasks it must perform. The Captain has got to be able to take control of the situation and create an environment where all the multitude of trivial and important things can be sorted out and completed.

SOME COMMENTS ABOUT CRM
*(CRM = Command Resource Management also called
CLRM = Command Leadership Resource Management)*

The check-person will be looking for you to take an active role in creating the cockpit climate for a smooth transition during moments of high stress and activity. This is a great opportunity for the flight crew to get some brownie points. CRM or Crew Resource Management is observed throughout the check and will account for much of the Check persons opinion at the conclusion of the check. The technique involves developing a communication and listening toolkit that requires a mature and consistent approach to leadership. This results in addressing the individual problems that will arise in a manner which utilizes each of the other crew members capabilities and input while resolving the problems in a manner that will accommodate all of the relevant pieces of information that are available or can be obtained. Consulting sources outside the cockpit such as the Dispatcher or ATC is considered favorable. It is not up to the Captain alone to have the solution to a problem. Using anger, excessive emotion, and other manipulative responses will definitely jeopardize your chances for a successful check-ride experience. So, it is my opinion that you should be prepared to enjoy the check-ride as much as possible and demonstrate your abilities to be a dependable and capable Captain to the best of your ability.

CAPT confirms DOOR STATUS

Generally this procedure is initiated by the **GROUND PUSHBACK PERSONNEL** contacting the cockpit with this report:

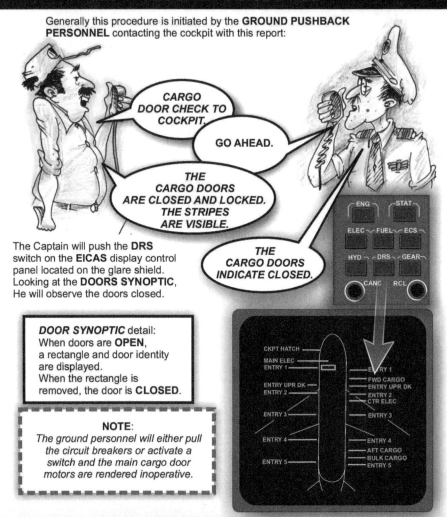

The Captain will push the **DRS** switch on the **EICAS** display control panel located on the glare shield. Looking at the **DOORS SYNOPTIC**, He will observe the doors closed.

DOOR SYNOPTIC detail:
When doors are **OPEN**, a rectangle and door identity are displayed.
When the rectangle is removed, the door is **CLOSED**.

NOTE:
The ground personnel will either pull the circuit breakers or activate a switch and the main cargo door motors are rendered inoperative.

CAPT verifies CABIN READY

When **ALL THE DOORS** are closed and the cabin is ready for departure:

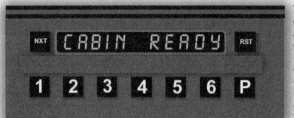

The **CHIEF PURSER** (also called the "A" or #1 Flight Attendant) will advise the cockpit when ready either by:
-**INTERPHONE** using voice,
or
-selecting **6P** on the **INTERPHONE** which creates a **CABIN READY** display on the **INTERPHONE CONTROL PANEL**.

CAPT - F/O   BEFORE PUSHBACK CHECKLIST

BEFORE PUSHBACK CHECKLIST

CHALLENGE (F)	RESPONSE
Doors	Closed (C)
Cabin Preparation	Complete (C)

CAPT - F/O   BEFORE PUSHBACK CLEARANCE

Obtain a **PUSHBACK CLEARANCE** (if required). The information regarding pushback requirements will be found on the **AIRPORT 10-7** pages.
The frequency for the controlling agency will be listed; however, **GROUND CONTROL** is normally the agency if no other information is listed.

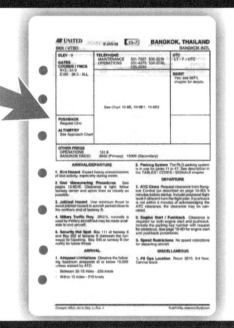

F/O   BEACON LIGHT SWITCH .... BOTH

The BEACON LIGHTS should be ON anytime the airplane is being moved or the engines are running or being started.

747-400 SIMULATOR TECHNIQUES ...

AFTER COMPLETING THE BEFORE PUSHBACK CHECKLIST

The **GROUND PERSONNEL** will initiate the **PUSHBACK SEQUENCE** dialog. Once voice contact is made, it is expected that the Captain will request the **"PUSHBACK CLEARANCE"** from appropriate authority ... usually **GROUND CONTROL** or **RAMP CONTROL** or authority as directed on the airport diagrm plate (Generally the 10-7 page).

GROUND / CAPTAIN

"Ground to cockpit: Pre-departure check complete. Ready for ___ person pushback."

    "Standby for pushback clearance."

"Standing by for pushback clearance."

    "Cleared to push, Brakes set."

"Roger, cleared to push. Release brakes."

    "Brakes released."

"Cleared to start engines.."
The command: "Cleared to start engines" (may also include which engine will be started) is to be given at the <u>**sole discretion of the ground crew**</u>.

    "Roger, cleared to start engines."

"Set brakes."

    "Brakes set, pressure normal."

"Tow bar disconnected."

    "Disconnect headset."
The Captain may desire for the ground crew to delay their disconnect, and they will not leave from under the nose of the aircraft until cleared to do so.

"Disconnecting, watch for salute."
Ground may indicate location of the salute: such as "check for salute out the right side of the aircraft". If a delay is expected or incurred, he will state "standby" and give the reason.

Ground guy stands at attention and then ... salutes.
    Cockpit crew acknowledges salute by flashing the
        RUNWAY TURNOFF light **ON - OFF**.

Ground guy gives the **"RELEASE FROM GUIDANCE SIGNAL"**.
And then He will extend both arms in the direction of the expected taxi.

© MIKE RAY 2014
published by UNIVERSITY of TEMECULA PRESS

ENGINE TECHNICAL STUFF

There are different engine configurations available for the 747-400. I am going to just focus on one specific type. All the other type of engines have similar performance and operating characteristics.

The 747-400 that we are considering is powered by four colossal **PRATT & WHITNEY PW-4056** engines. Each of these monster engines produces about 56,000 pounds thrust for a whopping total of 224,000 pounds of available thrust.

There are two rotating sections on this engine: an N1 rotor and an N2 rotor.

The N1 rotor is that huge fan section that you see when you look into the front of the engine. It also is connected to a low pressure compressor and a low pressure turbine section.

The N2 rotor is the set of blades that you would see if you look into the tailpipe of the engine. It is considered the "HOT" section of the engine and consists of a high compression section and a high pressure turbine section. This is the section where the EGT (Exhaust Gas Temperature) is measured. Interestingly, the N1 and N2 rotors are mechanically independent. It is almost like having two separate jet engines mounted in tandem. They create a very fuel efficient and powerful thrust team. The N2 rotor is used to drive the engine accessories.

	TIME LIMIT	COLOR	EGT (C°)
START	Momentary	RED	535
TAKE-OFF	5 Minutes	RED	650
MAX CONT	Continuous	AMBER	625
ACCELERATION	2 Minutes	RED	650

Min OIL PRESSURE	70 PSI
Min OIL QUANTITY	18 Quarts (Before engine start)
Min OIL QUANTITY (Eng at IDLE)	OK between 15 to 20 Quarts

Max OIL TEMPERATURE	177 °C
OIL TEMP (CAUTION ... up to 20 minutes)	164 to 177 °C
OIL TEMP (Continuous Operation)	163 °C
OIL TEMP (MINIMUM prior to take-off)	50 °C

N1 (MAXIMUM %)	114%
N2 (MAXIMUM %)	105.5%

The whole engine combination is "started" by a bleed air powered starter motor that can be engaged with the N2 rotor and cause it to spin with sufficient rotational velocity so that when the ignition cycle is started, the engine will operate on its own, and the starter will automatically disengage.

The start motor is normally powered by air from the APU; however, air from a ground cart or another running engine can be used. When using the air from another engine this is referred to as a "cross bleed start."

When the START SWITCH is pulled by the pilot, a solenoid will hold it open. The action opens two valves; the START VALVE and the ENGINE BLEED AIR VALVE. Both MUST be open to operate the starter. The light in the switch indicates that the START VALVE has opened. This activity requires AC power.

Ignition is supplied by two igniters in each engine. They require AC power.

At the end of the start cycle (52% N2), the start switch snaps in and the light goes out indicating that the 2 valves have closed and the starter has disconnected.

747-400 SIMULATOR TECHNIQUES ...

ENGINE START

DISCUSSION

There are basically two different types of **Boeing 747-400** start procedures. "Some" airplanes are equipped with an **AUTO-START** capability; others, retain the **MANUAL START** only

AUTO START (on the ground)

This automated system works great and is simple. Here are some of the features of the **AUTO-START** system.

       - Allows the **EEC** to automatically **ABORT** the start for:
           - **HOT START**
           -**HUNG START**
           -**NO EGT RISE**
But does **NOT** monitor **OIL PRESS** or **N1 ROTATION**.

       - Then it **MOTORS THE ENGINE FOR 30 SECONDS**, and makes another start attempt. The **EEC** is programmed to make **THREE AUTO ATTEMPTS** before aborting the start sequence.

AUTO-START also monitors **EGT** rise profile and if it detects a potential **HOT START** or **HUNG START**, it will shut off the fuel, adjust the fuel schedule, and re-applies the fuel flow for another start attempt.

If the **EEC** detects an impending **HOT START** or **HUNG START AFTER** starter cutout, **AUTO-START** sequence is terminated immediately. *In this case, the engine does not automatically motor for 30 seconds.*

Starter cut-out occurs at **50% N2 RPM**.

Operation of the **AUTO-START** is really intuitive and simple: When cleared to start ...
       - **PULL ENGINE START SWITCH,** and
       - **FUEL CONTROL SWITCH to RUN.**
       - **MONITOR** engine parameters.

MANUAL START (on the ground)

The tutorial will discuss the **MANUAL START** procedure in greater detail. Most airplanes are equipped with the **AUTO-START,** and since the monitoring and evaluation of the start is basically the same for both types of airplanes, it seems logical to cover the most complex procedure. There is always the chance that you will encounter a **MANUAL START** at some point in your 747-400 career.

After clearance to start engines is received:

F/O does four pre-start steps

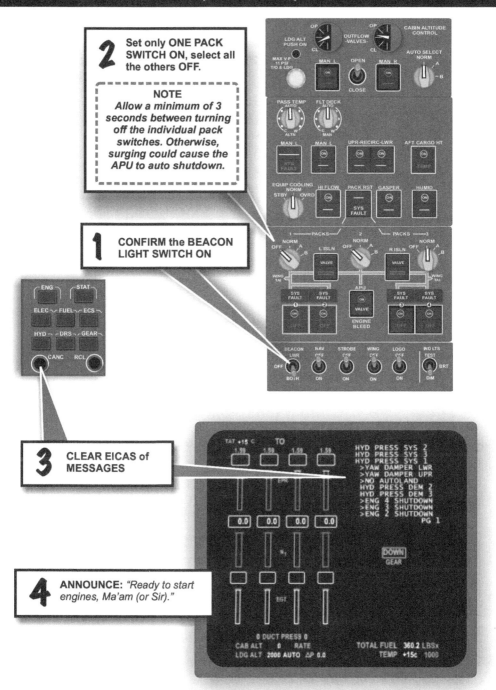

1 CONFIRM the BEACON LIGHT SWITCH ON

2 Set only ONE PACK SWITCH ON, select all the others OFF.

NOTE
Allow a minimum of 3 seconds between turning off the individual pack switches. Otherwise, surging could cause the APU to auto shutdown.

3 CLEAR EICAS of MESSAGES

4 ANNOUNCE: *"Ready to start engines, Ma'am (or Sir)."*

MANUAL ENGINE START SEQUENCE

Start sequence varies with different airlines.
I was taught to use the sequence: 4 - 1 - 2 - 3 when anticipating a 4 engine taxi.
Some airlines use other sequences such as: 4 - 3 - 2 - 1;
but as a general statement, NORMALLY expect to start engine #4 first.

> If the airplane weighs less than:
> ### 700,000 # GW
> Taxi with less than four engines should be considered.
> Start sequence for three engine taxi is: 4 - 1 - 2.
> Start sequence for two engine taxi is: 4 - 1.

The **CHECK CAPTAIN** will be really impressed to hear you suggest this; but before you do, consider what a three engine taxi entails, particularly if the weather in windy or wet and icy.

1 CAPT: states authoritatively, *"ENGINE START."*
or "start number 4," or "Let's do a three engine taxi, start number 4."

2 F/O: **PULL** appropriate **ENGINE START SWITCH**.
Solenoid will hold switch out.

3 CAPT - F/O: **OBSERVE** Start valve light in switch illuminates.
The light indicates that both **START VALVE** and the **ENGINE BLEED AIR VALVE** have opened.

4 CAPT - F/O: **OBSERVE DUCT PRESS** initially rise.

5 CAPT - F/O: **OBSERVE N2** increases.

6 CAPT - F/O: **OBSERVE OIL** pressure increases.

> **NOTE:**
> A significant drop in **ENGINE OIL QUANTITY** after **ENGINE START** and during **GROUND OPERATIONS** is *NORMAL!*

ENGINE START

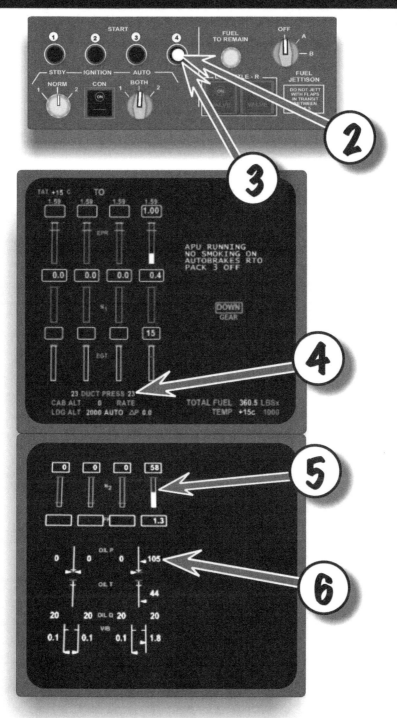

page 201

747-400 SIMULATOR TECHNIQUES ...

ENGINE START

7 At 25% N2 or MAXIMUM MOTORING, whichever comes first, but not less than the **FUEL-ON-COMMAND** indicator:

DEFINITION
The "**FUEL-ON-COMMAND** indicator" is the little horizontal tic mark on the N2 indicator of the lower **EICAS**. It indicates the minimum **N2 RPM** at which the **FUEL CONTROL** switch may be moved to **RUN** during start.

8 F/O: FUEL CONTROL SWITCH; LOCK in RUN (UP) position.

TECHNICAL INFO:
When the **START SWITCH** is pulled, the selected ignitor on each engine energizes when the respective **FUEL CONTROL** switch is in the **RUN** position and **N2 is less than 50%**. The selected ignitor de-energizes when the **FUEL CONTROL** switch is placed in **CUT-OFF**.

9 CAPT - F/O: OBSERVE some FUEL FLOW

NOTE:
If the **ENGINE FUEL FLOW** indications are abnormally high immediately after **FUEL** is commanded **ON**
..there is **NO PILOT ACTION REQUIRED IF:**
  - **FUEL FLOW IMMEDIATELY** thereafter stabilizes at "**NORMAL**" start fuel flow within **15 SECONDS**,
    AND
  - All other indications are **NORMAL**.
    IF ENGINE FUEL FLOW remains **HIGH**, be prepared for possible **HOT START** and/or **ENGINE TAILPIPE FIRE!**

10 CAPT - F/O: OBSERVE EGT rise within **20 SECONDS**.

11 CAPT - F/O: OBSERVE some **N1 ROTATION** prior to **40% N2**.

ENGINE START

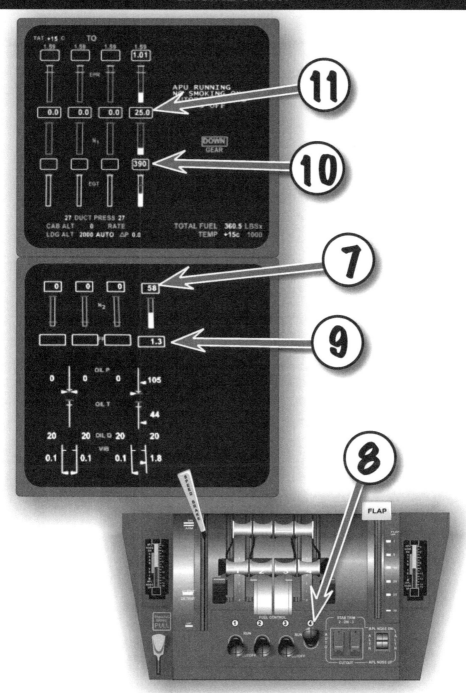

ENGINE START

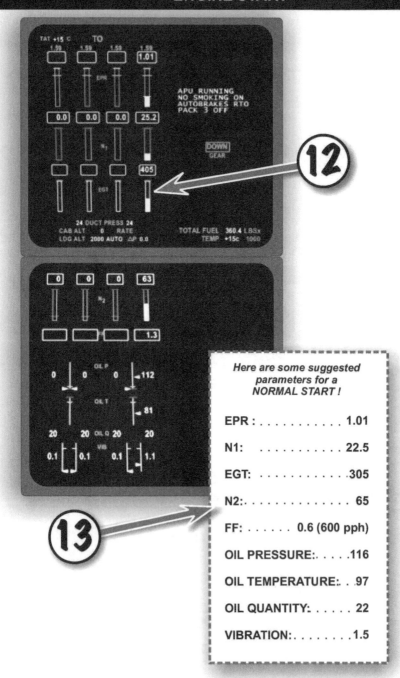

Here are some suggested parameters for a NORMAL START !

EPR : 1.01

N1: 22.5

EGT: 305

N2: 65

FF: . . . . . . 0.6 (600 pph)

OIL PRESSURE: 116

OIL TEMPERATURE: . . .97

OIL QUANTITY: 22

VIBRATION: 1.5

... and PROCEDURES FOR STUDY and REVIEW ONLY!

ENGINE START

12 **CAPT - F/O:** verify **EGT START LIMIT RED LINE** disappears when the engine has stabilized.
This should have occurred within **2 MINUTES** of moving the **FUEL CONTROL SWITCH** to **RUN**.

13 **CAPT - F/O:** Monitor the engine indications and **EICAS** for irregularity messages.

CAPT - F/O must be on the alert for an
ABNORMAL ENGINE START.

HERE ARE
THE ABNORMAL CONDITIONS
to look for:

NO EGT rise within 20 seconds after moving FUEL CONTROL SWITCH to RUN.

HIGH INITIAL FUEL FLOW.

EGT RAPIDLY approaching EGT start limit of 535 degrees C.

NO N1 increase by 40% N2.

NO OIL PRESSURE by 40% N2.

EGT start limit of 535 degrees C exceeded.

N2 fails to reach stabilized idle within 2 minutes after moving FUEL CONTROL SWITCH to RUN.

PNEUMATIC AIR interruption.

ELECTRICAL POWER supply interruption.

EICAS engine display disruption.

If you encounter these conditions:
The first step of the QRC is a memory item:

FUEL CUTOFF SWITCH .... CUTOFF

FOOTNOTE: *There have been upgrades to the **EEC** (Electronic Engine Control) units that can affect the **ENGINE IDLE** speeds as well as the **ACCELERATION** profile.*

AFTER START STUFF

At Captain's discretion ... **AIR CONDITIONING PACKS** may be turned on prior to the **RELEASE TO TAXI** salute if all engines are started.

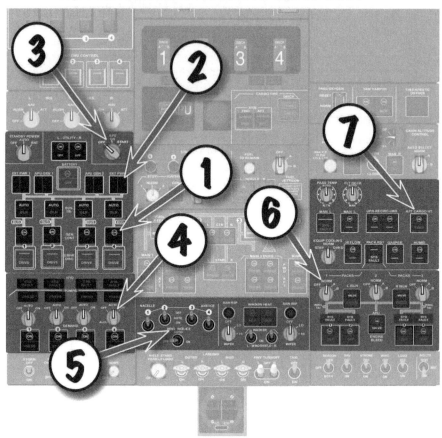

CAPTAIN DOES 5 THINGS:

1. Verify ELECTRICAL TRANSFER.
2. Verify EXTERNAL POWER DISCONNECTED.
3. APU SELECTOR.
4. HYDRAULIC DEMAND PUMP SWITCHES ... ALL AUTO.
5. NACELLE ANTI-ICE as desired.

F/O DOES 2 THINGS:

6. PACK CONTROL SELECTORS ... NORMAL.
7. AFT CARGO HEAT SWITCH ... ON.

AFTER START STUFF ... in detail.

1. CAPT: ELECTRICAL TRANSFER .... VERIFY ON.

If indication is **"ON"**, then the GENERATOR FIELD is closed and the GENERATOR CONTROL BREAKER is allowed to close automatically ... when system logic permits. If the indication is "OFF," then **BOTH** the GENERATOR FIELD and GENERATOR CONTROL BREAKER are OPEN.
When an engine is started, with APU generators or external power sources powering the AC system and the respective engine generator control switch is ON and the bus tie switches are in AUTO, the respective IDG **AUTOMATICALLY** powers its side of the tie bus when the voltage and frequency are within limits.

NOTE: *The previous APU or EXTERNAL power source is disconnected **AUTOMATICALLY**.*

2. CAPT: EXTERNAL POWER .... DISCONNECT.

The indication that the external power is connected is AVAIL being illuminated in the EXTERNAL POWER switches.

NOTE:
It may be necessary to prompt the ground people when external power is no longer needed.

3. CAPT: APU SELECTOR .... OFF

Once the GEN CONT OFF lights are extinguished, shut down the APUs.

4. CAPT: HYDRAULIC DEMAND PUMP SELECTORS............................. ON

The DEMAND PUMPS are operated by PNEUMATIC air.

NOTE:
If the switches are ON, the ENGINE DRIVEN pumps operate continuously.

5. CAPT: NACELLE ANTI-ICE............................ ON / OFF

If you need anti-ice for taxi or take-off, turn on now.

DISCUSSION:
Each engine supplies its our source for bleed air, which circulates ONLY in the nacelle. It, DUH, only operates when the engine is running. When selected, NAI is displayed on the EICAS to indicate minimum N1 limits if TAT below +10C. Also, continuous ignition is energized by this switch.

6. F/O: PACK CONTROL SELECTORS .... NORMAL

7. F/O: AFT CARGO HEAT switch .... ON

more AFTER START STUFF

CAPT and F/O:
CHECK the FMS
CDU take-off speeds.

Verify take-off speeds are indicated on the **TAKEOFF REF** page of the **CDU**. There are two things to look for.

1. If the speeds are in "small font." It is SOP to have all the speeds indicated as large dark fonts, even if there is no change to the value. To change from small font to large, simply push the button next to the speed number on the CDU.

2. Check the speeds indicated on the CDU display and determine if they differ from the speeds calculated by "hand." If there is a discrepancy, it MUST be resolved and the proper values placed in the appropriate position on the CDU.

F/O:
EICAS .... RECALL,
RESOLVE, CANCEL

Push the **RCL** switch and check the **EICAS UPPER DISPLAY** for alert messages. These can be warning (**RED**), caution (**AMBER**), advisory (**WHITE**), memo (**WHITE** at bottom of the display). All the **WARNING** and **CAUTION** messages must be appriopriately addressed! If an **EICAS** alert message appears, do what is necessary to extinguish the message. Reconfigure the aircraft, accomplish appropriate **IRREGULAR** procedure, contact **SAMC** (24 hour on-call maintenance facility) to determine if **MEL** relief is available.
Make appropriate logbook entry if applicable.

> **NOTE:**
> THE EICAS SCROLL SHOULD
> BE COMPLETELY RESOLVED
> PRIOR TO TAKE-OFF.

... and PROCEDURES FOR STUDY and REVIEW ONLY!

BEFORE TAXI OUT

WARNING:
THE CAPTAIN MUST NOT REQUEST A TAXI CLEARANCE NOR RELEASE THE PARKING BRAKE UNTIL BOTH OF THE FOLLOWING SIGNALS ARE RECEIVED AND ANNOUNCED:

SALUTE SIGNAL,
and
GUIDANCE SIGNAL (or RELEASE FROM GUIDANCE SIGNAL!)

NOTE:
If the signalman is NOT in a position to be seen by the Captain, it is permissible for the First Officer, with the Captain's concurrence (of course), to view the SALUTE and advise the Captain. **ONLY** then may the Captain acknowledge the receipt of the salute by
FLASHING A RUNWAY TURNOFF LIGHT!

THERE ARE SIX DISTINCT STEPS.

1 (C) "I HAVE A SALUTE" .... announce

2 (C) announce:
"AND GUIDANCE" or "RELEASE FROM GUIDANCE."

WARNING:
IT IS CRITICAL TO THE SAFETY OF THE OPERATION THAT NEITHER THE CAPTAIN NOR THE FIRST OFFICER SET THEIR RESPECTIVE AUDIO PANEL CONTROLS UNTIL THE CAPTAIN ACKNOWLEDGES THE MECHANIC'S SALUTE. DOING SO MAY RESULT IN MISTAKENLY ASSUMING TAXI CLEARANCE HAS BEEN RECEIVED AND MOVING THE AIRPLANE PRIOR TO ALL PERSONNEL / EQUIPMENT BEING CLEAR FROM BENEATH THE AIRPLANE.

3 (C - F/O) AUDIO PANELS .... SETUP.

4 The Captain will request a **TAXI CLEARANCE** when She / He is ready to listen and assimilate the information.

5 (F/O) TAXI CLEARANCE .... OBTAIN.
Here's the DEAL!
Do not even think about getting a clearance until AFTER the Captain requests it. Ensure that both pilots are monitoring the frequency and free from other duties.

6 (C) PARKING BRAKE .... RELEASE.

THIS IS A SERMON, PAY ATTENTION !!!

DO NOT ...

NEVER-NEVER-NEVER

IT IS **EXTREMELY** IMPORTANT that you DO NOT MOVE the jet without both:
- **SALUTE** and
- **RELEASE SIGNAL**

I am devoting a whole page to this topic because it is SO IMPORTANT !.

DISCUSSION: On this huge moving apartment house it is very easy to get all pre-occupied with some event in the cabin or cockpit and fail to observe the SALUTE-RELEASE from the ground man.

Perhaps you can see the tug disappearing into the distance and assume that all personnel have left and are clear ... DANGER - DANGER - DANGER.

There is no way to determine if some unsuspecting ground guy isn't waiting, still plugged in, for a release from you. There is NO WAY to see what's going on down there.

CAUTION

BEFORE you release the brakes and start to move this monster ... be ABSOLUTELY CERTAIN that there is no one under the airplane. The ONLY way to know this is to get a SALUTE and a RELEASE FROM GUIDANCE !!!!!

DO NOT RELEASE PARKING BRAKE TOO SOON.

NO - NO !

NOTE: I just want to EMPHASIZE the point that this is a VERY dangerous place and you can easily KILL or MAIM someone.

Even though you are wanting to beat Brand X, or get a taxi clearance during a lull in the Ground Control chatter, or you just want to "get things going," BUT ... This IS NOT the place to make up time. TRUST ME, you do not be known as a ROCKET SCIENTIST!

If you do this ... You will be talking to strange men in black at a long green table. You will be writing letters and making reports until you get your job back.

HOW TO TAXI

STEP ONE
GET A CLEARANCE.

It is ESSENTIAL, particularly when traveling to destinations where English is not the primary language, that the WHOLE crew assist in interpreting and understanding the complete taxi instructions.

Both the Captain and the First Officer should have copies of the airport diagram available for reference.

STEP TWO
DON'T HIT ANYTHING.

NOTE:
First Officers should keep their feet near the brake pedals and feel free to stop the airplane without the Captain's permission if they feel it is necessary.

YIPE:
Minimum pavement width required to complete a 180 degree turn is 153 feet.

1: The official flight manual has a whole page discussing the fact that this is a really "BIG" airplane and you can not always see what is going on down there. **2:** They chat in great detail about how important it is to have ground personnel help when close to other stuff.
3: They state that the loss of the body gear steering makes turns more difficult.
4: They want you to know that the rudder is so huge that there is this big weather cocking problem at taxi speeds with strong cross winds.
5: The winglets are way out there, and even though the wings are pretty high, don't taxi with them over something, taxi around it.
6: During turns, the wingtip swings out further than the cockpit end. Inboard pods are only 4 feet above the surface.
7: Use groundspeed readout on the ND for taxi speeds:
Here are the recommended taxi speeds:

Straight ahead: 25 KTS
45 degree turns: 15 KTS
90 degree turns: 10 KTS.

BIG !!!

Tires get HOT ... so taxi using "periodic" brake applications to keep speed under control. Do not "ride" the brakes. Take turns slowly to avoid sideloading.

AVOID EXCEEDING 40% N1 !!!!

Allow time for the application of **TAXI THRUST** to take effect. The airplane can be really heavy and it takes time to get it to start to move. Idle thrust is usually enough to keep the big bird moving.

WARNING:
DO NOT MOVE FLAPS and FLIGHT CONTROLS UNTIL THE AIRPLANE IS MOVING UNDER IT'S OWN POWER AND IS CLEAR OF GROUND EQUIPMENT AND PERSONNEL.

DURING TAXI PROCEDURES

747-400 SIMULATOR TECHNIQUES ...

After selecting the desired FLAP POSITION, OBSERVE the flap position indicator on the upper EICAS move to the selected position.

CAUTION:
Moving the flaps REQUIRES that the flap movement should be stopped when passing through the detents; and placed positively in the next detent. If you don't, one or more control units may disconnect.

REALLY TERRIFIC TIP:
If a flap control disconnects, here is what to do:
CYCLE THE ALTN FLAPS ARM SWITCH TO ALTN, THEN OFF.

(F/O) select the EICAS STATUS PAGE

This will bring up the "Flight Control Indications" for the FLIGHT CONTROL CHECK.

(C,F/O) check the FLIGHT CONTROLS

When on the taxiway clear of congestion; the Captain will call for the "**CONTROL CHECK**."

CAPTAIN: CHECK RUDDER MOVEMENT.
Hold the nose gear steering bar in a centered position
and push the rudder full throw left and right
while observing the indices respond properly.

CAUTION:
IF THE RUDDERS ARE KICKED TOO VIOLENTLY;
FAILURE OF THE RUDDER COULD OCCUR.

F/O: CHECK AILERONS and ELEVATORS.
Move the controls through their full movement.
Avoid violent movement, and
always be aware of the other pilot's hands and arms.

(F/O) select the EICAS ENG PAGE

NOTE:
After engine start, it is NOT necessary to check status messages. The reason is that any message that will have an adverse effect on the safety of flight or requiring crew attention will appear as an EICAS ALERT MESSAGE.

DURING TAXI PROCEDURES

(C) arm LNAV and VNAV (as required)

Normal SOP calls for
BOTH **LNAV** and **VNAV** to be armed.

(C - F/O) complete BEFORE TAKEOFF CHECKLIST

Complete the checklist down to the MANIFEST CHANGES LINE.

BEFORE TAKEOFF CHECKLIST

CHALLENGE (F)	RESPONSE
Flaps	_____ Planned, _____ indicated, detent (C)
Control Check	Complete (C,F)
LNAV, VNAV (as required)	Armed (C)
Nacelle anti-ice	On / off (C)

--------------- *MANIFEST CHANGES* ---------------

IMPORTANT NOTE:
Ensure that the flaps are **_SELECTED_** to the **_PLANNED_** position.
You will get **_NO TAKE-OFF WARNING HORN_** for the mistake.
For Example: If you PLAN 20 on the CDU but take-off with 10, you
will get **NO WARNING** but you will have the wrong V speeds.

(F/O) check FINAL WEIGHT MANIFEST

The "**FINAL WEIGHTS**" are obtained by ACARS from the company. If there is anything that is more frustrating, I don't know of it. As you are taxiing towards the takeoff end of the runway, it seems that everything comes to a jarring halt while you "wait for the weights." It is particularly critical if you are #1 for takeoff and have to taxi clear because you don't have your final weights.

NOTE:
NEVER start your take-off without your updated "weights." No matter how inconvenient or difficult, you **_MUST_** have the updated weight check and a **PASSENGER** count that matches with the data from the company load planner.

(F/O) confirm TAKEOFF PERFORMANCE

On the **TAKEOFF REF** page of the **CDU/FMC**, confirm (and change if necessary) the V-speeds and any other items that have changed.

(F/O) set STABILIZER TRIM

Position the stabilizer to the **TAKEOFF** trim setting. Observe that:
the stabilizer trim indicator is in the green band, and
STAB TRIM indicator **OFF** flags are out of view.
The trim setting setting should reflect the same as
the latest information from the "load planner".

747-400 SIMULATOR TECHNIQUES ...

DURING TAXI PROCEDURES

(C - F/O) complete BEFORE TAKEOFF CHECKLIST

Once we have received the updated "WEIGHTS" ..
Complete the checklist down to the FINAL ITEMS LINE.

BEFORE TAKEOFF CHECKLIST

CHALLENGE (F)	RESPONSE
Flaps	____Planned, ____indicated, detent (C)
Control Check	Complete (C,F)
LNAV, VNAV (as required)	Armed (C)
Nacelle anti-ice	On / off (C)
--------------- MANIFEST CHANGES ---------------	
Trim	____, ____, ____, Set (C)
Weight, speeds	Checked (F), set (C,F)
FMCs, radios	Programmed, set for departure (C,F)
EGPWS/RADAR DISPLAYS	SET (C,F)
Thrust	____ Reduced / max EPR, set (C)
MCP	V2____, heading____, altitude____, set (C)
-------------------- FINAL ITEMS --------------	

(PNF, PF) display appropriate CDU PAGES

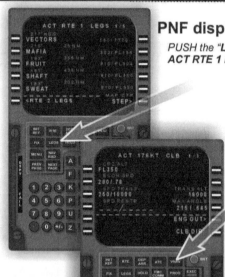

PNF display LEGS PAGE

PUSH the **"LEGS"** key to display
ACT RTE 1 LEGS page.

PF display VNAV ACT XXXX CLB

PUSH the "VNAV" key to display
ACT XXXX CLB page.

These are company "**CANNED**" suggestions. Obviously there are many different and unique situations. These are just suggestions and every Captain can coordinate with the other crewmember to setup whatever (qwirky or unique) display alternative she or he may desire.

(C or F/O) accomplish CABIN NOTIFICATION

At least **TWO MINUTES** prior to takeoff, notify the Flight Attendants via the PA:

"FLIGHT ATTENDANTS, PREPARE FOR TAKEOFF."

... and PROCEDURES FOR STUDY and REVIEW ONLY!

Some remarks about a confusing but important part of the airplane ...

THROTTLE (thrust lever) REVIEW

AUTO-THROTTLE SELECTOR SWITCH

TO/GA SWITCH
The TAKEOFF / GO-AROUND switch.

If the **TO/GA** is selected **AFTER** the airspeed gets above 50 KTS, the **AUTO-THROTTLE** will not engage and therefore it **WILL NOT** move the throttles to the takeoff range.

AUTO-THROTTLE DISCONNECT SWITCH

NOTE:
When we use this switch to disconnect the **AUTO-THROTTLE**, there are two things to be aware of.

*First, the auto-throttle switch on the **MCP** will remain "armed."
Second, we need to select it twice to disarm the warning horn that will sound if we select it only once.*

☠ DANGEROUS DISCUSSION:
*When you sit perched 5 stories above the earth, it is very difficult to accurately determine just how fast this big Momma is moving across the earth below. In fact, if you are cleared for a "rolling take-off" and you push up the throttles in anticipation;
Or have a strong headwind coupled with a low rolling speed, it is* **EASY** *to exceed* **50 knots** *waiting for the engines to stabilize before you push the TO/GA switch.
If you do this, the throttles* **WILL NOT** *move automatically towards takeoff power. You will be attempting to takeoff with some unknown wimpy power setting and could wind up in the canal.*

Let me EMPHASIZE; This is REALLY BIG!!!

ANOTHER IMPORTANT NOTE:
*If we screw up somehow, and at 65 KTS we notice that **"HOLD" DOES NOT ANNUNCIATE** on the PFD (YIPE!),
THEN:*
DISCONNECT AUTOTHROTTLE
(push thumb switch twice),
and
SET DESIRED THRUST MANUALLY!
(You do know what the desired thrust should be, don't you?)

50 KTS WARNING

© MIKE RAY 2014
WWW.UTEM.COM

747-400 SIMULATOR TECHNIQUES ...

BEFORE YOU BEGIN
THE TAKE-OFF EVOLUTION

We must take a moment at this point in the checkride to evaluate the situation. It is more than likely that the check person has sneakily introduced something into the environment which might affect the takeoff in a negative way. One of those would be to have the weather below **LANDING MINIMUMS** but above the **TAKE-OFF MINIMUMS**. It would be necessary to indentify a "suitable" **TAKE-OFF** alternative, notify **ATC** and **FILE** that alternative, and then "**SPECIFIED IN THE OPERATIONAL FLIGHT PLAN**". I interpret that to mean that the Captain must ensure that the TAKEOFF alternate is actually written on the onboard official flight plan.

If you can not return and land at the same airport, then you must designate a

TAKE-OFF ALTERNATE

T/O ALT REQ
When:

DEP AIRPORT is BELOW LANDING MINIMUMS

for
747-400
T/O ALT must be within

850 NM

or **NOT more than 2 HR**
from the departure airport
at **NORMAL CRUISE** in still air
with
1 ENG INOP

- NEW or HIGH MINIMUM CAPTAINS must use their minimums in this determination.

- If you decide to designate a TAKEOFF alternate AFTER filing your release; you MUST have the concurrence of the DISPATCHER and write that alternate on your original copy of the FLIGHT RELEASE DOCUMENT.

- If after dispatch, an ALTERNATE or a different or additional alternate is added to the flight release, that alte**rnate must be written on your original FLIGHT RELEASE DOCUMENT**.

© MIKE RAY 2014
published by UNIVERSITY of TEMECULA PRESS

some legal reasons
NOT TO TAKEOFF

- **MICROBURST:** The #1 reason **NOT TO TAKE-OFF** is a MICROBURST report.

- **TAILWIND EXCEEDS 10 Knots.**

- **DO NOT TAKE-OFF if:**
  STANDING WATER ... over 1/2 inch
  SLUSH .... over 1/2 inch
  WET SNOW .... over 1 inch

	TAKEOFF NOT PERMITTED	SUSPEND OPERATIONS (except emerg)
SLUSH	OVER 1/2"	OVER 1/2"
WET SNOW	OVER 1"	OVER 2"
DRY SNOW	OVER 4"	OVER 6"
STANDING WATER	OVER 1/2"	OVER 1"

RUNWAY CLUTTER CHART
(FOM page ALL WX-14)

- **ICING and FREEZING PRECIPITATION:**

  MODERATE RAIN:
  HEAVY FREEZING RAIN:
  HEAVY FREEZING DRIZZLE:

- **IF BRAKING ACTION NIL:**
  Takeoff NOT RECOMMENDED.

- **WEIGHT of AIRPLANE TOO GREAT** for the existing runway conditions.

- **AIRCRAFT SYSTEMS NOT READY:**
  Either checklist NOT completed, or
  Warning light or horn,
  Flight Attendants NOT ready,
  Ambiguity in clearance or routing,
  Other NO BRAINERS!

- Captain is to make the first 10 take-offs and landings after IOE. Captain or F/O under 100 hours and Captain under 300 hours have restrictions outlined on that page. There are also some EXEMPTION 5549 stuff and FAR PART 121.438 limitations.

KNOW THIS !

Here is the OFFICIAL definition of **CLUTTER** (a form of runway contamination):

STANDING WATER of 1/8 inch
SLUSH of 1/8 inch or greater
WET SNOW of 1/4 inch or greater
DRY SNOW of 1 inch or greater

Further, less than (the amounts listed in the definition above) are not considered clutter and no weight or V speed restrictions are required.

If clutter exists, there are pages in the Flight Manual to figure out what adjustments are necessary.
NOTE 1: using CLUTTER CRITERIA REQUIRES that you put the new clutter airspeeds on the AIRSPEED INDICATOR.
NOTE 2: Captain is supposed to make the takeoff.

KNOW THIS CHART!

	LIGHT	MODERATE	HEAVY
FREEZING RAIN	OK	NO-OP	NO-OP
FREEZING DRIZZLE	OK	OK	NO-OP
SNOW	OK	OK	OK

ICING and FREEZING PRECIPITATION CHART

PC ORAL QUESTION!

TAKE OFF:
Captain will ALWAYS make the take-off if the **TOUCHDOWN RVR is LESS THAN 1000 or ROLLOUT RVR LESS THAN 1000.**

LANDING:
Captain will always make the landing it the **VISIBILITY LESS THAN 1/2 MILE (1800 RVR).**

WARNING
It takes a LAWYER to read these pages!

Therefore, expect the checkguy to ask questions about this stuff!

747-400 SIMULATOR TECHNIQUES ...

WIND

Before we can start the takeoff, we have to make certain that the airplane is operating within the limitations of the **WIND** component restriction. The wind limitations for the airplane are listed in this table. Notice that the **CROSSWIND** is listed as "**demonstrated**". That means that it is a limitation that was derived by actually flying the airplane during certification. This "limitation" can be exceeded, particularly in an emergency.

The **LANDING TAILWIND** has some flexibility also, as the exceptions are listed below the table.

Since we are only human pilots, and it is really difficult to figure out what the actual headwind component is if wind doesn't exactly align with the runway ... there is a useful **WIND COMPONENT CHART** available that will help out.

MAXIMUM WIND (knots)

CROSSWIND (Demonstrated)	TAKEOFF	30
	LANDING	30
TAILWIND	TAKEOFF	10
	*LANDING	10

*If the **TAILWIND** is greater than **10 KNOTS** during landing, up to a maximum of **15 KNOTS** then that airplane can be landed provided these restrictions are complied with:
- Minimum RUNWAY 9500 Feet.
- MANUAL landing ONLY!
- BRAKING ACTION GOOD or better.
- AUTO-BRAKE 3 or greater used.
- Maximum Landing weight within limits.

Using the "**WIND COMPONENT CHART**"

Example problem:
RUNWAY 31
WIND: 260/30

What is the **CROSSWIND COMPONENT**?

STEP 1: Since the example has us taking off on **RUNWAY 31**, that means that we are going to be heading 310 degrees, so we will imagine that (310) is the orientation of the chart.

STEP 2: The wind is **COMING FROM 260** degrees, we can determine the difference between the airplane and wind heading (310 minus 260 = **50 degrees**).

STEP 3: Place a wind vector extending from the "0" (Center of the chart) to a point equal to the wind speed (**30 KNOTS**) along the **50 DEGREE** "angle line.

STEP 4: Drop a line from that point straight down to the "**CROSSWIND COMPONENT**" line at the bottom of the chart.

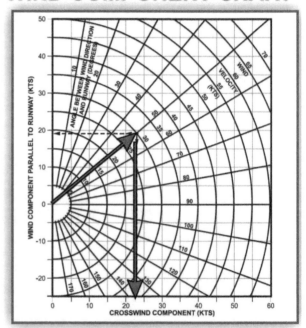

WIND COMPONENT CHART

STEP 5: Interpolate the value to be **23 KNOTS**. We can also determine that the HEADWIND component is 19 Knots.

Since 23 Knots is less than the 30 Knot limit on the **MAXIMUM WIND** chart we are not limited by crosswind. In like manner, if the wind was from the rear, we could determine both the TAILWIND and the HEADWIND components.

... and PROCEDURES FOR STUDY and REVIEW **ONLY!**

EOSID
(Engine Out Standard Instrument Departure),
Also called **T page, Escape Route,** or
whatever else you may know it as!

A long time ago, some airline managers took a hard look at the airports their airlines operated out of and evaluated the possible outcome should their departing airplanes experience an engine failure on take off. At some of those airports, they acknowledged that there existed the very real potential for a safety issue and so they developed special routing to assist their pilots in avoiding potential problems; and these "**ESCAPE ROUTES**" became required briefing material at these airlines. Good stuff. Then national aviation operating agencies (FAA, JARS, CAA, CSAS etc) looked at what these airlines had done and also thought it was a good idea for certain airports to have pre-figured "**ESCAPE**" routes created in case an airplane experienced an engine failure event on takeoff. These "special escape"routes were to be airport specific and be designed so as to take into account local conditions such as terrain, obstacles, and so forth. Their end-game goal was to provide a pre-thought-out way to get the airplane back to the airport without hitting something. Great idea!

Of course, every airport has the mandate that *"The **EXTENDED RUNWAY CENTERLINE** must have **TERRAIN** and **OBSTACLE CLEARANCE** up to 1500 FEET."* This "**ESOID ESCAPE ROUTE**" we are talking about here is something else in addition to that. It is interesting each participating airport and airline has the privilege to develop their own unique escape procedure and routing if they feel the published EOSID is inadequate or faulty. So, airlines have also published their own special engine out routing ... for example, over at United, they have "T" pages, Lufthansa has published its LIDO standards, and so forth.

Today, if the airport departure pages dictate a secondary departure routing, it will be necessary for the pilot to brief for and confirm that this escape routing to be useable. By that I mean, both the standard SID as well as the EOSID MUST conform to minimum standards such as convective activity (Clouds and weather) as well as landmark intrusion such as a construction tower or temporary event that would be indicated as a NOTAM or SPECIAL BULLETIN.

DISCUSSION:
If you are confronting a possible "**ESCAPE ROUTE**" then you have to also brief that departure as if you were going to actually have to fly it. By that I mean, for example, if there is significant weather along the pathway for the **EOSID**, then even though there may be a clear pathway along the **PRIMARY ROUTING**, you maybe should consider delaying departure until **_BOTH ROUTES ARE CLEAR_**.

Not All airports or runways have this requirement and generally speaking, the "T" page or **EOSID** routing is a simple turn to a heading or tracking to a waypoint or fix. Sometimes, the routing is more complex and establishing an alternate **ROUTE (RTE 2)** or additional **LEGS** segment on the **CDU** would be appropriate. Here is how to do that.

LOADING THE ESOID ROUTING.

If the "**T PAGE**", **EOSID**, or **ESCAPE ROUTING** is complex ... it is probably advisable to "pre-enter" in into the **FMC** using the **CDU**. There are basically **TWO WAYS** to do this.
- Use the **<RTE 2** feature, or
- Create a "**DISCONTINUITY**" at the end of the **LEGS PAGE** queue and enter the **ESCAPE ROUTING** below (or after) the Primary **RTE** on the **LEGS** page.
Let me make comments regarding both procedures.

Using the <RTE 2 technique

A lot of pilots become very proficient using the **RTE 2** feature because it can also be used as a utility to place **SIGMETS** and **TURBULENCE** reports on the **ND** as well as using the **RTE 2** to examine and pre-select **ALTERNATE** routing such as the **EOSID**. I personally seldom use this feature **BECAUSE** there is the real possibility of screwing up or erasing the whole **ROUTE 1** while you are inputting the information if you are not very careful. So, here is my suggestion ... unless you know what you are doing, I would not use the **<RTE 2** feature *on the checkride*. This is just me ... however, that being said, if you know what you are doing and don't get too rushed, the **<RTE 2** method is the technique recommended!

An ALTERNATE method

Entering the **EOSID** or **ESCAPE ROUTING** is based on the assumption that if we have an engine out event on take-off, we will "probably" not continue the flight on to our destination. To do that, an easy way is to establish a **DISCO** at the end of the **LEGS** queue and build the escape routing from there so it can be selected for use with a couple of quick key-strokes.

Here is how to place a **DISCONTINUITY** (also called a **DISCO**) at the end of the **LEGS PAGE** queue.

Let us assume that the Escape Routing was something like this ... In the event of an engine failure during take-off "*Proceed to directly to CLIFF and HOLD at 3000 FEET*."

STEP 1: Scroll the end of the **LEGS PAGE** queue using the **NEXT PAGE** key on the **CDU**.

STEP 2: Type the **FIRST FIX** of the **ESCAPE ROUTING** in the **SP** (Scratch Pad) using the keypad of the **CDU**. For us that would be "**CLIFF**".

STEP 3: **LS** (Line Select) to the -------- (dashed **ENTRY** line). Notice that there may be an **AMBIGUOUS FIX** message **POP-UP**. The **FMC** is asking the pilot to choose from among the alternatives that it has in the database. As a general rule, the top-most waypoint will be the most likely choice; however, this requires verification of correctness.
This will place the first way-point of the escape routing at the end of the queue.

STEP 4: **EXEC** (Execute).

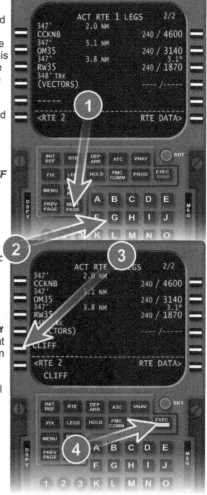

... and PROCEDURES FOR STUDY and REVIEW ONLY!

There is a little "trick" or secret to all this. It involves the use of a "pseudo" or "throw away" fix. This must be a waypint that exists in the database of the **FMC**. I use **KLAX** or **AAA**. You can make up your own.

STEP 5: Type the "*PSEUDO FIX*" (**AAA**) into the **SP**.

STEP 6: **LS** (Place) that pseudo fix AAA "*ON TOP OF*" the first waypoint of the escape routing which is now the last waypoint in the **LEGS** queue. That would be "**CLIFF**" in this case.

STEP 7: **EXECUTE**.

STEP 8: Select the **HOLD** key. This will place boxes into the **SP** (Scratch pad) for us to enter the location where we want to hold. In our case, we want to hold at the entry labeled "**CLIFF**" that is at the end of our **ESCAPE ROUTE**.

STEP 9: Scroll to the end of the queue. This will display the **DISCO** above the way-point **CLIFF**, isolating it from the rest of the **PRIMARY ROUTING**. And we can now complete the routing. Ours will include a it will include a holding fix, that is where we can dump fuel, sort things out, and prepare for the return to land scenario.

STEP 10: Select the way-point **CLIFF** to the **SP** and then to the **HOLD BOXES** and accomplish an ordinary garden variety holding pattern at **CLIFF** which could include the assigned altitude (**3000 FEET**) as well as a potential **HEADING** to the **LANDING RUNWAY** defining way-point. That would conclude the **ESCAPE ROUTING** setup.

When you have the "**V1 CUT**" and engine failure on takeoff event, you can call for the "**T PAGE**" or escape routing and the **PNF** can click-click-click have it up and staring you in the face on the **ND** and **LEGS** page in a flash.

How to activate the **ESCAPE ROUTING**.
At the point where you will need it, it is fairly intuitive for the **PNF** to scroll to the bottom part of the **LEGS** queue where the **DISCO** (discontinuity) is, and **LS** the first waypoint of the **ESCAPE ROUTING** (**CLIFF**) to the **SP**; then scroll to the first page and place the waypoint at the top of the queue.
Then **EXECUTE**. Once it is there, the rest of the primary routing is gone and you can concentrate on proceeding on the escape routing that you have prepared.

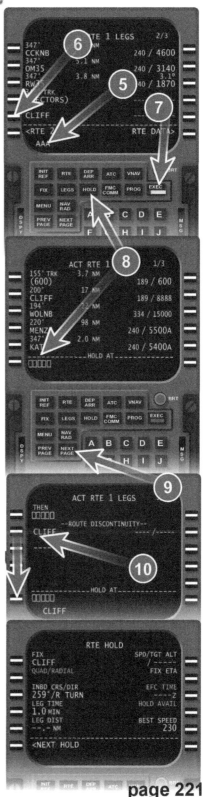

© MIKE RAY 2014
WWW.UTEM.COM

page 221

747-400 SIMULATOR TECHNIQUES ...

TOGA QUICKIE REVIEW of TAKE-OFF MODE

If you new to using the **TOGA** (Take Off Go Around) feature that is installed on every airliner that I know of, then you are in for a real treat. This is a fantastic automation that resolves many of the issues that arise during the power-up and take-off as well as the go-around phase of a flight.

Here are the two "**TOGA**" switches on the **747-400**. Depressing either one will activate the system.

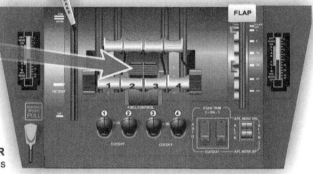

ON THE GROUND ... with **NO AUTOPILOT** engaged and **BOTH FLIGHT DIRECTORS** switches **OFF**, the **TOGA** system is **UNARMED**!

The system is **ARMED** when the "first" **FLIGHT DIRECTOR** is positioned **ON**. Three things occur:

- **FD** is displayed on the **PFD** as the **ENGAGED AFD**S mode.
- **TO/GA** is displayed twice on the **FMA** (Flight Management Annunciator) in the **ROL**L and **PITCH** modes.
- **PITCH** and **ROLL COMMAND BARS** are displayed on the **PFD**s.

During **TAKE-OFF**: At **100 knots**, the **FMC** selects a reference "**BAROMETRIC**" altitude that is used to:
- **ENGAGE LNAV** at **50 fee**t,
- **ENGAGE VNAV** at **400 feet,**
- **ENABLE A/T** (Auto-throttle) engagement (if not already engaged).
- **COMMAND AIRSPEED** acceleration for **FLAP RETRACTION,**
- **SET CLIMB THRUST** (if an altitude has been selected).

NOTE: *IMPORTANT!* When the airplane is less than **50 kts**, pushing the **TOGA** switch, engages the **AUTO-THROTTLE** in the **THR REF** mode. *If the AUTO-THROTTLE is NOT engaged by 50 kts, it cannot be engaged until 400 feet.*

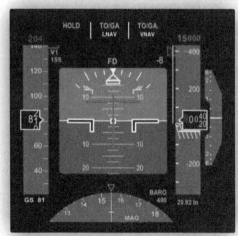

After the **TOGA** switch is depressed, and prior to **65 KTS**:

- **PITCH BAR** commands a fixed pitch of **8 degrees**.
- **ROLL BAR** comands **WINGS-LEVEL**.
- **AUTO-THROTTLE** engages in the **THR REF** mode.
- **THROTTLES AUTOMATICALLY ADVANCE to TAKEOFF THRUST**.
- **FMA** on **PFD** annunciates: **THR REF / TOGA / TOGA.**

At 65 Kts:
- **AUTO-THROTTLE** mode changes to **HOLD**.

Here is the **PFD** at **80 KNOTS**.

... and PROCEDURES FOR STUDY and REVIEW ONLY!

"CLEARED INTO POSITION"

BOTH CAPT and F/O move a switch

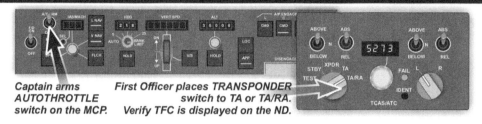

Captain arms AUTOTHROTTLE switch on the MCP.

First Officer places TRANSPONDER switch to TA or TA/RA. Verify TFC is displayed on the ND.

(F/O) PUSH RECALL/CANCEL for the EICAS

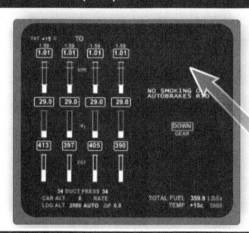

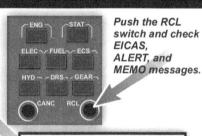

Push the RCL switch and check EICAS, ALERT, and MEMO messages.

The idea is to see if there are any 'hidden" messages that need attention. Once they are resolved, then the EICAS should be set up to receive new messages.

(C,F/O) complete BEFORE TAKEOFF CHECKLIST

BEFORE TAKEOFF CHECKLIST

CHALLENGE (F)	RESPONSE
Flaps	____Planned, ____indicated, detent (C)
Control Check	Complete (C,F)
LNAV, VNAV (as required)	Armed (C)
Nacelle anti-ice	On / off (C)

-------------- *MANIFEST CHANGES* --------------

Trim	____, ____, ____, Set (C)
Weight, speeds	Checked (F), set (C,F)
FMCs, radios	Programmed, set for departure (C,F)
Thrust	____ Reduced / max EPR, set (C)
MCP	V2____, heading____, altitude____, set (C)

-------------- *FINAL ITEMS* --------------

Cabin notification	Complete (F)
Transponder	TA / RA (F)
Autothrottle	Armed (F)
EICAS	Recalled, cancelled (F)

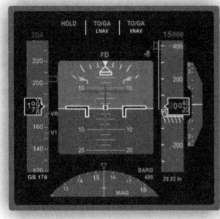

At **LIFTOFF**:
- **PITCH BAR** commands **V2 +10**.
- If the airspeed remains above target airspeed for **5** seconds **PITCH BAR** changes to **V2 + 25** kts.

If **ENGINE FAILURE** occurs:
- **PITCH BAR** commands **V2**, if the airspeed is below **V2**.
- If airspeed between **V2 and V2 + 10**, the **PITCH BAR** will show the existing airspeed.
- If airspeed above **V2 + 10**, then **V2 + 10** will be displayed.
- **ROLL BAR** commands bank to maintain **GROUND TRACK**.

After **LIFTOFF**:
At **50 feet**:
- **LNAV** engages. **ROLL BAR** commands track to active route.
At **400 feet**:
- **VNAV** engages (if armed).
- **PITCH BAR** commands current airspeed until reaching **ACCEL** (acceleration) height.

- **AUTO-THROTTLE** remains in **HOLD** mode at **TAKEOFF THRUST** until **VNAV** or **FLCH** is engaged.
The **AUTO-THROTTLE** sets the selected **REFERENCE THRUST**.

- **FD** (Flight Director) **AFDS** engaement terminates when an **AUTO-PILOT** is engaged.

- **TOGA** is terminated when another **ROLL** and/or **PITCH** mode is selected.

At **ACCEL HEIGHT**:
- **PITCH** commands **REDUCED CLIMB RATE** and
- **PITCH** commands airspeed **250 Kts, Vref + 100 Kts, or Speed Transition speeds**.

At **THRUST REDUCTION HEIGHT** or **FLAPS 5**:
- **FMC** reduces thrust to **(ARMED) CLIMB THRUST**.

With the airplane on the ground, with no autopilot engaged and both flight directors switches off, the takeoff mode of the TOGA is armed when the first flight director switch is positioned on.

Now, I am going to say next is something for which I have no reference for. At the gate, when you have the engines operating, and you **ARM** the **AUTO-THROTTLE**s, there exists the chance that should you inadvertently touch or depress the **TOGA** switch, the engines will go to **TAKE-OFF THRUST**, so ...

> **DO NOT ARM THE AUTO-THROTTLES AT THE GATE**
> **I RECOMMEND THAT WE DO NOT ARM THE AUTO-THROTTLES**
> **UNTIL JUST PRIOR TO TAKING THE RUNWAY FOR TAKE-OFF.**

Not selecting the AUTO-THROTTLES until just before take-off is **SOP** anyway, so I am not suggesting anything new; but to re-inforce my argument, if the engines are running, do not arm the auto-throttles and select a flight director on without thinking about the **TOGA** switch.

"FLAPS 10 -10 -10"

Here is a **"UNOFFICIAL GOUGE"** that many old-timers use. Just prior to take-off they check three places for their take-off flaps and make it a litany. In our example, we have elected to takeoff with 10 degrees of flaps.
Here is how it goes:

"FLAPS 10"

Check **FLAP HANDLE** in the **10 DEGREE** notch. Some pilots bump the handle with the palm of their hand to see if it is solidly in the notch.

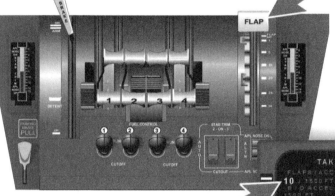

"FLAPS 10"

Check the **TAKEOFF REF** page to see that **FLAPS 10** is displayed.

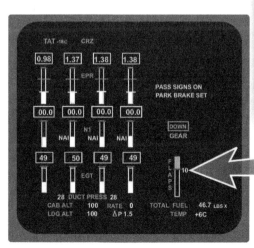

"FLAPS 10"

Check the **FLAP INDICATOR** on the **EICAS** at **10 FLAPS**.

This is not an official gouge, it seems like a good idea.

747-400 SIMULATOR TECHNIQUES ...

TAKING THE RUNWAY

When the airplane crosses the *"HOLD SHORT"* line:

BOTH CAPT and F/O turn on 2 lights

CAPTAIN:
Turns on
the **LANDING LIGHTS**
according to this schedule;
On EVEN numbered flights
use the INBD lights

CAPTAIN:
Turns on BOTH
**RUNWAY
TURNOFF LIGHTS**.

FIRST OFFICER:
Turns on:
STROBE LIGHTS and WING LIGHTS.
NOTE:
During restricted visibility, reflected flashback may preclude the use of the strobes.
If the Captain yells, "Shut those ***** strobes off!" this is usually a clear indication that they are not appropriate for conditions.

SOME TAKEOFF NOTES:

DISCUSSION:
REDO THE CHECKLIST ...
If any flight control is re-positioned (or not in the proper position) following the completion of the checklist,
or
There is a lengthy delay prior to takeoff.

---

2 HOURS and 5 MINUTES:
To prolong engine life, if the engines have been shut down for more than **TWO HOURS**, it is recommended that the engines run for **FIVE MINUTES** prior to application of takeoff thrust.

WHOOPS!
If a delay is encountered "in position" ...
Do not rely on holding the brakes with your feet. This big mother will creep. It will be a huge surprise to look up from "head down" and see the edge of the runway in front of the airplane.

Even with the **BRAKES SET**, there is so much residual thrust on this jet, that the airplane could (easily) begin to slide on a wet or icy runway.
FURTHER SURPRISE !
If you push up the throttles, this airplane can actually take-off with the brakes set
!!! **YIPE!**

"GEE, DUH ... What's that horn?"

© MIKE RAY 2014
published by *UNIVERSITY of TEMECULA PRESS*

... and PROCEDURES FOR STUDY and REVIEW ONLY!

TAKEOFF WARNING HORN
"Gee, What is that noise! DUH!!!"

If TAKE-OFF WARNING HORN sounds before the airplane is commited to takeoff, the takeoff should be discontinued and the problem corrected.

What causes the **TAKEOFF WARNING HORN** to sound?

**A/C ON GROUND,
FUEL CONTROL SWITCHES IN RUN,
ENG 2 or 3 in TAKEOFF RANGE,
AIRSPEED LESS THAN V1,
AND**:

1. ■ *PARKING BRAKE is SET*,
2. ■ *STAB TRIM NOT in the GREEN BAND (T/O band)*,
3. ■ *SPEED BRAKE lever NOT in the DOWN DETENT*,
4. ■ *FLAPS NOT in the T/O RANGE*,
5. ■ *BODY GEAR NOT CENTERED*.

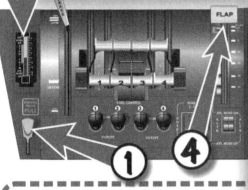

DISCUSSION about the
RTO (REJECTED TAKEOFF AUTOBRAKE).

*The **RTO** is a marvelous piece of engineering ... and it applies brakes **IMMEDIATELY** and **FULLY** when:*

**THROTTLES are retarded to IDLE and
"WHEEL SPEED" above 85 knots.**

*It is **NOT** a decelerate rate modulated event, it is **MAX BRAKING IMMEDIATELY!***

DO NOT EVEN TOUCH THE BRAKES if you want RTO.
*This mode is not like the regular "**RAMP**" on normal braking where you push the brake pedal and the deceleration mode operates giving you increasing braking the harder you push. Just slightly pushing the brakes for 2 seconds causes the **AUTOBRAKE (RTO)** to deselect!i*

*Also problematic: you could be looking at more than 85 knots on the **AIRSPEED INDICATOR**, but due to a headwind, you may get "**NO RTO**" when "**Wheel**" speed is **below 85 KTS**. This would occur when you pull the throttles to idle and expect **AUTOMATIC MAXIMUM BRAKING**!!! Confusion might occur because there is slight delay (about 2 seconds) **before** RTO activation after thrust levers pulled to idle.
If you don't get **RTO BRAKING** ... **APPLY MAXIMUM BRAKES MANUALLY.***

THE CHECK-RIDE BEGINS

...all that other stuff we just did was preparation.

You are prepared ... well, at least you have read this book ...
so relax and enjoy the next four hours. I realise that there is
a lot of pent up emotion and concern for the unknown; but
remember that you are as ready as you will ever be.
It is time to show 'em what you got!

... and PROCEDURES FOR STUDY and REVIEW ONLY!

TAKING OFF

Once you are "**CLEARED FOR TAKEOFF**"...

PNF

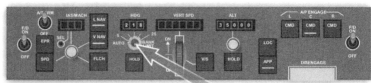

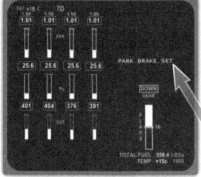

PNF (Pilot Not Flying): Checks and if necessary sets appropriate heading on the **HEADING SELECTOR** on the **MCP**. Consider: **RUNWAY HEADING**

Verify that the **PARKING BRAKE IS RELEASED**!

The **ONLY** valid indication that the parking brake has released is the **EICAS** indication. **PUSH** the **RCL** button to ensure that there are no "hidden" references to the parking brake.

IMPORTANT NOTE:
CAPTAIN sez, "I (you) have the aircraft, the parking brake is set (released)."

Verify **BRAKES RELEASED**,
ADVANCE THROTTLES MANUALLY,
and **SET TAKEOFF THRUST** by depressing the **TO/GA BUTTONS** on the **THROTTLES**.

TECHNIQUE:
PF advance throttles smoothly to about 1.10 EPR and allow EGTs to stabilize,

then, BEFORE 50 KTS, push the TO/GA switch

and observe the throttles advance to and stabilize at the TAKEOFF REFERENCE BUGS by 65 KNOTS.

NOTE:
THR REF will be annunciated on the PFD until **65 knots**, then **HOLD** is annunciated above **65 knots**.

REALLY SCARY NOTE:

IF EITHER TO/GA SWITCH IS NOT PUSHED **BY 50 KTS**, then the autothrottle **CANNOT BE ENGAGED** until above **400 feet AGL**, HOLD will be annunciated, and the throttles will stay **WHERE YOU PUT THEM**. In this situation, the **AUTO-THROTTLE** will **NOT AUTOMATICALLY** set take-off thrust !!!! **OMIGOSH!!!**

© MIKE RAY 2014
WWW.UTEM.COM

As the jet starts to move, be prepared to ...

STOP THE TAKE OFF!!!

If you get **ANY PROBLEM OR WARNING** just after you release the brakes and start the TAKEOFF roll, it would be considered acceptable to stop the take-off and (possibly) get off the runway and set the parking brake so as to address the issue. Low Speed aborts do not represent a problem for the jet ... and a low speed abort will (probably) not result in a **BRAKE COOLING** problem.

ON THE OTHER HAND ...

HIGH SPEED ABORTS are considered to one of the **MOST DANGEROUS** events in operating any airplane. And you can bet the farm that a **HIGH SPEED ABORT** event **WILL** be a part of your check-ride. You must make a plan for yourself and stick with it. Here is the criteria established at a major airline ... and I will repeat it for you few times in this document to emphasize the importance:

The official position is, and I quote:
"In the high speed regime, especially at speeds near V1, a decision to reject should be made ONLY if the failure involved would impair the ability of the airplane to be safely flown."

ALERT - ALERT - ALERT

During the course of the checkride, you are going to be making a **MAX GROSS WEIGHT TAKE-OFF** ... be alert. Don't let your mind wander ... and be prepared. This is a tip-off that you are about to experience:
- **HIGH SPEED ABORT**, or
- **FAILURE OF AN OUTBOARD ENGINE AFTER V1!**

In this phase of the take-off, it all boils down to these three decisions that you will have to make:

1. Make a **LOW SPEED ABORT** for almost anything.

2. Make a **HIGH SPEED ABORT** only if the ability of the airplane to fly is compromised.

3. After the **HIGH SPEED ABORT** ... assume you have "**HOT BRAKES**" and consult the "**BRAKE COOLING CHART**.

This will get you boo-coo brownie points early in the check-ride!

... and PROCEDURES FOR STUDY and REVIEW ONLY!

At least one of 5 exciting things is about to happen to you!

- **LOW SPEED ABORT**
- **HIGH SPEED ABORT**
- **V1 CUT**
- **V2 CUT**
- **NORMAL TAKE-OFF**

Let's talk about these in greater detail. On the following pages we will make some observations and suggest some techniques to cope with these critical maneuvers.

Let's break these options out and look at them one at a time: →

SIMULATOR TECHNIQUES ...

GET READY ... ONE of these
FIVE TAKE-OFF SCENARIOS
is about to occur.

1. LOW SPEED ABORT

A **LOW SPEED ABORT** can be triggered by **ANYTHING** that is not normal. A low speed abort is **"NO BIG DEAL."** And it is true in the real world, anytime you are in the power up phase or just beginning your take-off roll; if **ANYTHING** doesn't look exactly right, **STOP THE TAKE-OFF**. Terminating the take-off at this point is going to get you **BIG BROWNIE POINTS**; and you can expect that during a check-ride, you will be given some annoying low speed meaningless indication. Here's the rub, the airplane accelerates very rapidly, and you will very quickly transition into the **HIGH SPEED ABORT** zone ... and trying to stop for some annoying piddly problem above about **80 KNOTS** is a **MAJOR BOO-BOO**. So, make up your mind that you are going to stop for about anything below about **80 KNOTS**. Check-guys **LOVE** to hear this statement in the **"BEFORE TAKEOFF BRIEF."**

> **NOTE:**
> 1. There will be NO RTO.
> 2. IF an engine fails during the initial spool up and low speed transition, *DO NOT USE REVERSE*.

2. HIGH SPEED ABORT

> This is **THE MOST DANGEROUS EVOLUTION IN AVIATION**.
> However, there are events that could trigger it, and on your checkride I suggest that you focus and confine these to **ENGINE FIRE/FAILURE BEFORE V1**. I am, of course, sticking my neck out here, but normally on a check-ride one should **HIGH SPEED ABORT only for "BELLS and SWERVES BEFORE V1."**

You have to be dividing your attention between keeping the jet going down the centerline and monitoring the engine instruments for flickering guages, etc. A tough assignment!

OF COURSE, who didn't hear about the heroic crew who slammed them into reverse during the **HIGH SPEED REGIME** and avoided a head on collision in the fog. **WHEEEEEW!!!** And if you get the check-pilot from **HELL**, I guess that they could give you something like that in the sim, but ... I think you should expect swerves and bells.

DO NOT DO A HIGH SPEED ABORT for:
- Stall shaker at rotation
- door light
- side window popping open
- insignificant warnings

CHECK GUYS LOVE THIS !

... And it is a FAVORITE ploy to display some meaningless message on the forward display during the takeoff roll.

> **CHECKRIDE SUICIDE!**
> DO NOT TRY AND PUT THE BIRD JET BACK ON THE RUNWAY IF YOU HAVE ALREADY STARTED ROTATION!

... and PROCEDURES FOR STUDY and REVIEW ONLY!

3. ENGINE FAILURE AFTER V1 ("V1 CUT")

I think I can say without fear of contradiction that this is the **MOST BUSTED** event on the check-ride.

> At a point about 5 KNOTS prior to V1 ... The Captain <u>WILL</u> remove her (his) hand from the THRUST LEVERS.

When the Checkguy sees the captain's hand come off the thrust levers ... He (She) is spring-loaded to ruin your life. At that point, the "ENGINE FAIL" button is pushed on the Checkguy's secret problem panel! I <u>GUARANTEE</u> THIS WILL HAPPEN on your checkride! The event, depending on the selection made by the Checkguy, will take a couple of seconds to develop to a spooldown, and then a couple of eyeblinks for you to notice it ...

NOTE:
in the simulator; *DO NOT PUT YOUR HANDS BACK ON THE THRUST LEVERS AND TRY TO ABORT*!

The profile for the V1 CUT presented here is one that will keep you from busting the ride right off the bat ... but, like everything else, it is NOT engraved in stone. Modify it as you go to meet the needs of the problem.

One recommendation: If you have a FIRE INDICATION, do not be in a hurry to shut the engine down. Continue your take-off profile and fly on up to about 500 feet and pushover before you get all involved in the QRC items. You can use the thrust to your advantage.

4. ENGINE FAILURE AFTER V2 ("V2 CUT")

This is normally NOT a part of the check-ride, but sometimes, if the Checkguy screws up and the V1 cut occurs too late and you are already rotating ... it does get hairy. Regardless, you will be expected to control the airplane and not hit the earth. This is also a potential failure introduced during the **MISSED APPROACH**!

*The **BIGGEST PROBLEM** in this situation is the pilot pushing the **WRONG RUDDER!!!** YIPE! So, here is the gouge.*

> **ROLL THE WINGS LEVEL**, and then
> **PUSH RUDDER PEDAL UNDER LOWER YOKE HORN.**

The airplane will feel a little (lot) sloppy at this point, but it will fly with all that aileron and NO rudder without falling out of the sky. Then, as soon as you are **<u>ABSOLUTELY CERTAIN</u>** which rudder to push, feed it in and trim it up. Transition to the V1 CUT profile.

5. NORMAL TAKEOFF

Believe it or not, they have to observe you do a "normal" takeoff and climbout. So, don't be surprised when it happens, just enjoy it.

see more details

747-400 SIMULATOR TECHNIQUES ...

comments about
ABORTS ... in general.

RULE #1:

> **CAPTAIN IS EXPECTED TO MAKE ALL ABORTS !!!**

CAPTAIN DOES THIS:

Whether the Captain is the PF or PNF, she(he) should take control of the airplane, and announce loudly and authoritatively:
"I HAVE THE AIRPLANE ... ABORTING !"

▬ **THROTTLES RAPIDLY TO IDLE**

▬ **AUTOTHROTTLE RELEASE SWITCH ON THROTTLE ... DEPRESS TWICE.**

▬ **REVERSE THRUST** ... *apply maximum allowable reverse thrust consistent with directional control.*

IF LOW SPEED (less than ~85 knots)
... DO NOT USE REVERSE THRUST !

> IF BELOW 85 KNOTS,
> APPLY BRAKING MANUALLY.
> *There will be NO RTO below 85 KTS.*

▬ **SPOILER LEVER** ... *If the lever doesn't extend with the deployment of the reversers,*
Manually pull the lever AFT to the UP position.

FIRST OFFICER DOES THIS:

▬ **VERIFY SPOILERS EXTENDED** *(ONLY if the reversers are used).*

▬ **CALL TOWER** .... *Call the Tower and tell them what you are doing.*

▬ **PA** .... *Tell FLIGHT ATTENDANTS and passengers something appropriate; such as:*
"REMAIN SEATED."
If you do not do this, you can expect the Flight Attendants to evacuate the airplane when it comes to a stop.

DISCUSSION:

Use of the thrust reversers are, of course, up to the Captain: but, consider that below about 80 knots, very little reverse thrust is supplied and the engines will be nothing more than HUGE vacuum cleaners and suck up huge amounts of debris
... AND THEY WILL CATCH FIRE !!!
EITHER DO NOT DEPLOY THE REVERSERS OR
STOW THE SPOILERS IF BELOW 80 KNOTS.

... and PROCEDURES FOR STUDY and REVIEW ONLY!

ABORTS continued.

DISCUSSION:
Once the jet is stopped, evaluate your situation and determine whether you can move the airplane and taxi clear of the runway. Reference to the Flight Handbook (**REJECTED TAKEOFF BRAKE COOLING TABLE** in the **LIMITS CHAPTER**) is advised if speeds greater than 80 knots were achieved.

If **cooling time required exceeds 70 minutes**; THEN:

- DO NOT SET PARKING BRAKE.
- APPLY BRAKES MANUALLY ONLY TO STOP AIRPLANE.
- CONSIDER ENGINE SHUTDOWN TO AVOID BRAKE USE.
- DO NOT APPROACH MAIN GEAR.
- DO NOT TAXI.
- REQUEST NOSE GEAR BE CHOCKED.
- CONTACT MAINTENANCE PERSONNEL (SAMC).

CAPTAIN and FIRST OFFICER DO THIS:

"Reduce reverse thrust so as to reach idle by 80 knots, and move the reverse levers to forward idle by 60 knots after the engines have decelerated."

Once the airplane has stopped,
Complete the
NORMAL LANDING ROLL PROCEDURE ITEMS:

If *EVACUATION HAS STARTED* or is *DESIRED*:

■ NOTIFY ATC that evacuation is in progress.

■ SET PARKING BRAKE.

■ RETRACT SPEEDBRAKES.

■ FUEL CONTROL SWITCHES to CUTOFF.

■ DO THE QRC for "EVACUATION."

When time permits:

F/O OPEN BRAKE TEMPERATURE SYNOPTIC
and *MONITOR BRAKE TEMPERATURES.*

F/O OPEN FLIGHT HANDBOOK *to the*
BRAKE COOLING TIME TABLE.

CAUTION:
Conditions permitting, do not set or hold brakes until referencing the BRAKE COOLING TIME TABLE in the WEIGHTS SECTION of the LIMITS AND SPECIFICATIONS CHAPTER.

NOTE:
For your information, it takes about 12 minutes for the brake temperatures to be accurate enough to use for cooling calculation.

747-400 SIMULATOR TECHNIQUES ...

LOW SPEED ABORT

Let's talk about the ... **TAKE-OFF WARNING**
WARNING light and Aural warbling two-tone SIREN!

DEE - DULL - DEE - DULL - DEE - etc.

The **TAKE-OFF WARNING** occurs when:

Aircraft is on the ground, and
Take-off thrust set on ENG 2 or 3 ...and either

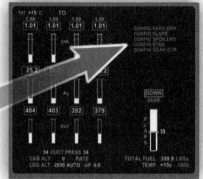

PARKING BRAKE SET, or
FLAPS NOT IN TAKE-OFF POSITION, or
SPOILERS NOT DOWN, or
STAB TRIM NOT IN TAKEOFF RANGE, or
either BODY GEAR NOT CENTERED.

DO NOT START TAKE-OFF until the situation is resolved!

☠ BIG-TIME PROBLEM
ENGINE FAILURE BELOW 80 KNOTS

DISCUSSION:

It is incredible how much thrust is developed. Probably few people on earth are able to command such a vast power resource, and it is an absolutely unbelievable experience to push up the throttles on a light 747-400. It accelerates like a drag racer.
BUT ...
What happens if a sudden engine failure occurs at low speed and high power settings. Without the airflow over the rudder, it does not have adequate authority to overcome the yaw motion. The tiller bar is virtually useless, and differential braking is inadequate. The situation develops in less than a second and the airplane **WILL LEAVE THE CONFINES OF THE RUNWAY** unless expeditious action is taken.

1 THROTTLES *IMMEDIATELY* TO IDLE
2 USE MANUAL BRAKING

NO RTO because the system is not armed until ground speed is greater than 85 kts.
DIFFERENTIAL BRAKING can't hurt but has little effect.
DIFFERENTIAL REVERSE AT A LOW SPEED IS NOT RECOMMENDED !

USE OF REVERSE THRUST
NOT RECOMMENDED !!!

...but, hey, do what you gotta do.

... and PROCEDURES FOR STUDY and REVIEW ONLY!

HIGH SPEED ABORT

Technically, any problem encountered "**IN THE HIGH SPEED RANGE** from **80 KNOTS to VR**" is in the **HIGH SPEED ABORT** category. However, the most likely event to occur, particularly on the check-ride, is an engine failure. For this discussion we will only look at the **ENGINE FAILURE** problem.

ENGINE FAILURE AFTER 80 KNOTS

DISCUSSION:

You MUST MAKE YOUR GO/NO-GO DECISION and stick with it.

If you have taken your hands OFF the throttle 5 KTS before VR, it is difficult to imagine the situation where it would be a good idea to make a "late" decision to abort and return to the thrust levers, particularly in the simulator.

The official position is, and I quote:

"In the high speed regime, especially at speeds near V1, a decision to reject should be made ONLY if the failure involved would impair the ability of the airplane to be safely flown."

This is a **BIG MAMMA-JAMMA** and once it gets trucking down the runway, it obtains incredible inertia. While the brakes are really fabulous, and the reversers really effective ... the **MOST** important factor in stopping on the runway is the rapidity with which the throttles can be retarded and the reversers deployed. Every part of a second you delay, the faster the airplane goes and the less runway that remains.

CAVEAT: *I think we are all aware of the courageous pilots who made a successful late abort because another airplane entered the confines of the runway on a foggy day ... correct assessment, no doubt ... and heroes by any measure. BUT on the check-ride, I would think twice before attempting a **LATE ABORT**!*

SYSTEM REVIEW: RTO

When **RTO** is selected;
The system provides **MAXIMUM BRAKING** when:
THROTTLES ARE RETARDED TO IDLE, with
AIRSPEED ABOVE 85 KTS.

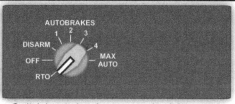

Switch located on lower console, right panel.

SYSTEM DISARMS WHEN:

BRAKE PEDALS ARE TOUCHED
    Even a slight pressure will cause the RTO to release.
THROTTLES ARE ADVANCED
SPEED BRAKE LEVER moved to the DN
    after speed brakes have been deployed on the ground.
A FAULT IN THE SYSTEM.

and at **LIFTOFF**.

> **WARNING:**
> When you depress the brake pedals, the **RTO** will disarm **BUT** incredibly, the switch will remain in the **ARM** position. It does **NOT** go to **OFF** on it's own. If you should inadvertently release the RTO, apply **MAXIMUM MANUAL BRAKING** for the same effect.

747-400 SIMULATOR TECHNIQUES ...

Once you have successfully managed to slow everything down ... and hopefully exitted the runway ... the very next concern will be the "*HOT BRAKES PROBLEM*".

HOT BRAKES

After any kind of an "**ABORT**" situation, the Check-guy is going to want to see you address the issue of "**HOT BRAKES**". Here is the situation: you have either emerged from a **LOW SPEED ABORT** or a **HIGH SPEED ABORT**. The cutoff between the two is a little hazy, but if you aborted before 80 Knots, then that is considered to be a **LOW SPEED ABORT**. Remember that the **RTO** (Reject Take Off) setting on the **AUTO-BRAKE** will be armed and will actuate at speeds above **85 Knots**. It is generally considered a **HIGH SPEED ABORT** anytime **RTO** braking occurs. Once we have stoppped the airplane after an abort, then we have to consult a specific chart; the "**BRAKE COOLING TIME**" chart. The Check-guy will *absolutely* expect you to get out this chart and interpret the data. You will need "**WEIGHT of AIRPLANE**" and "**SPEED AT TIME ABORT**".

This **CHART** will tell us how long the airplane has to be at rest before any subsequent take-off can be attempted. Let's take a look at a typical brake cooling time chart.

BRAKE COOLING TIME

Weight 1000 Pounds	SPEED at time of INITIAL BRAKE APPLICATION								
	80	90	100	110	120	130	140	150	160
875.0	29	43	58	MZ	MZ	MZ	MZ	MZ	MZ
870.0	28	42	57	MZ	MZ	MZ	MZ	MZ	MZ
850.0	27	40	55	68	MZ	MZ	MZ	MZ	MZ
800.0	22	38	51	64	MZ	MZ	MZ	MZ	MZ
750.0	20	33	46	60	MZ	MZ	MZ	MZ	MZ
700.0	14	29	41	55	67	MZ	MZ	MZ	MZ
650.0	0	25	37	50	63	MZ	MZ	MZ	MZ
600.0	0	19	33	45	57	68	MZ	MZ	MZ
550.0	0	15	28	39	52	63	MZ	MZ	MZ
500.0	0	10	22	32	45	54	62	MZ	MZ
450.0	0	05	20	26	37	45	56	65	MZ

> If **MAXIMUM BRAKING** or **RTO** is used; a **MAINTENANCE** check is required.

If the calculated time is either more then **70 MINUTES** or in the "**MZ**" is the "**THERMAL FUSE PLUG MELTDOWN ZONE**"; then the following criteria should be applied:
- *DO NOT SET PARKING BRAKES*
- *APPLY BRAKES MANUALLY ONLY WHEN NECESSARY TO STOP THE JET.*
- *TO AVOID USING THE BRAKES, CONSIDER SHUTTING DOWN THE ENGINE(S).*
- *DO NOT APPROACH MAIN WHEELS.*
- *DO NOT ATTEMPT TO TAXI UNTIL COOLING TIME ELAPSES.*
- *REQUEST NOSEWHEEL CHOCKS, and THEN RELEASE BRAKES.*
- *CONTACT APPROPRIATE MAINTENANCE PERSONNEL.*

... and PROCEDURES FOR STUDY and REVIEW ONLY!

INTRODUCING the dreaded, loathed, and feared ...

... MOST POPULAR SCREW-UPS

SCREW-UP #1: ABORTS AFTER V1!

The Captain attempts to abort after the airspeed of the airplane has exceeded V1.
When you are charging down the runway, and the PNF (Pilot Not Flying) calls out V1 at 5 knots before V1 ... Take your hands **OFF THE THRUST LEVERS** and **RESIST THE TEMPTATION TO REACH OVER AND GRAB THE THRUST LEVERS** in an attempt to **ABORT AFTER V1**.

The official position is, and I quote:
"In the high speed regime, especially at speeds near V1, a decision to reject should be made ONLY if the failure involved would impair the ability of the airplane to be safely flown."

SCREW-UP #2: PUSHES WRONG RUDDER

Since the **NOSE WHEELS** of the airplane are still on the runway, **THE RUDDER** which controls the nosewheels is initially the most important directional control. However, the absolutely **HUGE** rudder on this airplane becomes even more effective as the speed increases . It is (theoretically) not necessary to know which rudder to push ... rather it should be intuitve to keep the nose of the jet tracking down the runway. Even as big as the rudder is, it will take considerable extension to counter the effect of losing an outboard engine at take-off power.

Technique: **DO NOT "WALK"** the rudders. Get the airplane going where you want it and **LOCK YOUR LEG** ... allowing only teensy-tiny movements to fine tune the heading. Steer the airplane with your feet. This is a significant problem area and pilots frequently will "get behind" the airplane and by "**JABBING**" the rudders from side to side creating a wing oscillation that could "**DRAG A POD**"!

LOGICAL TIP: *Right rudder makes the nose go right and the left rudder makes the nose go left.*

SCREW-UP #3: ROTATES BEFORE Vr!

As sure as there is a tooth fairy, I guarantee that when the PNF yells, "**ENGINE FAILURE**", your first instinct will be to pull back on the yoke. If you do that, the airplane will be in the edge lights before you can slam the nose back on the concrete. So...

DO NOT RELIEVE PRESSURE ON THE NOSE-GEAR OR START TO ROTATE THE JET AT THE POINT WHERE THE ENGINE FAILURE OCCURS OR THE PNF YELLS.

Keep the nose firmly planted on the runway. It is my opinion that keeping some forward pressure on the yoke for a few knots beyond Vr will GREATLY assist in successfully completing this maneuver. There is **NO REQUIREMENT** to have the nosewheel coming off the concrete at Vr. The **MAXIMUM TIRE SPEED** is **204 KNOTS!**

DISCUSSION: *There is an "artificial" 35 foot barrier that the airplane must clear during the certification process, but during your check-ride that only thing you have to do is not hit anything on your way out of town.*

SCREW-UP #4: BAD ROTATION TECHNIQUE

When you begin to gradually pull back the yoke for the rotation, the directional control gradually shifts form the **RUDDERS** to the **AILERONS**. So as the airplane becomes more completely airborne, the relationship between the **RUDDER** and the **AILERONS** becomes critical. Excessive **RUDDER** will induce a significant **ROLL** so you have to reduce the **RUDDER** pressure and let the **AILERONS** take effect.

One technique is to use the **AILERONS** to maintain the wings level as you gradually **REDUCE RUDDER PRESSURE** while **TRIMMING THE RUDDER** so as to achieve a **LEVEL YOKE** requiring no control pressure. It is essential that you **DO NOT EXCEED 15 DEGREES BANK** as control of the jet could be jeopardized. Also be aware if excessive **AILERON** displacement is induced then the **FLIGHT SPOILERS** may be extended. At the same time, you will have to **GENTLY** pull back on the yoke to control the **AIRSPEED**. The target is somewhere between **V2** and **V2 + 10 Knots**. This translates to a **PITCH OF ABOUT 12 DEGREES**. The **TAIL** will strike the ground at 11 degrees so this rotation is a challenging task in itself.

---

(1) If you rotate too "smartly" you will drag the tail. 2 to 3 degrees per second is the "suggested" amount. We are wanting to get the nose to a *12 DEGREE PITCH*, BUT the tail will strike the earth at 11 degrees ... so we have to let the airplane fly off the ground "while" we rotate accordingly. *A TAILSTRIKE IS NOT GOOD!*

(2) If you delay rotation to less than **12 degrees**, the jet will not climb, but stay at a level altitude, or worse, be in a descent. This is **NOT GOOD**. It could result in **CFIT** (Controlled Flight into Terrain). *CFIT IS NOT GOOD!*

(3) If you get the nose too high, the speed will decay and this loss in airspeed will result in the rudder losing effectiveness and the nose of the airplane slewing into the "**BAD**" engine. If the pitch remains too high, then airspeed will decay and the airplane will **STALL! YIPE!** *A STALL IS NOT GOOD!*

Generally speaking, keep your eyes ouside on the far end of the runway, and as the airplane rotates and the visual cues disappear, go to the **PFD** and watch:
- **FOLLOW PITCH BAR** for **12 Degree** target (at MAX WEIGHT take-off)
- **@ POSITVE RATE** ... call for "**GEAR UP**".
- As the **GEAR DOORS** open, the increased drag will cause the airspeed to decay.

RIGHT HERE IS A PROBLEM AREA.
DO NOT CHASE THE AIRSPEED
... stay focused on the PITCH BAR.

12°

The **747-400** has a nice feature that is useful. If you have the presence of mind, you can use this to assist in trimming the rudder.

The little "**SAILBOAT**" at the top of the **PFD ATTITUDE INDICATOR** has two parts, and when they are aligned the airplane is in trim. You can "**STEP ON THE SAILBOAT**", thereby using your feet on the **RUDDER PEDALS** while you "**TRIM THE SAILBOAT TO THE SAIL**" using the **RUDDER TRIM** knob. This will be about **7 DEGREES of RUDDER TRIM**.

If the sailboat gets "**TOO FAR OUT OF ALIGNMENT**", the jet could require excessive aileron displacement to maintain heading control.

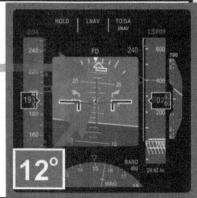

The whole idea is to **TRIM THE RUDDER SO THAT THE YOKE** is level before you engage the **AUTO-PILOT**, while maintaining a smooth and consistent rate of rotation.

... and PROCEDURES FOR STUDY and REVIEW ONLY!

SCREW-UP #5: DOESN'T USE AUTO-PILOT

@ 800 FEET AGL ...
Use of the AUTO-PILOT strongly recommended!!!

The **AUTO-PILOT** is a sensitive and temperamental device. It sometimes seems to have a mind of its own and this is particularly evident when you are trying to engage it during the "**ENGINE OUT**" procedure. The problem is that you must have the airplane in trim *BEFORE* attempting to engage the **AUTO-PILOT**. If the airplane "too far" out of trim, the **AUTO-PILOT** will not engage.

Before you attempt to engage the **AUTO-PILOT**, you should be able to release all the **YOKE** and **RUDDER** pressure and the airplane will continue to fly the attitude in which you left it. Now it will tolerate some slop, so I would say that it should be "about" in trim because during this rather busy moment you simply don't have the time to obsess over minor trim adjustments.

RUDDER TRIM with AUTO-PILOT DISCUSSION.
With the **AUTO-PILOT** engaged, you may continue to trim the **RUDDER** and the **AUTO-PILOT** will remain engaged. However, **PITCH** and **ROLL** inputs using the **YOKE** or **TRIM CONTROLS** will

SIDEBAR DISCUSSION
Here is a discussion about a feature of the **AUTO-PILOT** that you should be aware of ... and has nothing to do with take-off but seems to fit into the narrative and helps us understand the **AUTO-PILOT**.
NOTE: The **AUTO-PILOT** does **NOT** move or control the **RUDDER** ... unless the airplane is using **MULTIPLE AUTO-PILOTS** such as during the **CAT II** or **CAT III AUTOLAND APPROACH** mode. Then, once the airplane has descended below 400 feet AGL, and subsequently executes an **AUTO-PILOT CLIMB** or a **GO-AROUND**, the **AUTO-PILOT** will automatically input the necessary **RUDDER** input to maintain the airplane in trim. WOW!! Terrific!

HOWEVER
Once the airplane climbs to 400 feet or another roll mode is selected, the **AUTO-PILOT RUDDER** input will be instantly removed and what ever trim inputs it had are removed and the trim will revert to the position it had before the **MULTIPLE AUTO-PILOTS** were selected. What does this mean to us as pilots?.

IF THE PILOT FLYING DOES NOT HAVE THEIR FEET ON THE RUDDER PEDALS AND IS PREPARED TO "COUNTER" THE LOSS OF TRIM INPUT AND MAINTAIN THE TRIM INPUT WITH THEIR FEET ... THE AIRPLANE "COULD" ENCOUNTER A SIGNIFICANT UNEXPECTED ROLL MOMENT AND POTENTIALLY INDUCE A LOSS OF CONTROL! OMIGOSH!!!

SCREW-UP #6: NO PREPARED PROCEDURE !

There is probably nothing more pathetic than watching a hapless pilot trying to "invent the wheel" as the engine-out scenario progresses. There simply **MUST** be some thought given as to what you are going to do when this event occurs. It is virtually impossible to "think" when your brain turns to putty ... so a pre-thought out sequence of events is extremely useful. The book outlines a potential response, and I have added to that with some thoughts of my own. You will find "MY" V1 CUT profile on the next pages.

Remember that it is NOT an official construct and it will be up to you to create your own profile.

Here is a "suggested V1 CUT" profile for you to consider.

747-400 SIMULATOR TECHNIQUES...

"V1 CUT" ENG FAIL after V1 — Aircraft is on the ground

*This **V1 CUT** profile is **NOT** an official or **SOP** construct. It is just my idea and it is up to you to make up your own profile.*

@ 800 FEET

6 Reduce pitch, increase airspeed to CMS. Use IVSI ... **"THINK LEVEL"** PUSH...PUSH...PUSH.

- AUTO-PILOT
When in trim, select autopilot.

5 **PRST**

P — **PUSH** Reduce pitch, increase airspeed to **CMS**.

R — **ROLL** Turn the **YOKE** so as to **ROLL** the wings level.

S — **STEP** on the **RUDDER** under the lower **YOKE HORN**.

T — **TRIM YOKE** level with the **RUDDER TRIM**.

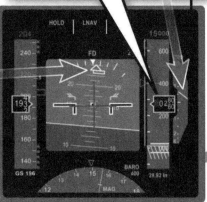

CLIMB at CMS

4 **@ POSITIVE CLIMB:**
"GEAR UP
HEADING SELECT
STEER ME TO C/L
or "T" heading"

12°

3 **ROTATE** 12 degrees

V2 to V2 + 10

2 While on GROUND CONTROL HEADING WITH RUDDER

If desired ...
"MAX THRUST"
add thrust slowly

© MIKE RAY 2014

1 **KEEP NOSEWHEEL GROUNDED** **TIP!**
until VR (DO NOT ROTATE TOO SOON)
a good technique is to initially rotate VERY slowly

MAXIMUM TIRE SPEED = 204 Kts ground speed.

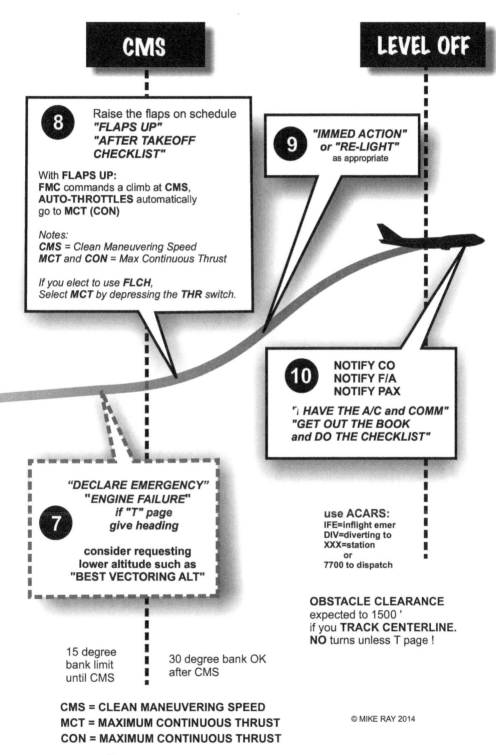

747-400 SIMULATOR TECHNIQUES ...
10 STEPS to GREAT "V1 CUT"

1 — **STEP 1**: Do not rotate prematurely. Keep some positive forward pressure on the nosewheel. You are not required to rotate the airplane exactly at Vr, and I have found that a few extra knots helps greatly in the control after the nosewheel breaks contact with the runway. There are TWO restrictions:
- **MAX TIRE SPEED is 204 Kts**.
- Don't hit the lights on your way out of town.

2 — **STEP 2**: Use of the RUDDER while on the runway it **CRITICAL**. The best technique seems to be to put in the required rudder, and '**LOCK YOUR LEG**". Make only teeny-tiny corrections as necessary to keep the airplane on the runway. Pushing back and forth too much sets up a wallowing roll that could lead to a **"POD STRIKE"**.

3 — **STEP 3**: The rotation target is about 12 degrees, and it is essential that you keep the rate of pitch as 2 to 3 degrees. Not only is it a **TAIL-STRIKE** problem, but the airplane is easier to control in the transition to airborne flight.

4 — **STEP 4**: Once the airplane has left the earth, ensure that you are in a climb and get the "GEAR UP". The PFD ROLL will indicate the heading of the airplane when it got airborne, and that could be slewed some so have the PNF use the HDG SEL and provide a heading that will compensate for the wind and correct back to the extended centerline of the runway. The 1500 Foot obstacle clearance is only valid if you are on the runway centerline.

5 — **STEP 6**: At 800 FEET ... **PFA**.

P=PUSH the yoke forward so that th airplane will start to accelerate. It takes a lot of pressure so you have to concentrate on trying to get almost level. *DON'T DESCEND!!!*

R = ROLL YOKE so as to keep wings level.

S= STEP on the **RUDDER PEDAL** under the **LOWER YOKE** horn.

T = TRIM RUDDER to remove pressures on the controls.

6 — **STEP 5**: **ENGAGE AUTOPILOT:** The technique that works best seems to be to get the heading of aircraft where you want it with the **YOKE**, and then trim the rudder so the the **YOKE** is **LEVEL**. You must trim out the pressures or the **AUTO-PILOT** will not engage.

7 — **STEP 7**: **DECLARE AND EMERGENCY**! If you are going to fly a "T" page, EOSID, or ESCAPE ROUTING .... you must tell ATC. Remember that you are only cleared to 1500 FEET with guaranteed obstacle clearance if you remain on runway centerline ... so it would be useful to have the cobntrolling agenmcy provide "terrain clearance" during your subsequent maneuvers. On you check-ride they won't tell you diddly

8 — **STEP 8**: Once you have the flaps up, and you have the airplane under control ... *DO NOT FORGET THE "AFTER TAKE-OFF CHECKLIST"*.

9 — **STEP 9**: Now is the time to ask, "*WHAT HAPPENED*" and consider a re-light or do the **IMMEDIATE ACTION** checklist. One pilot *MUST FLY THE AIRPLANE*!!!.

10 — **STEP 10**: Depending on the nature of the problem, once I had completed the **IMMEDIATE ACTIONS**, I would wait for a break in the activities so that we could ensure proper completion of the checklists associated with the problem(s).
We need to tell everyone else about our situation. That would be the **FLIGHT ATTENDANTS**, the **PASSENGERS**, and the **COMPANY** and anyone else that should be in the loop. Use the communication tools at your disposal.

... and **PROCEDURES FOR STUDY and REVIEW** ONLY!

LIST OF AWARENESS ITEMS

Brief the **EOSID**, "T" Procedure, or **Escape Routing** before starting the take-off.

There are at least two ways to set-up the Escape routing in the **CDU**.
- Use **RTE 2**, or
- Place routing at the end of the **RTE 1** queue after a **DISCONTINUITY**.

Plan for **LNAV/VNAV** takeoff.

Once the Takeoff roll has started ... look at the far end of the runway and control the heading of the airplane using the **RUDDERS**.
Once the **ENGINE FAILURE** occurs, use the **VISUAL SIGHT PICTURE** of the runway to keep the airplane headed in the right direction. Once you have the airplane going down the runway "**LOCK YOU LEG ... FREEZE IT**" and only allow teensy-tiny movements to tweak the heading.
Use **AILERON** as necessary to keep the wings level.
Keep some forward pressure on the yoke to help the nosewheel maintain "bite".

At **VR**:
Don't start your rotation immediately, but let the airplane accelerate a few knots beyond.
Remember that the **MAXIMUM TIRE ROTATION SPEED is 204 Kts GROUND SPEED**.
Keep your view of the end of the runway until the nose of the airplane rises and obscures it; THEN transition to the **PFD**.
Monitor the rotation speed to about 3 degrees per second,
make **12 degrees** your **INITIAL TARGET** and then follow the **PITCH BAR** (It will set the pitch for the desired V speed).

At **POSITIVE RATE** call for:
- "GEAR UP"
- "STEER ME TO CENTERLINE"

> **TERRAIN and OBSTACLE clearance is assured to 1500 FEET AGL on the extended runway centerline.**

- Expect **GEAR DOORS** drag to affect airspeed ... don't chase the airspeed, follow the **PITCH BAR**.
- Use the **RUDDER** to keep the **YOKE** level.
- Generally speaking **DO NOT** perform any **IRREGULAR** or **EMERGENCY QRC** until the airplane is **UNDER CONTROL** and **CLEANED UP**.
- If a **FIRE EMERGENCY** ... leave engine running and don't attempt a shutdown until airplane is stabilized in the climb (about **500 FEET AGL**).
- Treat the **PITCH** as if you are milking a mouse ... teeny-tiny 1/2 degree increments.
- Call for Escape Heading when appropriate.

PNF should declare an emergency and inform the tower of your actions. Example "XYZ Tower, this is ABC123 declaring an emergency with an engine failure. Coming to a heading of 234 degrees."
- **DO NOT** accept turnsuntil cleaned up and above**1500 FEET AGL** unless **ATC** Controller provides terrain and obstacle clearance.

When in trim and above **800 FEET**, Engage the **AUTO-PILOT**.
When ready, confirm the irregularity or emergency, "**WHAT HAPPENED?**"

During all this distracting activity, remember to RETRACT the flaps on schedule.

After clean up:
- Complete **CHECKLISTS**.
- Check THRUST for MAX CON and airspeed for CMS (Clean maneuvering Speed).
- Suggested **RUDDER TRIM** settings
  7 degrees in CLIMB.
  4 degrees in CRUISE.

© MIKE RAY 2014
WWW.UTEM.COM

747-400 SIMULATOR TECHNIQUES ...

"V2 CUT"
ENGINE FAILURE AFTER V2 or AIRBORNE

IF AN ENGINE FAILS BETWEEN LIFTOFF AND ALL ENGINE ACCELERATION ALTITUDE AT MAXIMUM TAKE-OFF PITCH ATTITUDE, TAKEOFF POWER,
AND AIRSPEED BETWEEN V2 AND V2+10 KNOTS ...
THE IMMEDIATE YAWING AND ROLLING OF THE AIRPLANE WILL SIGNIFICANTLY CHALLENGE THE PILOT'S ABILITY TO CONTROL THE AIRPLANE !

Do NOT be deceived ... this is an extremely dangerous situation. The pilot's immediate initial response is **CRITICAL** in the successful recovery and resolution of this situation. The airplane transitions to a possibly unstable control situation very quickly; and the possible potential for an **UPSET AT LOW ALTITUDE** is **VERY REAL**. However, as soon as control of the airplane is assured, the remainder of the procedure is fairly routine.

IMMEDIATE INITIAL ACTION

The **KEY** to a successful outcome is to **IMMEDIATELY** do these three steps:

P - **PUSH**=IMMEDIATELY LOWER THE NOSE.
(Pitch target: 12 degrees or V2)

R - **ROLL**=LEVEL THE WINGS WITH THE AILERONS,
and at the same time

S - **STEP**=APPLY RUDDER to counteract the yaw.
(Push rudder pedal under LOWER YOKE HORN).
This will cause the yoke to assume a level position.

T - **TRIM**= Roll the TRIM TAB to remove the pressure.

THEN

Once you have the airplane under control:
- "POSITIVE CLIMB,
  GEAR UP (if not yet raised)"
- ADJUST PITCH to maintain V2. Initially shoot for 12 degrees and adjust as needed.
- TRIM PITCH and RUDDER
- CORRECT HEADING.
- If a "T" page, revert to that heading, or continue to climb out on the runway centerline.

Once the airplane is out of immediate danger of loss of control .... continue with the V1 Cut procedure as outlined earllier.

"FLY THE AIRPLANE!"

© MIKE RAY 2014

published by UNIVERSITY of TEMECULA PRESS

... and PROCEDURES FOR STUDY and REVIEW ONLY!

Let's make some comments about an *"ENGINE FIRE DURING TAKE-OFF"*.

ENGINE FIRE ON TAKEOFF.

DO NOT SHUT DOWN THE ENGINE
... until the airplane is at an altitude that is clear of obstacles (Above at least 500 feet) and the airplane is under control. It might make sense to be above 800 feet with the AUTO-PILOT engaged before you initiate the engine shut-down.

Unless there is an *"ENGINE FAILURE"* associated with the *"FIRE"* the accepted technique is that you should not attempt to shut the engine down until you get to an altitude where **CFIT** (Controlled Flight Into Terrain) or **LOW ALTITUDE/LOW SPEED** upset is not a problem.

SHUT OFF THE FIRE WARNING BELL

You can do shut off the bell by pushing either **WARNING/CAUTION** light on the glare-shield panel. Continue the takeoff and climb while allowing the engine to continue operating even though it is indicating that it is on fire. This will result in a *"NORMAL"* climb profile to the altitude you seelct to shut down the engine. Generally speaking, it makes sense to be above **500** feet as a minimum; although I think that **800** feet is a good altitude since you can hook up the **AUTO-PILOT** and use it during the shutdown.

SOME COMMENTS REGARDING AUTO-PILOT

The **AUTO-PILOT** has some limitations that we should discuss.
FIRST: The **AUTO-PILOT** wil remain engaged with an **ENGINE OUT**; however, unless the **RUDDER** is trimmed, it cannot handle the aerodynamic forces encountered with the airplane flying in an excessive slip or skip. The **RUDDER** must be trimmed for the **AUTO-PILOT** to be able to accomplish "normal" attitudes for flight; such as turns.

SECOND: If the **AUTO-PILOT** should un-expectedly **DISCONNECT** while in this "abnormal" trim condition, there is the possibility that the airplane could quickly flip into an upset attitude. The pilots should be aware of this and have their feet *"ON THE RUDDERS"* in order to quickly restore proper airplane trim conditions.

THIRD: The *"SAILBOAT"* (Slip indicator) on the **PFD** must be trimmed for the **AUTO-PILOT** to operate properly and safely. If this is done, normal **AUTO-PILOT** ability to control the airplane is restored. If the airplane is flown around with the airplane *"NOT IN TRIM"*, it can become unstable enough at any moment so that the **AUTO-PILOT** could unexpectedly trip off..

This is the "SAILBOAT" (SLIP INDICATOR) I am talking about. Even with the AUTO-PILOT engaged, you can still trim the RUDDER and get the bottom part of the slip indicator aligned with the top. This will cause the YOKE to move to a more "level" position. The alternative, if you are hand flying, is to ROLL the wings level with the YOKE and while keeping the wings level on the PFD ATTITUDE INDICATOR, PUSH the RUDDER PEDAL under the "lower" yoke horn until the yoke is level and then TRIM out the pressure. Then you can re-engage the AUTO-PILOT and you should be good to go.

747-400 SIMULATOR TECHNIQUES ...

This airplane is so powerful that even at **MAX TAKE-OFF GROSS WEIGHT**, the loss of an engine procedure is the same basic profile as the four engine, normal take-off procedure.

REGULAR 4 ENGINE TAKE-OFF, and also ENGINE FAILURE ON TAKEOFF.

> A **NORMAL** take-off is with **LNAV** and **VNAV ARMED**.
> Take-off with **LNAV** and **VNAV ARMED**, particularly on the checkride. **VNAV** will give you **MAXIMUM CONTINUOUS THRUST** automatically when there is an engine failure.

20 KTS — BODY GEAR locks on Take-off roll.

50 KTS — If TO/GA button has not been depressed by the time the airplane has reached 50 kts tire speed, the autothrottles will not move to select T/O power. Instead, whatever throttle setting you have will remain.

65 KTS — CHECK and see if **HOLD** is annunciated on FMA...
IF IT IS NOT: manually set T/O power.

> **NOTE**: IF **HOLD** is not annunciated, then autothrottle will NOT be available until 400 feet and must be reselected at that time.

80 KTS — *"80 KTS-THRUST SET"*
Airspeed indicators are checked for validity and concurrence.
Also, **MASTER CAUTION** inhibited until 400'

85 KTS — RTO available.
ABORTS below this speed WILL NOT have RTO.
USE MAXIMUM MANUAL BRAKING!

V1 — The suggested GO/NO-GO point.
PNF callout made 5 kt before indication on the airspeed indicator.

CAPTAIN REMOVES HAND FROM THRUST LEVERS.

VR — At VR start ROTATION. The suggested technique is to apply back pressure on the yoke to achieve a 3 degree per second rotation. Aim for the **PFD PITCH** indicator (should be about 15 - 17 degrees). If **ENGINE OUT** make target pitch **12 degrees**.
This should give about V2 +10 kts initially.

V2 — Four engine climbout is V2 + 10 normally; However, climbout is made at V2 (if engine failure occurs at or before V2) or up to V2 + 10 (if speed attained prior to engine failure).

POSITIVE CLIMB GEAR UP — When BARO ALT and VSI agree
"GEAR UP"

THRUST REDUCTION

THR REDUCTION altitude on the **TAKEOFF REF** page is where the reference thrust setting is changed to the armed climb thrust. The throttles reduce automatically if auto-throttles are armed.

ACCEL HT
(default 1500 ft)

Let airplane ACCELERATE, and RETRACT FLAPS (on schedule).

At **ACCEL HT altitude**, or **ALTITUDE CAPTURE BELOW ACCEL HT**:
VNAV commands the **PDF COMMAND SPEED BUG** on the airspeed tape to slew to 250 KTS (or 30 REF + 100 KTS), and the nose pitches over to accelerate.

800 FEET AGL — ENGAGE AUTOPILOT (specific airline restriction)

During FLAP RETRACTION, move the flap handle to the next position **ONLY** when AT or ABOVE THE MANEUVERING SPEED for the new position.

Some of you pilots out there are moving the flap handle BEFORE it is appropriate.
REMEMBER: DO NOT move the handle to the next setting until the SPEED BUG is at or above the green indicator for that NEW flap position.

For example: If you want to raise the **FLAPS** to 1; **DO NOT** select FLAPS 1 until the speed bug is at or above the FLAPS 1 indicator.

400 FEET AGL

"VNAV" ENGAGES
(If pre-selected BEFORE TAKE-OFF)
SPEED INDICATOR on PDF goes from V2 to V2 + 10
THRUST MODE goes to **THR REF**

250 FEET AGL

AUTOPILOT COMMENTS
The Autopilot is certified for engagement above 250 feet. Most airlines, however, are more conservative and use higher figures such as the 800 feet suggestion used in this diagram.

50 FEET AGL

"LNAV" ENGAGES
(If pre-selected BEFORE TAKE-OFF)
ROLL bar swings to indicate heading to on course.

TAKEOFF

Possibly the most potential for things coming unglued occurs during the take-off ... and the check airman is planning to take advantage of the situation. Here is where the check-ride actually begins, and it will be to the benefit of the pilot to be aware that seldom does a check-ride go smoothly during the take-off.

This is particularly true if the airplane set up to accomplish a *MAXIMUM GROSS* take-off. It should come as no surprise that there are only three things about to happen:
- *A NORMAL MAX GROSS TAKE-OFF,*
- *A MAX GROSS REJECTED TAKE-OFF,*
- *A LOSS OF AN ENGINE DURING TAKEOFF AT MAX GROSS.*

I suggest ... a thorough brief as to what you are going to do in each of these events and an understanding by yourself of what you are going to do ... particularly in the event of an engine failure.

> On **ALL TAKE-OFFs** ... have an "Escape route", "T page", or EOSID (Engine Out Standard Instrument Departure) firmly in your mind. If there is weather, **DO NOT** take-off until **BOTH** the **PLANNED** routing and the **ESCAPE** corridor routes are free of convective activity (Storm Clouds).
>
> If the Engine Out routing is going to require **CDU/FMC** or special instrument inputs (such as tuning a VOR, selecting an NDB, or ...) accomplish that prior to beginning the take-off.

It is important to note that the aft fuselage clearance on rotation for takeoff is quite limited. This requires that the pilot be aware of that and use an initial pitch angle that minimizes the risk of a TAIL STRIKE. Good technique suggests that the rotation rate be limited to about 2 to 3 degrees per second.

Be aware that the potential for a tail-strike exists even *AFTER THE AIRPLANE IS ACTUALLY AIRBORNE*. The idea is that we should remain aware of the situation and even if we are the PNF, we should, call out excessive pitch input.

> IF PITCH EXCEEDS
> **11 DEGREES**
> WITH GEAR STILL ON RUNWAY
> THERE IS A POSSIBILITY OF A
> **TAILSTRIKE**

Initial pitch targets are less than 15 - 17 degrees and the target airspeed is V2 + 10 Kts.

DURING TAKE-OFF ROLL

PNF .... AIRSPEED CALLOUTS

As airspeed tapes pass 80 knots, PNF calls out,
"80 KNOTS, THRUST SET."
This call provides:
- Verification that desired thrust is set.
- Airspeed indicators have been cross-checked.
- An alert that the high speed phase of take-off has been reached where the GO/NO GO decision is critical.

CONFIRM:
1: **TO** is annunciated on the **TMA**
2: the numerical **READOUTS MATCH.**
3: the thrust tapes are at the **TIC MARKS**

4: **CONFIRM** the **CAS** on the **PFD** matches the **IAS** on the Standby Airspeed Indicator.

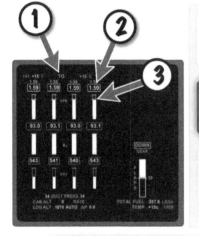

After the **TAKEOFF** thrust is set, the Captain's hand **MUST BE ON THE THROTTLES** until the **V1** call is initiated.

The **PNF** calls out "**V1**" 5 knots prior to the indication on the speed tape indicator.

At **V1,** the Captain's hand are to be removed from the throttles. It is highly unlikely, even in the real world, that an attempt to abort a take-off after V1 would be successful in remaining on the runway.

The **PNF** calls "**VR**" at the tic mark on the **AIRSPEED TAPE**. The **PF** initiates rotation.

The **PNF** calls "**V2**" at the V2 tic mark on the **AIRSPEED TAPE**.

ROTATION TECHNIQUE

PF............ ROTATION TECHNIQUE

Start with a slight forward pressure on the yoke and as the airspeed approaches **80 KTS**, start slowly relaxing this pressure.
At **VR**, rotate smoothly and continuously using about **3 DEGREES PER SECOND** towards your initial target pitch angle which will be indicated on the **PFD** (approx **15-17 DEGREES**) and which will produce about **V2 + 10 KTS**. If a greater pitch is required to maintain V2 + 10, **DO NOT EXCEED 20 DEGREES**. This is for passenger comfort.

CROSSWIND TECHNIQUE: While accelerating down the runway, apply **RUDDER** as necessary to maintain runway alignment and as the airspeed increases, introduce **AILERON INTO THE WIND** to maintain wings level.
Sometimes, in poor visibility, if the horizon is not distinct, use the **PFD** to maintain wings level.

As airspeed increases and the controls become more effective, displacement has to be reduced slightly. Be smooth. Hold your corrective displacement throughout the rotation, and as the airplane "slips the surly bonds of earth" smoothly return the control wheel and rudder displacement to neutral.

TAILSTRIKE

11° pitch up with main gear still on runway !

NOTE:
It is possible to get a tailstrike, even though the airplane may actually be airborne. AVOID excessive or abrupt pitch inputs; especially until well clear of mother earth.

NOTE:
In gusty or windshear conditions consider delaying rotation and increasing initial climbout speeds.

GETTING AIRBORNE

PF, PNF ....

Either pilot announce:
"POSITIVE CLIMB."

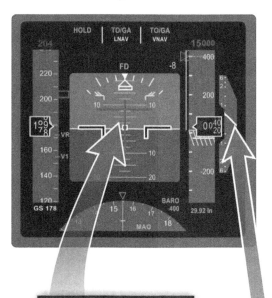

NOTE:
After liftoff, with all engines operating,
the FLIGHT DIRECTOR commands V2 + 10 or greater.
The ROLL COMMAND is the GROUND TRACK at liftoff.

NOTE:
"POSITIVE CLIMB" is identified by reference to **BOTH** the **PFD VSI** and the **BAROMETRIC ALTIMETER**.
They both **MUST** indicate a climb.

When BOTH pilots agree that POSITIVE CLIMB has occured:
PF calls, **"GEAR UP"**
in response PNF calls, **"GEAR UP"** and raises the gear handle.

FLAPS stuff

> **LIMITATION**
> Use of the Flaps is restricted to below 20,000 feet.

The flaps position is selected by the flap handle, which transmits the information to three **FLAP CONTROL UNITS (FCUs)**. They control the sequence of movement, monitor assymetry, control the flap relief system, and provide information to the **EICAS** and other systems.

Normally, **TRAILING EDGE FLAPS** are **HYDRAULICALLY** powered and **LEADING EDGE FLAPS** are **PNEUMATICALLY** powered. If part of the flaps malfunction, the **FCU** automatically shifts that group of the flaps and its opposite group to **ELECTRICAL** motors. This is called **SECONDARY** mode.

Secondary mode operation is much slower than primary (How much slower is it?)

   0-5 FLAPS TAKES APPROX. 4 MINUTES,
   0-25 FLAPS TAKES APPROX. 6 MINUTES.

Since this is an **"EICAS DRIVEN AIRPLANE"**; as such, the flap indicator is on the **EICAS** screen. 10 seconds after the flaps are raised the entire flaps display is removed.

Should any flap position be **"NON-NORMAL"** or you are operating the flaps electrically or **FLAPS ALT** control mode is armed ; then:

the **PRIMARY FLAP INDICATOR** will automatically be replaced with the **SECONDARY** or **ALTERNATE MODE FLAP EICAS** indicator.

It looks like this and provides information about the individual flaps and leading edge devices.

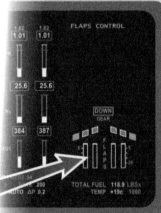

IMPORTANT NOTE ABOUT RAISING the FLAPS

The flaps MUST NOT BE RAISED until the little green numbers corresponding to that NEXT FLAP SETTING are AT or BELOW the ACTUAL AIRSPEED INDICATOR on the PFD.

For example, we CANNOT raise the flaps to 5 degrees until that little "5" is next to the pointer on the actual airspeed indicator.

Some Brain Surgeons posing as pilots have actually infiltrated our cockpits and continually raise the flaps **BEFORE** the indicator gets to the that flap setting number on the airspeed tape indicator.
"GEE ... DUH, What's that vibration?"

More boring FLAPS stuff

IMPORTANT NOTE ABOUT <u>EXTENDING</u> the FLAPS

Most airlines understand that pilots can get really busy sometimes and not be able to remember the **MAXIMUM FLAP SPEEDS**. So, they usually have a **MAX FLAP SPEED PLACARD** posted on the instrument panel.

Some airlines have even come up with a 10 kts company restriction to those limits to provide an additional buffer against the dreaded **FLAP OVERSPEED**.

FLAPS	1	5	10	20	25	30
V_{fe} -10 kts	270	250	230	220	195	170

THESE ARE MEMORY ITEMS !

If the FLAP HANDLE is moved too rapidly or there is not a short pause at the detents (little notches) then there is the good chance that the flaps will get all screwed up. You gotta be methodical and smooth in raising or lowering the flaps.

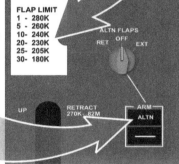

FLAP LIMIT
1 - 280K
5 - 260K
10- 240K
20- 230K
25- 205K
30- 180K

If the FLAP CONTROL UNITS should disconnect, and they will do so occasionally, here is a potential fix.
Reach up and cycle the ALTN FLAPS ARM switch to ALTN and then back to OFF.

The flaps MUST NOT BE EXTENDED until the actual airspeed is BELOW the restricted values on the chart above.
ALSO
The airplane should NOT be flown at speeds lower than those indicated on the PDF for the flaps selected.

In our example, we CANNOT fly at this speed without 1 degree flap, but we CAN extend flaps to 5 degrees.

The command to extend flaps involves three steps:

1. Ensure that the airspeed is below the limit for that flap setting.

2. Pilot Flying (**PF**) call for the desired flap using this terminology
"**FLAPS 5,
SET SPEED 191.**"
Get the speed from the **PFD** airspeed indication next to the little green flap setting number.

3. The Pilot Not Flying (**PNF**) will confirm the limit airspeed from the placard and set the speed for that flap setting from the Pilots Flight Director (**PFD**) on the Mode Control Panel (**MCP**) using **SPEED INTERVENE** knob.

AFTER TAKE-OFF

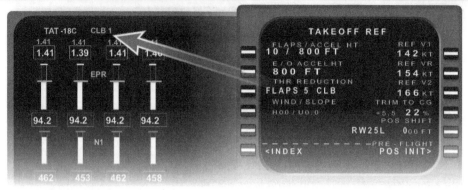

(PNF) MONITOR REDUCTION TO CLIMB THRUST.

The **PNF** is to observe the reduction to climb thrust. This will automatically occur @ **CLIMB THRUST** reduction altitude (**THR REDUCTION**) set on **TAKEOFF REF** page of the **CDU**, or

On **NON-VNAV** take-offs, by selection of the **THR** switch.

(PNF) LANDING GEAR LEVER .... OFF

This shuts off hydraulic system pressure to the landing gear.

(PNF) PACK CONTROL SELECTORS ... NORMAL

An automatic time interval inhibits multiple packs from starting simultaneously in the air. The auto protection is not available on the ground.

(C) SEAT BELT SIGN SELECTOR.....AUTO/ON

> **OFF** .... Turns OFF signs.
> **AUTO** .... The fasten seat belt signs are OFF when:
> Gear and flaps up and airplane above 13,000 feet.
> The fasten seat belt signs are ON when:
> Airplane descends below 13,000 feet OR
> when the gear are extended OR
> flaps are extended.
> The fasten seat belt signs come on when:
> The cabin altitude exceeds 10,000 feet.
> **ON** .... Turns ON signs.

(PNF) AFTER TAKE-OFF CHECKLIST .. COMPLETE

> ### AFTER TAKEOFF CHECKLIST
> (To be checked *ALOUD* by the pilot not flying)
>
> Landing gear lever... Off
> Flaps .... Up

... and PROCEDURES FOR STUDY and REVIEW ONLY!

HOW TO CLIMB and DESCEND

BRIEF and BORING DISCUSSION

There are a whole bunch of ways to operate the FMC on this airplane, but for the beginner, here are four commonly used ways to make this machine climb:

HANDFLY. This is always an option but a poor one and beyond the scope of this book. This airplane is a GREAT flying machine and a pleasure to hand fly ... but I don't recommend that you try to fly your check-ride by hand. Particularly in high traffic areas like the departure and arrival airport. One of the reasons is that when you are yoke-pumping by hand, you have to have the PNF (Pilot Not Flying) or "other guy" do all the MCP switch flipping and takes her/him away from doing other stuff.
When around other airplanes, <u>HOOK IT UP</u> to the auto-pilot and get your head out of the cockpit!

V/S ...... The Vertical speed knob is really flexible and easy to use, but has some serious shortcomings. Such as, it can fly the airplane into a catastrophic stall ... consider the situation where you select an altitude "**ABOVE MAX ALLOWABLE**" when you are attempting attempting to step climb when you are too heavy, **V/S** can fly you right into a stall. ***There is NO STALL or LIMIT AIRSPEED protection.***
It also can depart an the altitude without a target level off altitude and there is the very real possibility of flying into the ground.
For example; During a non-precision approach where you select the **MISSED APPROACH ALTITUDE** and leave the **MDA** in a descent using **V/S**.

TO RE-CAP THE MOST CRITICAL ELEMENTS OF A V/S:

- **V/S** will fly away from an altitude even though it is set in the **MCP** and captured. Vertical speed flies the airplane away from the altitude set in the altitude window on the MCP.

- **V/S** will fly to and **CAPTURE** an altitude if it is set in the **MCP**. And when a selected altitude is reached, the **PITCH MODE** changes to **ALT** on the **PFD**s.

- If in a descent or climb without a target altitude set in the **MCP**: **V/S** can climb into a stall or descend into the ground.

It seems contradictory, and according to studies conducted in such matters, pilots are confused by the **DUAL** function of the **V/S** knob.

FL CH .... This mode provides protection. ***It will not leave an altitude that is on the MCP***. It will only fly towards an **MCP** selected altitude and it will not climb or descend beyond that limit. It has a shortcoming, however. It relies on the airspeed set in the MCP for control and if you are making a large climb/descent, it does not take into account any factors that adjust for altitude. This could drive the airplane into an overspeed situation.
My advise: Use **FL CH** for climb/descents of only a few thousand feet or less.

> **SOP** dictates that the **AUTO-THROTTLES** should be shut **OFF at 50 feet** on all approaches including the visual ... unless the approach is a multi-auto-pilot auto-land (**CAT II/III**).

HOW TO CLIMB and DESCEND

WARNING ... "FLCH trap"

There is one area that I think deserves attention ... and pilots call it the **"FLCH TRAP"**. When descending in **FLCH** mode, the **AUTO-THROTTLES** will command **IDLE THRUST** and **HOLD** is annunciated on the **PFD**. When reaching the selected level off altitude, it is captured and the auto-throttle mode changes to **SPD**. The **THROTTLES** will **ACCELERATE** to **FMC** commanded settings and the airplane will fly level at the selected **MCP** airspeed. **HOWEVER** ... Since some airlines use the practice of placing either the **FIELD ELEVATION (QFE)** or **"0" (ZERO)** in the **MCP** "target airspeed" window for the **APPROACH DESCENT**, this means that the **AUTO-THROTTLES** will remain at idle since the airplane will never reach capture altitude. It will be up to the pilot to manage the airspeed using **MANUAL THROTTLE** manipulation.
If the **AUTO-THROTTLES** and the **AUTO-PILOT**(s) have been dis-armed, and the pilot is attempting to fly the **FLIGHT PATH** using the **YOKE** with no **THROTTLE** input ...
<u>**THE AIRPLANE WILL LOSE AIRSPEED AND EVENTUALLY STALL ... AND CRASH!**</u>

Here is my problem with that whole scenario.
First, I don't place **FIELD ELEVATION** or a **"0"** in the **MCP** during an approach. At **MDA**, I think it makes more sense to place the **MISSED APPROACH** altitude in the **MCP**. That way, since I am shutting off the **AUTO-THROTTLES** at **50 Feet** (as per SOP) ... I would have my hands on the **THROTTLES!** Setting the **MISSED APPROACH** altitude and manually pushing on the throttles are just one less thing that I would have to worry about during the rush of the **"HAND FLOWN" TOGA** response.

There are a coupla reasons why **QFE** (Field Elevation) might want to be placed in the **MCP** during a visual approach. One of these is the use of the "**PREDICTIVE ARC**" which is a green arc that appears on the ND and tell the pilot when the airplane is forecast to be at that altitude. The pilot simply flies the pitch that places the arc on the landing area of the runway.

There is one more tool in our arsenal of vertical pitch weaponry and that is the ...

VNAV. This complex mode is simply **NOT INTUITIVE** and so it is confusing and difficult to operate. If you get in the position where you simply don't know what the **VNAV** is doing ... or why, revert to the simple **FLCH** until you sort things out. I would say that 63% of the times when you ask, "What is it doing?" It is because of the **VNAV**.

Because of the complexity, I am going to spend a few pages talking about **VNAV**

HOW TO CLIMB and DESCEND

FOUR DIFFERENT VNAV MODES

As if there wasn't enough complexity to this **VNAV** thingee, now I tell you that there are **FOUR** different annunciated modes of operation.
Here are the four modes:
- **VNAV**,
- **VNAV SPD**,
- **VNAV PTH**, and
- **VNAV ALT**.

Let's take them individually and try to explain what they do and how and maybe even why.

VNAV: This mode is displayed if the airplane is **BELOW 400 Feet**, or if the **PERFORMANCE INITIALIZATION PAGE** was incomplete.

Definition: **AFDS** means Automated Flight Director System with **AUTO-PILOT** and **AUTO-THROTTLES ON**.

VNAV SPD: **AFDS** commands the **PITCH** to **MAINTAIN TARGET AIRSPEED** that was placed in the **MCP** window.

VNAV PTH: **AFDS** commands **PITCH** to maintain the **FMC** target **ALTITUDE** on the **LEGS PAGE**.

VNAV ALT: **AFDS** commands **PITCH** to maintain the **MCP** selected **ALTITUDE**.

NOTE:

With **VNAV ALT** engaged, when a conflict occurs between the **VNAV** profile on the **LEGS** page and the **MCP** selected altitude, the airplane levels at the first altitude encountered and the **PITCH MODE** on the **PFD** becomes **VNAV ALT** and the airplane remains at that altitude. To continue the climb or descent, either:
- **SELECT** another **PITCH MODE**, or
- **PUSH** the **ALT** selector again, or
- **INTERCEPT** the **VNAV PATH**.

What this does is effectively make the **MCP** the guardian of the altitude; that is, the **AFDS** will not violate any altitude that has been placed in the **MCP**. One could say that the airplane must have the permission of the **MCP** to climb or descend through a specific altitude.

This occurs when the airplane reaches **T/D** (Top of Descent); without a **LOWER** altitude set in the **MCP**, the **AFSD** will not begin the descent path programmed on the **LEGS PAGE** of the **CDU** and a message appears on the **CDU** scratch pad "**SET LOWER ALTITUDE**".

747-400 SIMULATOR TECHNIQUES ...

The mysterious VNAV enigma

The **VNAV** function is far too complex for a mere human airline pilot to understand so let's accept the fact that we will **NEVER** fully understand **VNAV**. However, that being said, we must be constantly aware of what it is doing and confirm that it complies with what we want it to be doing.

MAINTAIN SITUATION AWARENESS

According to piles of engineering reports by everyone from NASA and the FAA to Boeing and the airlines themselves; everyone of them agrees:

> "The **VNAV** function ... Accounts for the majority of reported human factor issues with cockpit automation."
>
> "63% of pilot-cockpit interaction issues were in the control of the ... **VNAV** function."
>
> 'The **VNAV** function is the most disliked feature of automated cockpit systems."
>
> "...73% of pilots used **VNAV** in the climb phase, while only 20% used the function in descent and 5% use the function in approach."

The heart of the VNAV problem !!!

If you are changing from a **NON-VNAV** pitch mode such as **FLCH** or **ALT** (for example after a go-around) or de-selecting speed intervene while in **VNAV mode;** and simply push the **VNAV** selector button, the **AIRSPEED COMMAND BUG** is likely to slew up or down to some airspeed that has no meaning to the human operator. In most cases, the engines either come on with a sudden burst of power, or worse yet, go to idle and the airspeed starts dropping below the selected flap speeds.

Anyway ... **NOT GOOD** and this unexpected event causes pilots all over the world to start pushing buttons, clicking off stuff, twiddling knobs, and generally getting all excited while trying to regain control of the jet. This is usually followed by some comment like:

"WHAT THE CAT HAIR IS IT DOING NOW????"

> **THIS COULD ALL BE AVOIDED**
> If we simply knew what airspeed the **VNAV** would annunciate as its target when selected!

The **VNAV**, however, wants to fly a secret algorithm predicated on stuff not told to the human operator in order to meet criteria usually hidden from display in the **FMC** (Flight Management Computer). It will even use modes that it will not annunciate on the **FMA** (Flight Mode Annunciator). There are literally hundreds of complex options open to the **VNAV** module computer.
What's a pilot to do? well, there is at least one thing to do ...

... and PROCEDURES FOR STUDY and REVIEW ONLY!

VNAV SECRETS

It may seem that I am saying that VNAV can only be mastered by an Albert Einstein ... this is simply NOT true. I think that the VNAV function is truly marvelous, BUT we have to get control of the rascal and understand why it selects the airspeeds that it does ... OR, and here is my suggestion, tell it what you want it to do for you.

The airspeed targets it selects are usually reasonable when we see what it is trying to do. They are airspeed from the bowels of its little computer heart that are imposed to protect the jet from exceeding some restriction or limitation.

Here is a way to input an airspeed request into the VNAV.

1. Select **VNAV** page on the **CDU** and Confirm that the speed listed as **MCP SPEED** is the same as the speed on the **MCP**.

2. Enter that speed or whatever speed you want the **VNAV** to maintain in the scratchpad.

3. Line select it to the **MCP SPD** line and observe the entry changes to **SEL SPD** and your speed is inserted.

4. Select illuminated **EXECUTE** button. The **FMC** now has the **VNAV** speed as your requested airspeed in its database.

5. On **MCP**, de-select the speed intervene or depress **VNAV** button to select **VNAV**.

6. Observe the **SPEED COMMAND BUG** on the **PFD** move to your desired speed.

Now, here is the place where the **VNAV** may or may not comply with your request. It has its own secret reasons.
If it does not, depress **SPEED INTERVENE** knob and set the desired speed manually.

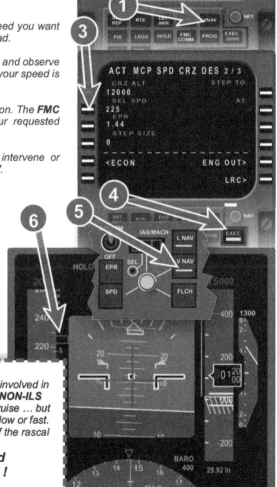

Here is the bottom line ... Don't get involved in the **VNAV** magic. It will fly a great **NON-ILS** profile, climb profile, descent, and cruise ... but do not be remiss and let it get you slow or fast. When it is necessary, take control of the rascal ...
**speed intervene and
FLY THE AIRPLANE !**

© MIKE RAY 2014
WWW.UTEM.COM

HOLDING!

There is something that strikes dread into the heart of even the most intrepid aviator ... it is that dreaded statement from the controller...

"I HAVE HOLDING INSTRUCTIONS WHEN YOU ARE READY TO COPY."

When we receive this message, there are three things we have to compute:

1 HOW MUCH FUEL DO I HAVE IN MINUTES.

2 HOW LONG CAN I HOLD?

3 WHAT IS MY "NEW" PLANNED ALTERNATE AIRPORT?

After receiving your clearance to hold, don't forget to:
CONTACT THE DISPATCHER with your assigned HOLDING FIX and EFC.

> You should do this via ACARS, because you can bet that the Dispatcher is really busy with lots of other pilots holding and diverting. The HOLDING/DIVERT scenario is particularly difficult because all the resources normally available to you for consultation will be occupied and not available to you. If unable to contact the Dispatcher or she (he) does not respond it is essential that you go through the same steps she (he) would in assessing your FUEL/TIME problem.

HERE IS HOW IT WORKS:

STEP 1: Use alternate designated on your FPF or go to FOM page APT-111 ff. Assume you are no longer going to your destination and select some realistic alternates. The selection process **MUST** involve the list on FOM APT-111 ff.

STEP 2: Determine how much fuel you want to have on board when you park the jet at THE NEW DESTINATION.

STEP 3: Determine how much fuel it will take to:
   (a) get from the holding fix to the initial approach fix at the new destination,
   (b) fly the approach and landing,
   (c) fuel for additional holding if appropriate,
   (d) taxi to the gate.

STEP 4: Subtract that from what you have on board; convert to minutes and decide when it will be time to depart holding.

STEP 5: DO NOT STAY AROUND beyond your limit. It has been my experience that these things can drag on and on and even when cleared to the next controller, additional holding may be required. **BE TOUGH!**

... and PROCEDURES FOR STUDY and REVIEW ONLY!

HOW TO SET UP THE HOLD

This machine is simply FABULOUS at holding. It is so intuitive that even I usually have trouble screwing it up.

STEP 1
Depress **HOLD** *key, then Either*
type it the designator for the holding fix, or line select the fix from the LEGS page. Another option is the **PPOS** key. If you select **PPOS**, then the airplane will immediately begin a holding pattern at the **PRESENT POSITION**.

NOTE
The selected holding fix does not have to be on your route of flight or even on an airway. Any fix may be used as a holding fix without entering it in the route of flight first.

STEP 2
Depress LS6L (next to the little row of boxes).

STEP 3
Depress the LS button next to place where you want to hold in the list of fixes. In our example, we want to hold after **MAGGI** so we pushed the button next to **BAMBO**.
Then, without prompting, the **MOD RTE 1 HOLD** page comes up for us to peruse and add to or change. Changes could include Inbound leg to the holding fix which is useful to align with your desired departure leg; or Right or Left turn, or Altitude, or Airspeed. Once it is the way we like it, we push the **EXEC** button

STEP 4
The "HOLD" page changes to add **<NEXT HOLD** at the bottom. At this time you may add additional holding patterns as needed. If the "**NEXT HOLD**" is already displayed, select it until you get to a page without a HOLDing pattern displayed.

STEP 5
Select the LEGS page and close up the discontinuity. When the jet gets to the holding pattern fix, it will automatically enter holding.

NOTE
Airspeed will have to be controlled by the pilot. It is suggested that you obtain a clearance to slow to holding speed from Air Traffic Controller on your way to the fix.

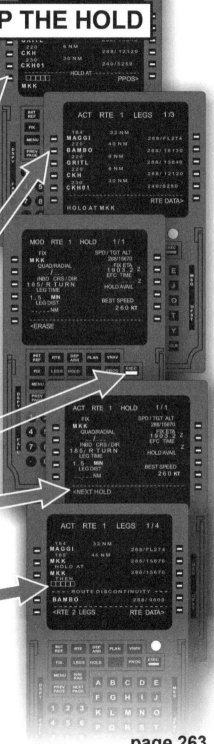

© MIKE RAY 2014
WWW.UTEM.COM

747-400 SIMULATOR TECHNIQUES ...

HOW TO GET OUT OF HOLDING

Pilots can always dream up some unique and esoteric way to do things ... work-around or just creative thinking. However, here are four ways to get out of holding; two bad ways and two good ways. First, the so-called "BAD" ways.

THE TWO BAD WAYS

OPTION 1: There is a brute force selection. Shut off the auto-flight system, grab ahold of the yoke and **HAND FLY** the airplane clear of the pattern. Always an option.

OPTION 2: Use the Heading Select (**HDG SEL**) and steer the airplane somewhere else. The big problem with these solutions is that once we get clear of the holding pattern remains in the **CDU** and the the airplane cannot be reconnected to the **LNAV** until the routing problem has been resolved.

THE TWO GOOD WAYS

OPTION 3: EXIT HOLD PROMPT. This the most clean and efficient way to exit a holding pattern ... and probably the one that the check pilot will be looking for. No matter where you are in the pattern, the airplane will make the most expeditious way to the holding fix and depart along the established routing.

OPTION 4: Select any **WAYPOINT/FIX** below the Holding fix in the **CDU LEGS** page queue. Then line select that to the top of the routing queue, and **EXECUTE**. The airplane should turn immediately towards that fix and the **HOLD PAGE** should drop from the routing queue.

OPTION 5: A variation on this is to pick any other fix and type it into the scratchpad and then place it at the top of the routing queue. The major difference is that the original routing in the **RTE** page will remain in place.

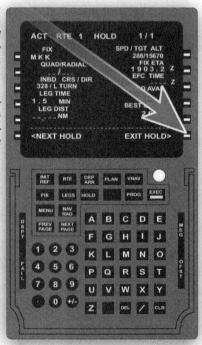

CHECK PILOT TRICK: There is unique situation (which could be a problem) if you have already loaded the approach for the landing runway and it contains a holding fix associated with the missed approach routing. Since the missed approach holding fix will appear on the HOLD page, if you are cleared to hold at that fix, you will have to select the NEXT HOLD in order to establish another holding pattern at that same fix. Otherwise, if you proceed directly to that fix, you will lose the complete approach and missed approach routing that you had already established.

HOLDING OP SPECS

Here are some of the things that you will be expected to know about holding. Even though the "MAGIC GLASS" does a miraculous job of figuring the entry, etc; there are some other things that we have to have at our fingertips.

"NORMAL" HOLDING SPEEDS

265 KTS MAX ———— 14,000 feet
230 KTS MAX ———— 6,000 feet
200 KTS MAX ———— Minimum holding altitude

DOMESTIC US ONLY

NOTE 1: If your airplane is "**TOO HEAVY**" to hold at that speed, you **_MUST_** obtain **ATC** clearance to hold at a higher airspeed.

NOTE 2: It is a common ploy for the check person to issue a clearance to hold when the **CMS** (clean maneuvering speed) is greater than the allowed holding speed. **BE ALERT!** Possible **STALL** danger. **YIPE!**

NOTE 3: Remember that if you are given holding right at the 14,000 feet boundary, then your holding speed **MAXIMUM IS 230 KTS.**

NOTE 4: Be aware of "Minimum Holding Altitudes." Usually the best indicator is the "**GRID MORA**." **MEA**s usually are not good choices as they require remaining too close to the airways. If the holding pattern is depicted, the little altitude inside it is the **MHA**.

NOTE 5: These "**NORMAL**" holding speeds do not count when flying to places that have published their own speeds; such as Military places, London, New York Area, etc. These places will inform you what their speeds are with some hidden note on the approach plates, charts, 10-7/20-7 page, etc.

NOTE 6: The recommended "**BEST SPEED**" annunciated on the **HOLD PAGE** does not take into account "compressibility." This means that holding **AT ALTITUDE MAY** necessitate a correction, otherwise the possibility of **LOW SPEED BUFFET** is enhanced. I recommend that at altitude, **_ADD 20 KNOTS TO THE FMC MINIMUM HOLD SPEEDS_**.

HOLDING INSTRUCTIONS

An ATC clearance to Hold will include:

1. Direction (NE, SW, etc)
2. FIX
3. Radial, course, bearing, etc.
4. Leg length
5. Direction of turns
6. EFC ... Complete holding instructions **MUST** include **EFC** !

Standard holding is:

Right Hand Turns
Inbound legs 1 minute at or below 14,000 feet
and 1 1/2 minutes above 14,000 feet
25 degrees bank (using autopilot)

IF ... THEN: Pilot Training 101. If you accept a clearance without an **EFC**, the Check Person will have an **UNCONTROLLABLE URGE** to fail your radios!

HOLDING REPORTS

ENTERING HOLDING: You MUST report:
1: FIX
2: TIME
3: ALTITUDE

IMPORTANT!
WHAT PILOTS SCREW UP: Most common deletion is for the pilots to forget to report their atitude.

LEAVING HOLDING: You MUST report:
departing the holding fix

You are then expected to resume normal operating speed.

© MIKE RAY 2014

747-400 SIMULATOR TECHNIQUES ...

After we have flown the agonizing long trip, we get up from the bunk to prepare to land in some exotic far-a-way place.

> The cockpit of the 747-400 is spacious, I would say "H-U-G-E". There is even room for a sleeping area with bunks separated from the rest of the cockpit by an enclosure that has a door.... I say it is spacious and huge except in the forward part where the pilots are seated. The way that the fuselage slopes, the pilot and co-pilot seats are quite cramped. The pilot's head is close to the side windows, which helps for visibility, but makes for difficulty in getting into the flight bag ... which is wedged between the seat and the side of the airplane.
> So ... I mention this because it is extremely useful to pre-select and arrange all the approach plates and charts that you will need ahead of time since it is virtually impossible to reach into the flight bag to paw through your stuff looking for the material you will edd while sitting in the seat.

SLEEPING ON THE JOB:
... AND RETURNING FROM ENROUTE CREW REST.

Unless you have experienced the whole whacky, unreal paradigm of actually being in charge of a huge airplane streaking through the stratosphere at near the speed of sound and yet having an assigned sleeping period; and then in addition having additional crew members on board and the idea of rotating personnel at the controls for the lengthy cruise portion of the flight, there can be some confusion as to just what has transpired during the time that you were on your "crew rest". My first time I was very uncomfortable and every time there was a change in pressure or engine sounds I was wide awake trying to think what was going on. Then, as time goes on, trip after trip, there comes a time when you accept and become at ease with the whole idea. Protocols vary with the different airlines, but it was my experience that a complete briefing as to what transpired while you were not at the controls was in order. It is my opinion that regardless of the assigned sleeping period, the Captain should be awake and take control of the airplane long enough before the commencement of the descent and approach phase so as to be able to emotionally prepare and completely brief the approach and landing. Occasionally, the airplane will have diverted during the Captains absence and a whole new set of weather and airport conditions may confront the crew.

DID WE "RA*"?

*
"RA" is short for "REDISPATCH ACCEPT." When flying long distances, particularly over-water, it may be necessary to file for a destination short of your desired landing airport for fuel purposes. Upon arriving at your "redispatch point," if the fuel situation is appropriate, you can accept a redispatch to continue to your destination.

© MIKE RAY 2014
published by UNIVERSITY of TEMECULA PRESS

This is probably the best memory gouge in this book.

The FABULOUS A.I.R.B.A.G. MEMORY GOUGE

HOW TO REMEMBER ALL THOSE LAST MINUTE THINGS TO DO TO PREPARE FOR THE APPROACH.

I don't know about you, but right here in the flight evolution, there is a lot going on ... and also, everything has to be done completely and accurately. In order to assist in doing everything from memory, there is a really great gouge; it is called "**AIRBAG**."

During the stress of the simulator check-ride, when things are coming apart ... revert to this gouge.

If you have a problem, just remember the gouge. The first thing you do after the immediate action items during an emergency is to try and get the jet on the ground. Any time you are preparing to land ... go to the gouge.

It is fairly easy to tick off the items in your memory and ask for them in a manner that makes you look like you know what you are doing. Particularly after an emergency or irregular event requiring a return to the field, it makes preparation to land fairly easy.

A = Get the ATIS,
I = Install the approach,
R = Tune the Radios,
B = Brief the approach,
A = Approach-Descent Checklist,
G = Go-Around.

Any questions?"

Before starting any approach, there is a bunch of stuff that we have to do. I have heard some pilots using the acronym *A-I-R-B-A-G*.

what is "A-I-R-B-A-G"

A - ATIS
I - INSTALL APPROACH
R - RADIOS
B - BRIEF
A - APPROACH-DESCENT CHECKLIST
G - GO-AROUND/GET OFF

On the following few pages, I will take each of the items in turn and treat them in some detail. The usefulness of this gouge (pronounced GOWJ) comes when you are confronted with either a normal landing ... or an irregular landing situation.

NOTE: This will set you up to do the **APPROACH DESCENT CHECKLIST** items.

... and PROCEDURES FOR STUDY and REVIEW ONLY!

get ATIS

ATIS - Airport Terminal Information Service

There are a lot of ways to get the weather at the destination airport. If you are within VHF range, of course, you can monitor the ATIS facility on the airport. The frequency is located at the upper left corner of the approach plate.

If you are out of radio range, you can monitor the HF radio, especially over the ocean. That works pretty good ... sometimes.

On this airplane, however, there is a third CDU unit on the console that is normally dedicated as the ACARS unit. It has the capability of receiving the weather and sending it to the printer for a hard copy readout. Very nice.

Here is the information that will be contained in the typical **ATIS** message:

AIRPORT NAME
ATIS INFORMATION PHONETIC DESIGNATION (alpha, bravo, juliet, etc).
TIME OF REPORT (Zulu time i.e. UTC)
WIND DIRECTION / SPEED
VISIBILITY
CEILING
TEMPERATURE
DEW POINT
BAROMETRIC PRESSURE
TYPE OF APPROACH IN USE
RUNWAY IN USE
NOTAMS (Notices to airman)

ATIS FORMAT

I'm telling you what to expect because if you are receiving the ATIS by listening over the radio, sometimes the reception can be garbled or more commonly, the operator giving the message is difficult to understand.

Since the international language is English and many of the persons making the transmission do not speak English, it can be very difficult for a non-English speaking pilot to understand. If we know what we are expecting, then it greatly reduces the challenge.

**FYI: THE ATIS IS ALWAYS
GIVEN IN THIS SAME FORMAT.**

INSTALL THE APPROACH

"T" = Install approach

Some really intelligent pilot thought up a memory gouge that I will pass on to you for your consideration. It is a simple way to remember how to install the approach and tune the radios.

The five "keys" involved make a "T."

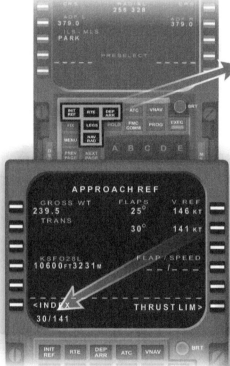

1 INIT REF page

This will bring up the "**APPROACH REF**" page. On there are displayed the V speeds and gross weight.

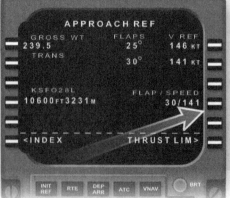

Decide whether you are going to use a 25 or 30 degree flaps setting for landing. Then Line select either to the scratch pad and place in the **FLAP/SPEED** line (**LS4R**). This will then tell the **FMC** what the target airspeed and flap setting will be for the approach and landing.

SOME THOUGHTS ABOUT FLAPS:
I have found that either works really well. The 30 degree setting causes more pitch over and so more over the nose visibility is available. That seems to be the choice during a low visibility/CAT III situation. The 25 degree setting seems to work a little better in a crosswind.

INSTALL THE APPROACH

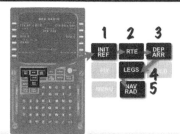

2 RTE page

The second key is the **RTE** key.

We should check that we have the proper destination inserted here.
If you are diverting or if you have changed your destination, here is the place where you would input that destination.

NOTE:
DO NOT place the runway of intended landing on this page. We place the runway information on the **DEP/ARR** page. Let me re-iterate this for you.

CAUTION:
DO NOT USE THIS PAGE TO INSERT THE LANDING RUNWAY!
USE THE DEP/ARR PAGE FOR RUNWAY UPDATES OR CHANGES.

If you are using an **UPLINK** system to install your routes, the data will be placed in the FMC automatically and you are just checking to see that it meets your company criteria.

LAND SHORT option

If you have decided to **LAND SHORT**, then you will have to tell the box where you are going so it can get all ready to help you out.

> Type the designator for the airport that you intend as your new destination and **LS1R** (push the first button on the right side of the **CDU**). Check for the correct entry and agree with the other pilot that we should really be doing this,
> Then depress the **EXECUTE** button.

Another place where you would use this option is if you are going to do multiple approaches to the same airport as during training. Simply place the departure airport in the departure slot as well. Now the airplane knows what we are doing.

INSTALL THE APPROACH

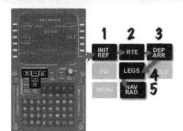

3 DEP ARR page

WHEN MORE THAN 400 MILES FROM YOUR DEPARTURE AIRPORT OR FURTHER FROM THE DEPARTURE AIRPORT THAN THE ARRIVAL,

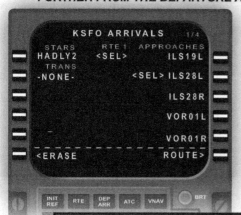

then the **DEP ARR** page will display all the approach, stars, and transitions that are in the FMC memory for the
DESTINATION AIRPORT SELECTED on the RTE page.
When the above criteria has been met, the **FMC** decides which airport to display predicated on the **DESTINATION** you have selected on the **ROUTES** page. Normally, we have selected the proper destination airport during the preflight phase; however, during a divert or land-short scenario, it is incumbent on the pilot to enter the "new" airport; but this must be done in the **RTE** page.

CAUTION:
Select the **APPROACH, STAR**, and **TRANSITION** in that order. If one of the items is changed or deleted, you must install all three items in the order I have just listed, even though one or more of the items may not have changed.

AIR-TURNBACK feature

Mr. Boeing built into the FMC a great feature. If you have some reason to immediately turn back and land, the quick and dirty solution to setting up the box for arrival is the "AIR TURNBACK feature.

If, after departure, you are:

LESS THAN 400 MILES FROM DEPARTURE AIRPORT, or NEARER TO THE DEPARTURE AIRPORT THAN ARRIVAL

*Then; all you have to do is depress the **DEP/ARR** key and the arrivals for the departure airport are automatically displayed.*

INSTALL THE APPROACH

4 LEGS page

On the legs page, you are going to see **ROUTE DISCONTINUITIES**. These would be represented by a row of little boxes. We must inspect the route (using the **ND and charts**) and see if "closing up" the discontinuity would be appropriate. Normally, the FMS desires a continuous path of linked legs all the way from departure to destination.

EXAMPLE 1

Looking at the **LEGS** page and comparing it with the **ND** and the **STAR** page, we can see that it is desirable to have a smooth linked transition between **MENLO** and **BRIJJ**.
To do this we would;

Line Select the FIX directly below the row of boxes to the Scratch Pad, and
Then Line Select the row of boxes.

This would "**CLOSE THE DISCONTINUITY**."

Don't forget to push the "**EXECUTE BUTTON**."

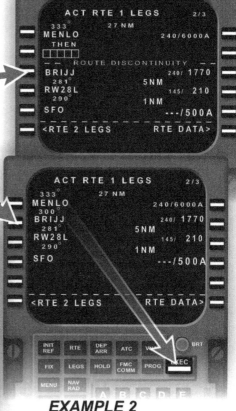

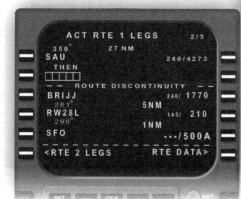

EXAMPLE 2

In this first example, we can see that it would be INAPPROPRIATE to close up the discontinuity. When we examine the route from SAU direct to BRIJJ on the ND, we see should expect an *LNAV- RADAR VECTOR-APPROACH* transition to the fix **BRIJJ**. So, we would consider leaving the discontinuity "open" and expect RADAR VECTORS from ATC.

Ensure that the IAF or FAF were NOT DELETED or the auto-tune on the navigation radios WILL BE INHIBITED !

RADIOS tune and identify

5 NAV RAD page

When you first start flying this airplane, you will ask, "Where are the nav radio knobs?" Well, there aren't any. You use the **NAV RAD** page on the CDU. This modern marvel uses **AUTO-TUNING** and it is fabulous, **BUT** ... we always have to be on guard that it is tuning the right radios and giving us what we want.

EXAMPLE 1

> **NOTE**:
> AUTO-TUNE only applies to the ILS and the VOR radios. It does **NOT TUNE THE NDB (ADF)** radios.
>
> AUTOTUNING will not override MANUAL tuning.
>
> Deletion of a corresponding manual frequency returns the system to autotuning.

The **VOR** has 4 tuning status modes:

A - Auto-tune. **FMC** has selected best frequency for position updatng.
M - Manual tuning is **PILOT** entered.
R - Route defining **VOR**s are being tuned by the **FMC**.
P - Procedure on active flight plan requires this **FMC** selected frequency.

Using the **EFIS**, the pilot can select the **VOR** or **ADF** displays on the **ND** (Navigtion Display).

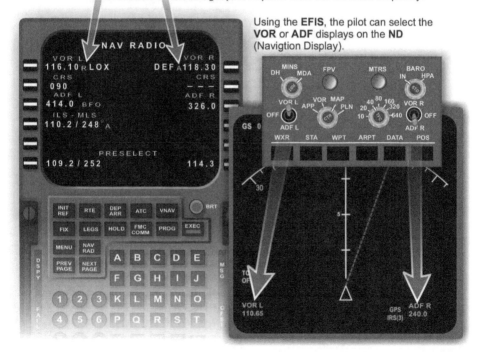

page 274

© MIKE RAY 2014
published by UNIVERSITY of TEMECULA PRESS

RADIOS tune and identify

5 NAV RAD page continued

On the **PFD**, the "**APPROACH REFERENCE**" will display either:
  IF IDENTIFIED by FMC, it will show the **ILS IDENTIFIER**; but
  IF NOT IDENTIFIED by FMC, it will display the frequency.
Once receiving it, it will display the **DME**.

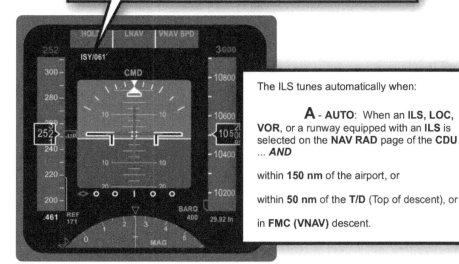

The ILS tunes automatically when:

A - **AUTO**: When an **ILS, LOC, VOR**, or a runway equipped with an **ILS** is selected on the **NAV RAD** page of the **CDU** ... **AND**

within **150 nm** of the airport, or

within **50 nm** of the **T/D** (Top of descent), or

in **FMC (VNAV)** descent.

PARK - On the **NAV RAD** page the **PROMPT CARET (< >)** and freq/crs will be displayed after **ILS** or **LOC** is selected, and

within **200 nm** of **T/D**, or

MORE THAN HALFWAY along the active route (whichever comes first).

Line selecting changes the tuning status to **M** (Manual).

M - **MANUAL**: Receivers tuned manually. Deleting a manually tuned freq/crs returns the **ILS** to the **FMC** autotuning.

NOTE:
AUTOTUNING WILL NOT OVERRIDE MANUAL TUNING.

ILS WILL NOT TUNE WHEN:
Autopilot engaged, **AND** either glideslope or localizer are captured.

A FLIGHT DIRECTOR is engaged, **either GLIDESLOPE or LOCALIZER is captured**, and the airplane is **BELOW 500 FEET RA**.

On the ground, with localizer alive, airplane heading within 45 degrees of the front course, and greater than 40 knots.

Auto tuning inhibited for 10 minutes after take-off.

747-400 SIMULATOR TECHNIQUES ...

BRIEF - APPROACH PLATE etc.

DISCUSSION

The Approach Chart brief is a somewhat difficult venue to define because every briefing situation and approach is different. Here is what we are trying to achieve: **SAFETY**. We want the approach to be as safe as we can make it, so we are interested in discussing anything on that approach that might compromise that goal. Mountainous terrain, towers,

We also want to make certain that the pilots all have the same approach plate and that the airplanes database conforms to what we are trying to do.

We want to know what we are going to do if we have to make a missed approach. And anything else you can think might be applicable.

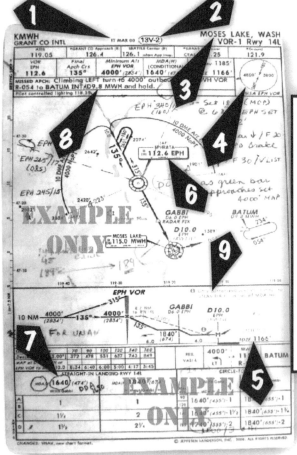

Here is some sample stuff. The list is NOT to be taken as definitive or complete. You will have to make up your own as the situation dictates.

1: Date and number of plate.
2: Chart number
   (a "V" indicates VNAV ready).
3: City (Airport name) and Runway.
4: MSA indicator
   (Minimum sector altitude).
5: TDZE
   (Touchdown zone elevation).
6: Frequencies of NAV facilities.
7: MDA
   (Minimum Descent Altitude).
8: Missed Approach
9: NOTES and comments.

So, when you do the approach brief, try to include as much IMPORTANT information as is available.

GOOD CAPTAIN HABIT

ALWAYS END THE BRIEF WITH THE QUESTION
...
"Are there any questions or comments?"

SOMETHING IMPORTANT THAT SOME PILOTS DON'T KNOW!

IF WEATHER GOES BELOW LANDING MINIMUMS DURING THE APPROACH

NON-PRECISION and **ILS CAT I/CAT II**: If you had adequate minimums to commence the approach at **FAF** (Final Approach Fix) and it subsequently goes below minimums during the approach, you are **OK** to continue the approach to **MDA** or **DH**; and if you "**VISUALLY ACQUIRE**" the runway (and meet all the other criteria for making a safe landing), you may land.

CAT III: ...a little Different. Once you have commenced the approach with acceptable weather at **FAF**, and the **RVR** subsequently goes below minimums, You may continue to **AH/DH**. *If weather at AH/DH is not above landing RVR minimums, you must Go-Around; EVEN IF YOU SEE THE RUNWAY. To put it another way, you MUST have landing minimums at AH/DA to continue.*

THIS IS THE CRUX OF THE PROBLEM!

FAQ (Frequently asked question)

If you are flying a **CAT III** approach and you arrive at **CAT II** minimums with the runway environment in sight **BUT** the **RVR** below **CAT III** minimums. That is, you started the approach with **CAT III** minima but you arrive at the **DH** with all **CAT II** landing criteria satisfied. **MAY YOU CHANGE TO CAT II CRITERIA AND LAND?**

*Some airlines allow the "**pre-briefed CAT II**" idea for every **CAT III**; That is to say, in your **CAT III AUTOLAND** approach brief you include the statement, "If the **AIRCRAFT SYSTEMS STATUS CHANGES** or **AUTOLAND ANNUNCIATION** goes to **LAND 2**, we will revert to **CAT II** criteria for landing at AH/DA … That is, **SEE TO LAND**."*

MIKE RAY's INTERPRETATION:
*If you arrive at the **CAT III DA** and the **RUNWAY RVR** is below **CAT III MINIMA** ... If you **VISUALLY ACQUIRE** the runway environment, you **MAY LAND**. You are changing the landing criteria to **CAT II** minima **DURING THE APPROACH**.*

---

*This is controversial territory and **MOST** airlines **DO NOT ALLOW** you to change the operating criteria **DURING THE APPROACH** ... By that, they mean, "If you started the approach using **CAT III** criteria, even though you arrive at the runway with **CAT II** minimums, you **MAY NOT LAND** if the **RVR** has dropped **BELOW CAT III MINIMUMS**."*

the A part of AIRBAG is
APPROACH DESCENT CHECKLIST

OBSERVATION:
The **APPROACH DESCENT CHECKLIST** has to be done expeditiously and accurately. I ABSOLUTELY DO NOT recommend memorizing checklists; however, I also feel that being familiar enough with the contents of this particular checklist to quickly and without faltering or stumbling be able to complete it succinctly and with total accuracy. **This is a particularly important checklist**.

THE APPROACH DESCENT CHECKLIST IS TO BE READ ALOUD AND COMPLETED BY THE PNF.

APPROACH DESCENT CHECKLIST
(To be checked *ALOUD* by the pilot not flying)

Approach briefing	Complete
FMCs, radios	Programmed, set for landing
EICAS	Recalled, cancelled
Airspeed	_____ Flaps, _____, set (Ref)
Autobrakes	Level _____, /Off
———— TRANSITION LEVEL ————	
Altimeters	_____, Set (In/hPa)

IMPORTANT:
The Altimeter setting MUST be cross-checked and confirmed correct ... This means:

BOTH ON THE SAME SETTING, AND
BOTH USING THE APPROPRIATE SETTING (HPA OR IN HG).

the G part of AIRBAG is
GO AROUND - GET OFF

GO AROUND

The approach will have several possible conclusions. You will land and go to some nice hotel and have a great dinner.
Or, you will not land but have to go-around. Here is where you discuss what will happen if the approach doesn't work out.
There are three possibilities to discuss:
RETURN TO LAND, OR
GO TO HOLDING, OR
PROCEED TO ALTERNATE.

It may be useful to include in your brief an estimate of the amount of time that you can remain in the area and what would be your divert plans. These may be predicated on fuel available, weather trends, traffic in the area, etc.

If holding after the go-around is desirable or even possible,
(1) check the hold page on the CDU and see if the routing and hold pattern are described properly, and
(2) indicate how you would get to the holding fix, and
(3) how long you could remain in the hold before you would have to leave for your alternate or divert airport.

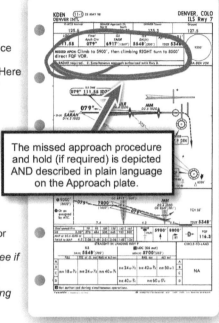

The missed approach procedure and hold (if required) is depicted AND described in plain language on the Approach plate.

GET OFF

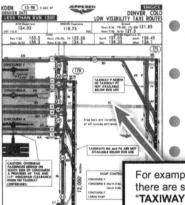

There are two things to be concerned with here. First, if there are a lot of airplanes ahead, such as at LAX and the high-speed turn-offs may be blocked, brief a reasonable plan for getting off the runway. VMC landings should have a "target" off ramp.

In limited visibility situations, particularly where "smegs" (Low Visibility Taxi Routes) are in effect, there are "usually" charts that address the situation. They will have specific instructions and notes.

For example: On the Denver low visibility taxi route chart there are several notes, one which says,
"TAXIWAY P NORTH OF TAXIWAY P7 NOT AVAILABLE BELOW RVR 600."
Notes like these should be briefed if they are applicable.

747-400 SIMULATOR TECHNIQUES ...

A BRIEF DISCUSSION ABOUT:

VNAV, CANPA, CROD, and CDAP

> The **Boeing 747-400** is certified to fly approaches using **VNAV** procedures; i.e. Using the autopilot to control the glide slope on approaches where the glide slope is **NOT** defined by a ground based transmitter (that is **ILS**). In the past several years there has been numerous attempts to use this capability in routine airline use. The "refinements" and suggested improvements have undergone a continual re-assessment due to skepticism on the part of airline personnel in charge. These guys have printed and promulgated their changes and revisions ... constantly ... and we are currently at a place where the now named "RNAV" series of approach criteria is reaching useage.
>
> However, this all started at the attempt to define the place (a virtual point in space) where the pilot may (should) start their final descent. Here is one of those procedures which is still used at some airlines. It is called in a generic way as **Constant Descent Approach Procedure**, or simply **CDAP**.

This approach segment discussion only applies to the **VERTICAL** portion of the **APPROACH PATHWAY**.

VNAV: **V**ertical **Nav**igation

CANPA: **C**onstant **A**ngle **N**on **P**recision **A**pproach

This is defined as flying the approach from the Final Approach Fix to the Missed Approach Altitude using a constant angle.

CROD: **C**onstant **R**ate **O**f **D**escent

This is defined as flying the approach from the Final Approach Fix inbound to the Missed Approach Altitude using a constant rate of descent.

CDAP: **C**onstant **D**escent **A**pproach **P**rocedures

This is defined as flying that part of the approach from .3 miles outside the FAF to MDA + 50 feet using a predetermined Vertical Rate of Descent.

These procedures only apply to that small but important portion of the approach from the FAF to the MDA.

The reality is, however:

1. We will still be using "**DIVE AND DRIVE**" techniques for those portions of the approach outside the FAF.

2. The only place where there are changes to our previous technique is in that portion of the approach from the **FAF** to the **MDA**. All the rest is the same as before.

3. Also, this will **ONLY** apply to Non-ILS (Non-Precision) approaches ... even though the airline reminds us that **ALL** approaches will use **CDAP**. How can this be???

4. Well, here is why. Precision Approaches (**ILS** based glide-slope) have always been descent-rate controlled by their very nature, so they will be flown using the same basic techniques.

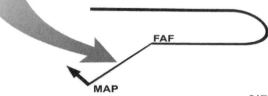

page 280

... and **PROCEDURES FOR STUDY and REVIEW** *ONLY!*

Let's talk about how we go about

flying ...

THE CDAP PROFILES

At least 90% of the **IMC** (instrument flight rules) approaches flown today use the **ILS** related approach evolution. While it is true that the state of the art is constantly in a transition to the more sophisticated **GPS** and "other" advanced technologies, currently the **ILS** glide-path is still considered the most commonly used approach in the world.

While the **NON-ILS** (formerly called **NON-PRECISION**) approaches are usually required to be demonstrated by the pilot in the simulator for training purposes ... we as pilots should gain proficiency in accomplishing these maneuvers and consider that there actually may be sometime-someplace when we may have to accomplish one of these **NON-ILS** maneuvers. So we will discuss them in detail also.

We will **NOT** include the **FMS** or other **GPS** related related approaches in our discussion; neither will we consider the **ASR**, PAR, GCA and other voice-command controlled glide-slope venues.

While it is true that the 747-400 series aircraft are certified by Boeing and the FAA to fly all instrument approaches to a point 50 FEET above the end of the runway using the **VNAV** (called **RNAV**)... and it works fabulously; we will be describing the **CDAP** (Constant Descent Approach Procedures) technique. Doing this adds a continuity to both the **NON-ILS** and the **ILS** approaches fleetwide that is lacking in using other procedures.

Just to reiterate: **"CDAP"** is different from the old **"Dive and Drive"** and from the **"VNAV"** (**RNAV**) technique described by Mr. Boeing. While from some viewpoints, it is not necessarily the "best" way to fly these approaches, but has been adopted for other reasons.

We will now attempt to describe the
CDAP technique
for flying the approaches.

It is appropriate to say that at some level, "**ALL**" approaches are actually **CDAP** approaches, even the **ILS** precision approaches.

747-400 SIMULATOR TECHNIQUES ...

SOME SPECIAL CDAP RULES
CONSTANT DESCENT APPROACH PROCEDURES

- All approaches are to be flown at a constant descent rate to a descision point (DA or decision altitude) where a decision is made to either land or go-around. This is called CDAP (Constant Descent Approach Procedures).

- For all non-precision approaches, if weather is less than 1000/3 they must be flown using the autopilot, disconnecting at no less than 50' below the published MDA.

- For Non-precision and VMC; If landing, the autopilot MUST be disconnected no later than 50' BELOW THE PUBLISHED MDA.
  I included this restatement in order to emphasize that this gives us 100' from DA to the MDA-50' disconnect altitude to tweak the final descent using the autopilot. This is the recommended technique. STAY ON AUTOPILOT if acquiring the runway for landing at DA.

- CDAP non-precision approaches are flown to a DA, where the decision to land or go-around is made.

- All non-precision approaches require one pilot (PF or PNF) to monitor raw data no later than the FAF or IAF for piloted constructed approaches.

- There is no longer a requirement to compute a PDP.

- There is NO ALLOWANCE for descent rates greater than 1000 fpm below 1000' AFE.

- DA is computed by adding 50' to the published MDA.

- All CDAP approaches are flown with the TDZE in the MCP. TDZE is computed by rounding up the published TDZE to the next highest hundred.

- A descent rate correction of NO MORE THAN +/- 300 fpm from the computed descent rate may be made during the approach.

BIG !

> IF more than +/-300 fpm correction is required, the approach is considered unstable and a go-around is **REQUIRED**.

EXPLANATORY NOTE: Momentary corrections exceeding +/- 300 fpm **DO NOT** require a go-around.
Frequent or sustained corrections DO require a go-around.

- The MISSED APPROACH altitude is to be set in the MCP **during** the go-around, **after** the "gear up" command.

ALSO BIG !

... and PROCEDURES FOR STUDY and REVIEW ONLY!

CDAP TALK THROUGH
CONSTANT DESCENT APPROACH PROCEDURES

The "old" dive and drive technique is still to be utilized outside of the FAF. Remember that the CDAP only applies to the vertical component of the approach INSIDE the FAF; however, there are some CDAP items that must be completed prior to reaching the FAF.

Even though there is NO PDP calculation, the PF will be **REQUIRED** to compute and include in your brief:

> Computed descent rate, and
> Computed TDZE, and
> Computed DA
> (and set on appropriate altimeter).

Once you level off at the FAF altitude, and ALT CAP is annunciated, you should set the TDZE in the **MCP EVEN THOUGH THERE MAY BE STEPDOWN FIXES INSIDE THE FAF**.
That means that we don't use "dive and drive" inside of the FAF.

Once established at FAF altitude,
select Vertical Speed and check for "zeros."

If a stepdown segment exists inside the FAS (Final Approach Segment) it is considered GOOD TECHNIQUE to prefigure a "howgozit" by adding 1 mile and 300 feet to the stepdown crossing fix criterium. You can write that right on the approach plate.

At .3 miles prior to the FAF on the non-precision approach, roll the Vertcal Speed selector on the MCP to the pre-figured descent rate figure. Start descent aggressively and don't delay as you **WILL** get high.

Once stabilized in the descent, you are allowed to observe and adjust the "green arc." It should rest approximately at the approach end of the runway.

The 1000 foot call should occur at 3 miles from the runway.

Vertical corrections using the Vertical speed knob should be carefully selected. I found that one "click" adjustments sould be adequate and it takes some time for the correction to be reflected in the green arc.

> *In any case, DO NOT SUSTAIN +/- 300 fpm deviation from planned!*

Once established in the Go-around it is necessary to set the MISSED APPROACH ALTITUDE on the MCP. The callout goes like this:

> "Go around Thrust"
> "Flaps 20"
> "Gear Up"
> "Set Missed Approach Altitude"

DO NOT FORGET!

THERE ARE **THREE KEY ELEMENTS**
TO THE CDAP BRIEF:

- **Computed TDZE**
- **Computed DA**
- **Computed DESCENT RATE**

Computing the TDZE

The definition of TDZE is Touchdown Zone Elevation, and we get that from the APPROACH PLATE PLANFORM DIAGRAM. The "COMPUTED" TDZE is that value rounded up to the next higher 100 feet.

For example: If the published TDZE is 301 feet, round up to 400 feet; and if the published TDZE is 399 feet, round up to 400 feet.

This computed TDZE is placed in the MCP (Mode Control Panel) once ALT CAP is annunciated at the FAF (Final Approach Fix) altitude and outside the FAF.

Computing the DA

The definition of DA is DECISION ALTITUDE, and we get that from the APPROACH PLATE PLANFORM DIAGRAM. The "COMPUTED" DA is the MDA (for non-precision approaches) value plus 50 feet.

For example: If the published MDA is 1060 feet, then the computed DA will be 1110 feet; that is 1060 + 50 = 1110 feet.

This computed DA is placed on the barometric altimeter for non-precision and precision approaches.

HOW TO CALCULATE THE
VERTICAL RATE OF DESCENT!

There are THREE suggested ways to figure the Vertical Speed:

METHOD 1.
USE THE APPROACH PLATE PLAN-FORM DIAGRAM.

Determine your approach GROUNDSPEED. This information can be taken right off the ADI (757/767), HSI (737) or ND (747-400).

NOTE
Technically, the FMC generated groundspeed information should not be used for glideslope computation UNTIL:
- the airplane is fully configured and
- inbound on approach airspeed.

A couple of notes:

1. We can only make simple adjustments to the V/S wheel anyway, so more "accurate" calculations are a waste of time.

2. If the airspeed is "off scale" in the chart (a common situation), make an estimate. It seems to be better to guess higher rather than lower.

3. If the wind-speed on the ground is low and the wind at altitude is dramatically different, be aware that it could affect the calculation significantly. The pilot must constantly be aware of the changes and adjust by "tweaking" the VERTICAL SPEED knob.

BELOW 1000 AFE (above field elevation) Op Specs DO NOT allow:
vertical speeds greater than 1000 fpm; or sustained
corrections greater than +/- 300 fpm

METHOD 2:

THERE IS A CHART IN THE FOM called:
 "**Descent angle and descent rate chart.**"
It seems to me, however, that it would be excessively complicated to be digging into my flight bag and thumbing through some bulky FOM to find a chart that would require lots of interpretation.

METHOD 3.

USE A RULE OF THUMB:

> ½ groundspeed X 10 + glideslope correction.
>
> *Note: Glideslope correction = +50 fpm for each 0.25 degrees that the G/S is greater that 3 degrees.*

Example: at 140kts G/S on a 3 ½ degree glideslope,
 the descent rate should be:
 140/2=70, 70 X 10 = 700 fpm + glideslope correction.
 since the G/S is 2 X .25 degrees greater than 3 degrees,
 then we add 100 fpm.
 Therefore, the computed descent rate should be
 700 fpm + 100 fpm = 800 fpm.

Wheew!!! That seems way too complicated for me, so I recommend the special MIKE RAY **Simple application of METHOD 3:**

> USE 800 FPM INITIALLY FOR A STANDARD 3 DEGREE GLIDE-SLOPE.
> IF GLIDE-SLOPE GREATER THAN 3 DEGREES,
> START OFF WITH 900 FPM.

IMHO *(In Mike Ray's humble opinion)*
It seems to me that there is only a small time window where the ground speed can be evaluated. It is from the point 3 miles outside the FAF until pushover at .3 miles from FAF. This is the only place where this observation can be accurately made. I thought to myself that this was a real time tight area where there is a lot going on and I would be hard pressed to concentrate on this.

I also observed that the descent rate solution was nearly always 800 fpm and also that a higher initial descent rate worked better than a shallower descent. That way, corrections requiring reducing the descent rate can be made without exceeding the 1000 fpm descent restriction; however, steeper descent corrections made to make descent milestones are SEVERELY restricted by the +/- 300 fpm limitation and the 1000 fpm restriction below 1000 FAE.

This applies particularly on approaches where the glide-slope is greater that 3 degrees. On those approaches, the required descent rate was around 900 FPM.

My advise, GET ON THE DESCENT QUICKLY, and BE AGGRESSIVE in your calculation and in starting down. The nose of the airplane **SHOULD** be coming over by the time you cross the FAF.

... and **PROCEDURES FOR STUDY and REVIEW** ONLY!

*C*ONSTANT *D*ESCENT *A*PPROACH *P*ROCEDURES SECRETS !

WHAT PILOTS SCREW UP !

THERE ARE (at least) 11 MAJOR THINGS PILOTS FAIL TO DO ON THE CDAP:

☐ 1. Failure to set next altitude on the **MCP** after altitude capture (**VNAV PATH**) when maneuvering *__OUTSIDE__* **FAF**. This is not to be confused with arriving at the **FAF** altitude inbound, in which case you would set the **TDZE**.
It is important to still use the "dive and drive" techniques when maneuvering outside of the **FAF**.

☐ 2. Failure to use the **CDAP** procedures and restrictions on the ILS and Visual approaches.

☐ 3. Failure to **BRIEF**:
　　　Computed Descent Rate (**FPM**)
　　　Computed **TDZE**
　　　Computed **DA**

☐ 4. Failure to set Computed **DA** on the barometric altimeters, setting instead the published **MDA**. Remember, computed *DA = MDA + 50 feet*.
Op specs still allow you to use the autopilot down to **50 feet below published MDA**.

☐ 5. Failure to set **COMPUTED TDZE** in the **MCP** at **ALTITUDE CAPTURE** (**VNAV PATH**) on the level off inbound to the the FAF.

☐ 6. Pilots tend to **OVER-CONTROL** the glidepath. Excessive reliance on the green arc and not allowing enough time for the arc to "settle down" after a correction is applied.

☐ 7. Pilots **EXCEED the +/- 300 FPM** restriction to the announced Computed descent rate without initiating a go-around. The approach is considered unstable inside the **FAF** if that restriction is exceeded for a sustained period of time.

☐ 8. If a step-down fix is depicted, the pilot *MISTAKENLY* sets "step down fix" altitude inside the **FAF** in the **MCP** instead of the computed **DA**. The suggested technique for determining if the restriction at the step-down fix is going to be met is to
add 1 mile and 300 feet to the fix altitude.

☐ 9. Failure to cross-check that the 1000 foot call-out occurs at 3 miles from touchdown. *This is presented as awareness technique and **NOT** as a requirement.*

☐ 10. Failure to initiate go-around at the "computed **DA**," Instead, allowing the airplane to descend to the "published **MDA**" before initiating the go-around.

☐ 11. Failure to set Missed Approach Altitude in the **MCP** after the request to raise the gear.

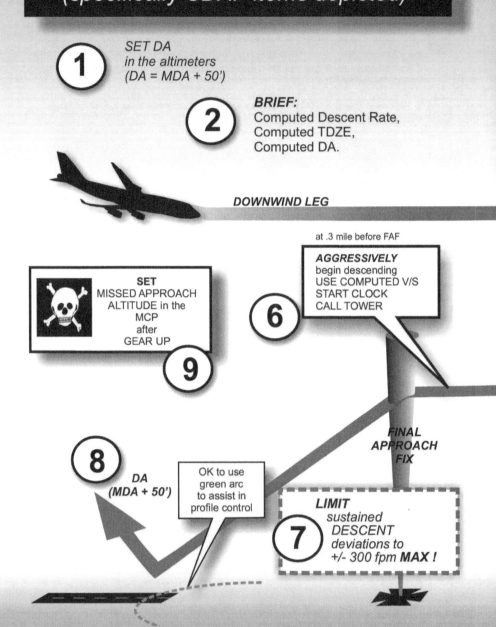

... and PROCEDURES FOR STUDY and REVIEW ONLY!

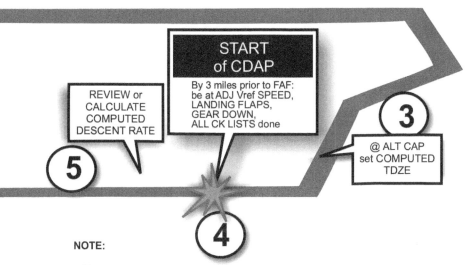

NOTE:

CONSTANTLY "MARRY THE BUGS." By this I mean; continually keep the HEADING INDICATOR (buckteeth) on the ND aligned with the TRACK INDICATOR (Triangle at the top of the instrument). *Boeing pilots quickly develop a habit of reaching up to the MCP and tweaking the HDG SEL knob contunually.*

CHECK for green dashed line on the ND during the VOR approach. This green line indicates that the VOR radios auto-tuned and are being received suitable for navigation.

CHECK FLIGHT PATH DEVIATION INDICATOR visible on the ND.

747-400 SIMULATOR TECHNIQUES ...

NON-ILS CDAP
Generic approach

APPROACHING PATTERN

A - ATIS
I - INSTALL APPROACH
R - RADIOs (tune & Ident)
B - BRIEF
A - APP-DESCENT CKLIST
G - GO-AROUND

SET DA
in altimeters
(DA = MDA + 50')

MISSED APPROACH

"GOING AROUND"
PUSH TO/GA button
VERIFY throttles go to G/A THRUST.
IF MANUAL, ROTATE (Follow F/D)
FLAPS 20°
@ POSITIVE CLIMB
"GEAR UP"
"SET MISSED APPROACH ALTITUDE"
"L-NAV" or **"HDG SEL"** @ 400 feet.
VNAV or FLCH @ 1000 feet.
A/P switch above 1000 feet on CMD.

after gear up...
SET M/A altitude

at DA (MDA +50')
with RWY IN SIGHT

Continue descent on AUTOPILOT.
DISCONNECT A/P before
50 feet below MDA
(takes about 10 secs)

PF or BOTH PILOTS
MUST display
appropriate
COURSE GUIDANCE
INFORMATION.

SELECT V/S
Check for zero's

A .3 miles prior to FAF

Using the V/S knob
AGGRESSIVELY
begin descending
USE CALCULATED
DESCENT RATE
START CLOCK (if req)
CALL TOWER

FINAL APPROACH FIX

OK to use
green arc
to assist in
profile control

+/- 300 fpm
MAX !

© MIKE RAY 2014
published by UNIVERSITY of TEMECULA PRESS

... and PROCEDURES FOR STUDY and REVIEW ONLY!

THE KEY to flying NON-ILS APPROACHES

WHEN DESCENDING:
@ **ALT**
"SET NEXT ALTITUDE"
or
"TDZE if at FAF ALT."
"OPEN V/S WINDOW
and CHECK FOR ZERO'S"

NOTE: *OF COURSE, DON'T SET IN THE NEXT ALTITUDE UNTIL YOU ARE CLEARED!*

TIP: FLCH OK when outside FAF; this keeps you altitude protected; but, INSIDE the FAF YOU MUST USE COMPUTED DESCENT RATE.

CROSSWIND LEG
FLAPS 10
speed as appropriate
If Engine failure;
Terminate fuel transfer

START CDAP
At least 3 miles prior to FAF:
- **GEAR DOWN,**
- **LANDING FLAPS,**
- **SLOW TO TARGET A/S,**
- **FINAL DESCENT CHECKLIST.**
ALL *"OTHER"* CHECK LISTS done.

INTERCEPT VECTOR
When "Cleared for Approach" and Course Deviation Bar begins to move ("CASE BREAK"),
OR you are within 10 degrees of the inbound course;
then you are cleared to begin descent to next altitude.

REVIEW or CALCULATE COMPUTED DESCENT RATE

@ **ALT** on FMA
1. SET TDZE
2. OPEN V/S
   CK zero's

DON'T FORGET

USE AUTOPILOT and AUTOTHROTTLE!

try to USE LESS THAN when tracking

USE V/S for descents inside FAF
SET COMPUTED DESCENT RATE @ .3 MILES FAF

BOTH PILOTS MUST MONITOR ADF POINTERS on ND if flying ADF

USE ND to continuously monitor LNAV and RAW DATA alignment during approach.

LNAV OK FOR TRACKING **BUT** IF CRS deviates from RAW DATA; GO TO HDG SEL and fly the RAW DATA.

AT ALTITUDE "ALT" on FMA COMPUTED TDZE select V/S, CHECK ZEROs

MAX DESCENT CORRECTION +/- 300 FPM

MISSED APPROACH at DA MDA plus 50 feet

FLYING THE VISUAL APPROACH

> *Flying the airplane VISUALLY does not necessarily infer that you should attempt to HAND FLY the approach. This is a surprisingly complex and challenging exercise and will demand that the pilot use every available tool at your disposal.*

Flying the **VISUAL** or **VMC** (old-timers called it **VFR**) approaches would seem at first glance to be the least challenging of all the pre-landing evolutions that you will have to perform during a check-ride ... but as it turns out, the visual approach is one of the top **FIVE** most screwed up items during the check sim-ride ... and, do I dare say, out on the line also. Generally when this procedure is given as a check-ride demonstration item, the check-pilot will use the opportunity to have you also display both your engine out landing and strong crosswind technique (or lack thereof). The whole process can quickly turn into a can of worms if you don't quickly make a judicious and useful plan. "**HIGH** and **FAST**" or "**LOW** and **SLOW**" are not places where you want to be as you approach the runway.

> **NUMBER ONE** item on the list of useful things to include in the setup is to select and use *"LINE SELECTABLE" WAY-POINTS* that have **ALTITUDE** and **DISTANCE** milestones by which you can gauge your progress during the visual approach descent. If these are available on the **CDU LEGS** queue ... use them.

While each situation is different, using the **AFDS** is a great help, and I recommend that you use those tools ... unless the check-person strictly forbids it. If you are left with only the option of hand flying the whole scenario, or you for some reason elect to manually operate the flight controls, keep in mind that doing so will create a demand for a continual stream of verbal commands to the **PNF** (or **PM**). This could be, I suppose, an opportunity for you to demonstrate your **CRM** (Control Resource Management) ... so in that case, I recommend that you take that as a challenge and use every available asset you can think of to assist you.

For this tutorial, we will be presenting the "**HIGH ENERGY VISUAL** approach with a downwind entry" as typical of the problems associated with the VISUAL approach. While this will be representative of some of my thoughts specifically regarding the KSFO slam-dunk type of profile, the information contained will also apply to most other airports. Each approach, however, presents its own separate set of challenging surprises so I suggest that you take what is being said for what it is ... a general set of parameters.

AGL versus MSL

This discussion is about the seemingly unimportant matter of **MSL** versus **AGL**. Sometimes in the heat of battle, we will slur and lose the importance of the difference between the two. Generally speaking, there is enough "slop" in the system to make the difference of no particular consequence. However, when flying into "**HIGH ALTITUDE**" airports such as Denver or Colorado Springs, the spread between **AGL** and **MSL** can be large enough affect your calculations.

MSL: MEAN SEA LEVEL. This is the altitude, that when corrected for the **LOCAL FIELD BAROMETRIC ALTIMETER** will represent the distance above the sea level at that point. *Below 17,999 feet MSL it is normally the altitude depicted on the altimeters.*
On Approach plates, **HAZARDS** to navigation such as towers, or terrain are depicted as **MSL**. Crossing altitudes and altitude criteria for the approach are depicted in **MSL**. The developers assume that pilots will utilize their **MSL altitude indicators** (such as the **PFD** tape) as their primary reference instruments.

AGL: ABOVE GROUND LEVEL. This is also depicted as **AFE (ABOVE FIELD ELEVATION)**. Cockpit indications of the **AGL** are generally limited to the **RAD ALT (RADAR ALTITUDE)** and only when below **2500 feet AGL** (it is also indicated on the **PFD**). However, in attempting to construct a totally **VISUAL** approach, the pilot **MUST** transition to an **AGL** mindset. Here is what I mean.

Approaching a high altitude airport, such as **KDEN** (Denver, Colorado, USA) if the pilot is planning to use a 10,000 foot **AGL** starting point for their calculations, then they would have to add about 5000 feet. Similarly, at **KLAS** (Las Vegas, Neveada, USA) an additional 2000 feet in necessary to account for the altitude of the runway.
If the pilot fails to make the correction, their calculations will place them **LOW** on their projected or desired flight path.

TECHNIQUE
Plan to have a **30 miles FLIGHT DESCENT PATHWAY** distance at **10,000 feet** above the runway at less than **250 knots**. If I don't have that situation, adjust the flight profile to compensate for the difference as far out as possible so as to avoid "**CLOSE IN**" radical flight maneuvering. Get yourself established in the descent early on in the profile.

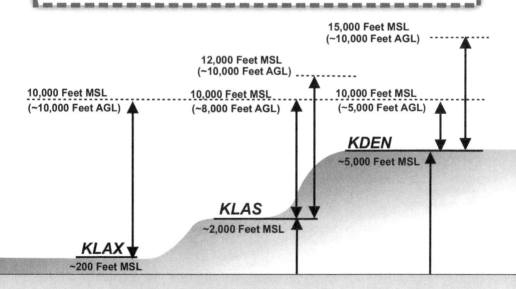

747-400 SIMULATOR TECHNIQUES ...
SAMPLE SITUATION.

Using an **APPROACH PLATE** for a **VISUAL APPROACH** makes sense if you understand what to look for. For our example, Assume **VISUAL** conditions that we are approaching the **DENVER INTERNATIONAL** airport from the **EAST**. We have been cleared directly to **DEN VOR** at 15,000 Feet (**MSL**) and to anticipate a download entry for a **VISUAL APPROACH** to **RUNWAY 8**. We can anticipate a **HIGH ENERGY DOWNWIND** entry. The altitude of the runway is ~5000 Feet **MSL**; therefore, we would want to cross a point about **30 "DESCENT PATHWAY MILES"** from the **RUNWAY THRESHOLD** at about **15,000 Feet** above the airport.

Right here is where pilots screw up! If you don't extend your "downwind" leg enough, when you start you turn inbound ... you will be "**HIGH and FAST**". Not good! So on your checkride ... place at least one "**MILESTONE**" fix or waypoint to evaluate your inbound leg. Here are two possibilities ... you could "create" a virtual point ... or you could use milestones from the "line selectable" approach menu on the **CDU (FMC)**. In this case, I have selected the **RWY 8 ILS** and see that there are two "crossing" points that would satisfy our requirements.

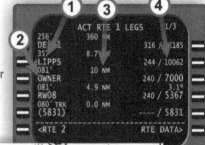

1. LIPPS
and

2. OWNER.

We can verify from the **CDU** display and **APPROACH PLATE** profile values for **LIPPS**:

3. ~15 miles (~10 + ~5) from the runway
4. 10,000 Feet MSL and ~5000 Feet AGL (10,000 - 5000).

> Notice that the **APPROACH PLATE** displays "**BOTH**" **AGL** and **MSL**; while the **CDU** displays **ONLY MSL**.

A quick 3:1 (5000 Feet X 3 = 15 Miles) estimate verifies the values.
Do the same analysis and assessment for the waypoint named **OWNER**.

The **ND** should look somewhat like this as you approach **DEN VOR** .... Fly the **MAGENTA LINE** and manage the descent so as to cross the **MILESTONE** waypoints at the planned target altitudes.

I am not suggesting that the **AFDS LNAV** "should" be used during this approach, but if you are set-up properly, there is no reason while the auto-flight could not be utilized as much as possible.
HANDFLYING or using the **HDG SEL** feature are generally used.

At a normal 3 degree glide slope, when at 500 feet **VDP** (Visual Descent Point), with the runway is insight and the airplane lined up laterally and established on the extended centerline, the runway should be **1.7 NM ahead**. Use **V/S** (**VERTICAL SPEED**) of **800 FPM** (Feet Per Minute).

REVIEW

... and PROCEDURES FOR STUDY and REVIEW ONLY!

It is generally accepted that a modern airliner will descend at a rate of 1000 Feet for every mile it travels across the ground. This is referred to as the "**3 to 1 RULE**". Now, of course, noone is

> ### 3 to 1 RULE
> Generally speaking, it will take a modern jet airliner 3 NM to descend 1000 Ft.

saying that this is a precise measurement that accounts for every situation and airplane; but what is true about this rule is that it makes the mental calculations and routine estimates fairly simple to approximate.

For example: If an airliner is cruising at 35,000 feet and intending to land at Denver (About 5000 feet in elevation), it should start its descent about 90 miles from the airport. As a quick rule of thumb, that is a useful piece of information ... however, it does not take into account any wind component, weight of airplane, or any of the other factors that affect the descent rate. It is just a quick estimate.

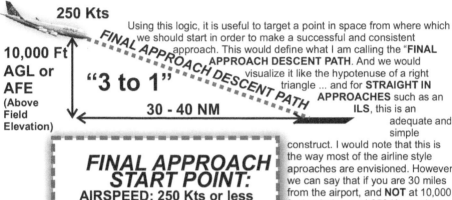

Using this logic, it is useful to target a point in space from where which we should start in order to make a successful and consistent approach. This would define what I am calling the "**FINAL APPROACH DESCENT PATH**". And we would visualize it like the hypotenuse of a right triangle ... and for **STRAIGHT IN APPROACHES** such as an ILS, this is an adequate and simple construct. I would note that this is the way most of the airline style aproaches are envisioned. However, we can say that if you are 30 miles from the airport, and **NOT** at 10,000 feet or below and 250 Kts or slower ... then you are likely to have **EXCESS ENERGY** and be **FAST** and/or **HIGH** during the approach.

> ### FINAL APPROACH START POINT:
> AIRSPEED: 250 Kts or less
> ALTITUDE: 10,000 Ft or less
> DISTANCE: 30 NM

DESCENT FROM ALTITUDE TO APPROACH STARTING POINT

For an example, let's look at a High Altitude airport such as Denver, Colorado. It would be more accurate to say that in order to make the approach we should plan to be at 30 miles from Denver at 15,000 feet and 250 Kts for the **FINAL APPROACH START POINT**. Since Denver is (about) 5000 FEET MSL, We have to add the 5000' into our calculation so that we will be descending to 15,000 Feet MSL in order to be at 10,000 Feet AFE (or AGL). Then when we can calculate the distance required to start our descent to 10,000 feet AFE using the "3 to 1" (3/1000) rule. If we were starting at 35,000 Feet MSL we would have to use the 15,000 Foot MSL for our calculation. So, 35,000 minus 15,000 = 20,000 times 3/1000 equals 60 miles.

There is another consideration, that is this. It will take the jet bout 9 - 10 NM to go from CRUISE SPEED to 250 Kts. We will have to add 9 Nm to our descent planning. So, the total disctance required for the descent to Denver goes like this:

60 = Distance to Descend to 10,000 Feet AFE.
 9 = Distance to slow to 250 Kts
30 = Distance to descend to Runway.
----
99 = Total distance

> ### "SLOW UP" RULE
> Generally speaking, it will take a modern jet airliner 9 NM to slow from about 300 Kts to 250 Kts.

SIMULATOR TECHNIQUES ...

Flying the
HIGH ENERGY VISUAL APPROACH
pilots call it ...
SLAM-DUNK

All over the world, there are airports where ATC may clear you for a **"VISUAL APPROACH"**, some with a **"DOWN-WIND ENTRY"** into the pattern. The slang term used by pilots to refer to these type of approaches is **"SLAM DUNK"**. These **"HIGH ENERGY"** approaches present the unwary pilot with the opportunity to get "high and fast" on the approach. If, for example, you are arriving into San Francisco from the North and West (Asia and Hawaii for example) you can anticipate that you may be cleared (weather permitting) for a "visual approach" with a downwind entry to Runway 28L or R. While on the surface, this may seem to be a relatively simple task; however, there are few more challenging activities associated with routine airline flying than making the visual approach. The whole evolution is even a more visceral experience for the pilot since they may only be using visual references for navigation and possibly manually controlling the airplane's approach descent pathway. Since the airplane is descending from cruise altitudes and airspeeds, the tasking is further complicated since the jet will have acquired "HIGH ENERGY". Put it all together and you will have a handful. While a late start in the descent can be a crucial mistake, it is only one of the many places where the approach can be screwed up with a potentially disasterous outcome.

While it would be correct to simplistically take the position that the every **"HIGH ENERGY VISUAL"** profile is flown precisely like a **"STRAIGHT IN"** approach and landing; since there is at least one significant difference. The pathway from the start of the final segment to the **RUNWAY THRESHOLD** may not be a constant heading during the descent. The **DESCENT PORTION** may include heading changes during the approach that will result in up to a 180 degree change in the airplane's direction. Particularly during the execution of a typical **"SLAM DUNK"** profile, the airplane will generally start the landing evolution over or abeam the landing runway and, more importanntly, on a heading that is the reciprocal of the runway of intended landing.

On some airports, the approach profile starts **BELOW 10,000 Feet**. That eliminates the potential of carrying excessive airspeed, since "below 10,000 feet" the 250 Kt restriction will already require the airplane to be at or below 250 knots. The approach we will be profiling in this tutorial will be the San Francisco (**KSFO**) visual approach from the north (**PYE**) to **Runway 28R or 28L**. This is a particularly challenging evolution since the starting altitude may be **10,000 Feet OR ABOVE GROUND LEVEL (AGL)**. This removes the 250 Knot restriction and allows for a higher insertion airspeeds.

... and PROCEDURES FOR STUDY and REVIEW **ONLY!**

THIS IS THE KEY TO THIS APPROACH !

Right here is where pilots can screw up the whole approach ... they fail to extend their downwind leg enough and start their turn inbound **TOO SOON!!!** Once they are committed, chances are the airplane is already **HIGH and FAST** when they start the inbound leg. The capabilities of the airplane to descend and slow up at the same time becomes problematic and deploying the flaps in a timely manner so as to be stabilized at 500 feet is marginalized. The potential outcome of the landing in the touchdown part of the runway is compromised. This is particularly important at a short runway such as Kahalui, Maui (**PHOG**) since the runway is only about 7000 feet long.

CAUTION - DANGER:
DO NOT START YOUR TURN TOO SOON.
It is important that you extend the downwind leg at least 1/2 the anticipated distance from your starting altitude to the runway.

DANGER!!!!

When flying the **FAS** (Final Approach Segment) it is **NEVER** a good idea to intercept the Glide-path (or Glide-slope) from above. This situation can rapidly develop by starting your turn inbound too soon and getting behind the airplane. The planning should always be to avoid getting **HIGH**.
We should plan to intercept the **FINAL APPROACH SEGMENT** or **GLIDE-PATH** from below ... That is, to "fly into" it.

AVOID ATTEMPTING TO INTERCEPT THE GLIDE PATH FROM ABOVE.

DEFINING PARAMETER FOR THIS APPROACH

Our discussion will start with a review of some of the rules that apply to this approach profile.

This is a place where the you can start your screw-up. If you maintain your descent airspeed into the abeam position, you will have to delay your descent until you have bled off the excess energy. I recommend that you start your speed reduction so as to be at the abeam position at **LESS THAN 250 Kts** or **CMS** (Clean Maneuvering Speed or Zero Flap speed). You also want to be slow enough to immediately start your descent and to initiate **FLAP** extension.

RULE
No airplane anywhere is allowed to exceed 250 Knots **BELOW** 10,000 feet MSL.
However, if the airplane is **AT or ABOVE** 10,000 feet MSL the 250 Knot restriction does not apply.
OK, so there are some exceptions ... such as air-shows or military operations, or operations that take place outside of controlled airspace.

© MIKE RAY 2014
WWW.UTEM.COM

Flying the "SLAM DUNK" VISUAL

747-400 SIMULATOR TECHNIQUES ...

Here are some of my thoughts about flying the "**HIGH ENERGY VISUAL**" approache that are in regularly usage. Clearly there are as many ways to fly this approach as there are pilot. The whole approach process starts well before you are in the airport area and so it is important that you get set up while you are still some distance from the airport.
Here are some of my thoughts.

FIRST VERY IMPORTANT ITEM: Get the local field barometric pressure setting and place it in the **ECU** (**EFIS** Control Unit) and the **STANDBY ALTIMETER**.

NOTE: You may use any "**LINE SELECTABLE**" approach available and utilize any of the waypoints in your planning. In our case, even though we do not expect to fly the ILS, we could load that approach and use the **CF** (Centerline Fixes) fixes as milestones in our descent planning on the **FAS** (Final Approach Segment).

POINT 1: Either select or create a waypoint or fix abeam the runway to start the approach. Generally speaking, you want to establish a distance abeam that will also make a comfortable radius for your base leg turn. Consider a 5 miles offset. Plan your distance by taking into consideration the high terrain or other obstacles. This will allow you to use **VNAV** during the first part of the approach. It is useful (and recommended).

POINT 2: For example, we could use 5 miles abeam the **KSFO RWY 28R**. Enter this as a waypoint into the **CDU** like this: **SFO011/05** and when cleared **LS1L**. This will place the new waypoint with a name such as "**SFO02**" at the top of the queue.

POINT 3: Expecting a clearance for "the visual approach," select an altitude and airspeed for starting the approach descent pathway.

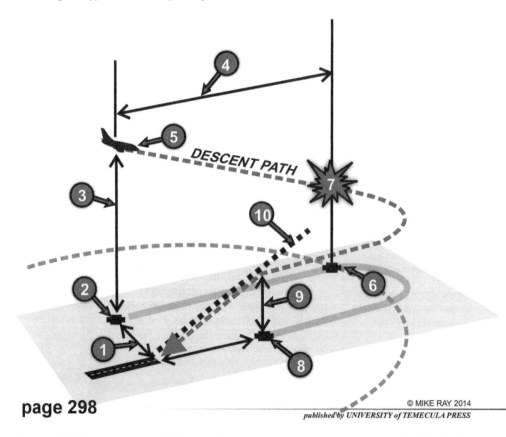

ARGUMENT FOR SLOWING TO CMS

Here are my thoughts on airspeed at the **APPROACH STARTING** way-point. Planning 10,000 feet (or above) allows you to "keep the speed up" and come slamming into town up to 300+ Kts, and you may legally do that if you desire; but a high initial speed would restrict the ability to start the descent and subsequently extend the downwind leg of the "**FLIGHT DESCENT PATH.**" Remember that it takes an additional **7-9 miles to slow to 250 Kts** or less in order to descend below the 10,000 ft **MSL** altitude restriction and start the approach. Generally speaking, a high insert speed crossing the abeam point is not necessarily desirable.

Plan to be abeam the runway at an airspeed at or below **250 Kts**, such as **CMS** (Clean Maneuvering Speed).

If we have the choice of selecting our own abeam point altitude, then the descent flight pathway distance will be appropriately shorter and so the distance to the downwind leg turn point will be less.

For example, at Kahalui, Maui (PHOG) the rapidly rising terrain to the East and West requires a starting altitude of 8000 Feet AGL, and that would place the turning point about 12 miles (8 X 3 = 24 divided by 2 = 12 miles)..

If we make this decision far enough out, we can even add that information to the **CDU LEGS** page and **FMC VNAV** can be used to assist in making the abeam point crossing. Type the desired altitude and airspeed in the scratch pad of the **CDU** and then line select it to the right side queue opposite the **SFO02** waypoint. This gives us a target fix with airspeed and altitude for the **VNAV**.

ABEAM POINT

AIRPORT

ONE HALF COMPUTED DESCENT PATHWAY DISTANCE FROM THE ABEAM POINT

START OF COMPUTED DESCENT PATHWAY MAXIMUM!
30 Nm
10,000 Ft
250 Kts

POINT 3B: Then select an outbound heading from SFO02 that will be the reciprocal of the runway. In our example, since we are using Runway 28, the reciprocal will be (281 -200 + 20 =101 degrees). Pre-select that heading using the **HDG SEL** on the **MCP** (Mode Control Panel).

FINAL APPROACH START POINT:

POINT 3C: Once **ALT ACQ** is displayed on the **FMA** (Flight Mode Annunciator across the top of the **PFD**) select the crossing altitude of one of the waypoints from the **ILS** or other approach selected on the **CDU**. I selected 3200 for the crossing altitude of **my** selected **CF** (**CenterField**) fix (**CEPIN**). You can create your own waypoint, say 3 miles from the end of the runway at keeping in mind that in that case the crossing altitude should be about 1000 Ft **AGL**.

POINT 4: Determining the distance to the point to start your turn inbound is the **"CRITICAL DECISION"** in this maneuver. Let's discuss this in detail.

Calculating the ***TOTAL FLIGHT DESCENT PATH LENGTH*** is absolutely essential. Making a quick estimate (3 to 1) ... at 10000 feet it will take about 30 miles to descend to **SFO RW28R** runway elevation. So my "**MINIMUM**" downwind leg length should be 15 miles (30 / 2 = 15).

> **TIP:**
> Before starting the maneuver, select the **FIX** page.
> Place the runway or co-located identifier
> into the "boxes" and add the distance to the turn (/12)
> to the dashed line list. This will place a **GREEN ARC**
> that will show you where you have calculated the turn point.

This will give us a "**GREEN DASHED ARC**" on the **ND** (Navigation Display) that is 15 miles from our landing runway. We should continue our downwind leg until we reach the "green arc" before we start our crosswind leg.

747-400 SIMULATOR TECHNIQUES ...

POINT 5: Crossing the point abeam the airport on the **DOWNWIND** leg, do the following things:

- Select desired **DESCENT MODE**:
  V/S on the **MCP** and roll in about **-800 to -1000 FPM** descent rate, or
  FLCH set lower **ALTITUDE and AIRSPEED**, or
  VNAV (requires a target waypoint), or
  MANUALLY FLY (disengage **AUTO-PILOT** and **AUTO-THROTTLES**).
- Select **SPEEDBRAKE** (OPTIONAL).
- Select **FLAPS 1** and set speed on the **MCP** for the next **FLAP** setting when airspeed allows,
- Select **HDG SEL**. Monitor the turn to the downwind leg heading.
- As the descent develops, select **MCP** speed and continually take flaps accordingly until you are at **15 degrees**.

Due to the constantly changing airspeed target for the **FLAP EXTENSION** schedule, the **THRUST LEVERS** have a continual tendency to want to move into forward thrust during the descent. There are two ways to counter this.
    1. Shut off the **AUTO-THROTTLES** and **MANUALLY** adjust the thrust.
    2. Use the **FLCH** and this keeps the **THRUST** commanded to **IDLE**. Keep **AUTO-THROTTLES** selected **ON** and set in an altitude **BELOW** your planned descent profile.

"THE FLITCH TRAP"

Here is the problem with that. This is what is called a **"FLCH" TRAP**. Let me explain briefly ... when flying without the **AUTOPILOT**; in this configuration with the **AUTO-PILOT DISCONNECTED**, you will be manually controlling the altitude and attitude pitch axis, and that will in turn control the **AIRSPEED**. It will probably be impossible to control the airspeed while maintaining the desired flight path with the **THRUST in IDLE**. Attempting to hold the airplane on **GLIDE PATH** without a complementary increase in **THRUST** will require a **PITCH UP** that will cause a resulting decay in the **AIRSPEED** that will eventually result in a **STALL** situation. **NOT GOOD!!!**
Of course, the solution is that you **MUST KEEP YOUR HAND ON THE TRUST LEVERS** and you **MUST CONTROL THE THRUST MANUALLY**.

POINT 6: Point of starting the turn: *This is the crux of the problem*. This is the main factor that will determine whether you will make a smooth and timely descent or be **HIGH** and **FAST** (*YIPE!!!*). Determining when to start the turn will dictate whether you will intercept the inbound bearing to the runway from a **BELOW GLIDE PATH** position or you will have everything hanging including the **SPEED BRAKE** trying to get back down on the **GLIDE PATH**.

POINT 7: As you approach the point where your path crosses the "**15 MILE GREEN ARC**", roll in a heading of 190 degrees (the crosswind heading).
        - **GEAR** down, **FINAL DESCENT CHECKLIST**
        - **FLAPS 25**, set **REF** speed
        - **FLAPS 30/40**
        - **SPEEDBRAKE** retract. Check for green "**ARMED**" light on instrument panel.

As you descend and continually set in the appropriate **FLAP TARGET SPEEDS**, the **THRUST LEVERS** will constantly wanting to add thrust as you approach the target **AIRSPEED** for the intermediate **FLAP SPEEDS** that are set in the **MCP**. Physically take your hand and **MANUALLY** "**OVER-RIDE and HOLD**" the thrust levers in the idle position where appropriate.
Manage the descent so that you can retract the **SPEED-BRAKE** before the thrust levers start to move forward to maintain airspeed as you complete the descent portion and intercept the **GLIDE PATH**.

It is useful to have either a "**LINE SELECTABLE**" waypoint or a virtual point to aid in evaluating your descent progress. At 3 miles, you should be about 1000 feet AGL. You can place an additional arc on the fix page for this information, or create a virtual point
        - Select **VOR LOC** or use **HDG SEL** or other heading guidance to cross the waypoint. Green arc should rest on the waypoint that corresponds to the altitude set in the MCP. If predictive ARC does not rest on that waypoint, adjust the **V/S** to move it to that position.
        - Before reaching the waypoint, place **MISSED APPROACH ALTITUDE** in the **MCP**.
        - Crossing that final waypoint attempt to be totally stabilized crossing that fix.

... and PROCEDURES FOR STUDY and REVIEW ONLY!

POINT 8: My recommendation is this: as you approach your final milestone waypoint on the final approach course, set an appropriate **MISSED APPROACH ALTITUDE** in the **MCP** that is **HIGHER** than the **FLIGHT DESCENT PATH** altitude. If you wait too long, the airplane will want to level off at that altitude and you will get high on the approach. Plus, if you must execute a **MISSED APPROACH**, then you will have already set the target for the missed approach into the **MCP**.

COMMENT:

Some airlines use **QFE** (Field Elevation Altitude) or "**ZERO**" in the **MCP**. If you elect to do that, then when you execute the missed approach, you will be required to set the Missed Approach Altitude in the **MCP** "*DURING*" the initial rotation for the procedure. Seems to me like there is a lot going on at that point and is just an unnecessary complication of an already cluttered operation. Particulatrly if you are dealing with an engine-out or a irregularity/weather/icing/visibility situation.

POINT 9:

@ 2000 Feet AGL:
- **TOGA ARMED**
- **THRUST APPROACH IDLE** set as new minimum thrust for idle.

@ 1000 Feet AGL:
- **SPEED BRAKE** must be stowed. If the use of the **SPEEDBRAKE** is necessary, the approach is to abandoned
- Airplane should be stabilized on glide-path, heading, and airspeed.

@500 Feet AGL:
- If airplane not in a position to make a landing, the approach must be abandoned. **TOGA** and get out of there.

@ 50 Feet AGL: AUTO-THROTTLES must be disconnected.

POINT 10: If there is available **HEADING** and **GLIDE-PATH** information available, it is **SOP** to have them turned and operating for cross-check information. So, even though we are "hand-flying a visual approach" it is *ALWAYS* useful to have a back-up source for checking our vertical and lateral progress.

WHEEW!!!

Flying the NON - ILS APPROACHES

Formerly known as
THE NON-PRECISION APPROACHES

VOR
NDB (ADF)
LOC ONLY (ILS G/S OUT)
VNAV (Not flown at some airlines)

*** THE 747-400 WILL NOT FLY A BCRS**

NOTE!

..AND I WANTED TO FLY A BACK-COURSE APPROACH, DANG IT!

There are several ways to fly the **FINAL APPROACH SEGMENT** of a **Non-ILS** approach. We will cover the **CDAP** method in this manual

... and **PROCEDURES FOR STUDY and REVIEW** *ONLY!*

LET'S SIMPLIFY
EVERYTHING ... including the non-ILS approaches.

WORTHLESS DISCUSSION:

The really good news about flying this incredibly complex machine is that it is pretty easy. The engineers at Boeing realized that ordinary, garden variety airline pilots would be operating the jet ... so they tried to keep it simple. The airline training people, however, took this to be a challenge and an opportunity to add complexity. So they did what they could to complicate things.

By a stroke of engineering genius and pure luck, the big bird is so powerful, that we are able to group the normal four engines approaches and the three engine approaches together. All it takes is a little rudder trim, and the procedures are the same.

Mr. Boeing failed to include that obnoxious "reverse sensing" doo-dad" that made us able to fly that goofy back-course procedure ... so, this particular airplane **CANNOT FLY A BACK COURSE** approach. Breaks my heart.

REGARDING THE NDB OR ADF APPROACH.

I regard it as a quasi-emergency anyway. I cannot think of anyplace in the world (other than some obscure place in China) where pilots are subjected to using this semi-dangerous piece of aviation memorabilia on a regular basis. It was clearly intended to be used when there was NO OTHER AVAILABLE APPROACH ! No matter how perfectly we can fly the "tail of the needle," there is ALWAYS gross errors in the line-up of the signal. Anyway, I am telling you, that if you have to do this maneuver, get your head out of the cockpit and don't get all tied up in flying the maneuver. The idea is to land the airplane safely.

The ADF, even perfectly flown, can place you in a position where a safe landing is in doubt. Be spring loaded to the go-around position. DO NOT accept a dangerous transition.

This approach belongs in a museum and was NEVER intended to be flown by a high performance jet the size of a 747-400.

I am going to make the bold assertion that since we fly all the non-precision approaches similarly, and since there are only three types; ADF (NDB), VOR, and ILS with GLIDE SLOPE OUT (Localizer only); let's emphasize the similarities, then the things that are different become simple to see.

NON PRECISION APPROACHES
OP SPECS

GENERAL STUFF:

In weather less than 1000/3, or published minimums, whichever is greater ... **USE THE AUTOPILOT**.

LNAV OK on VOR and NDB approaches IF the approach is built on the LEGS page and the RAW DATA information is monitored.

If LNAV does not agree with the RAW DATA ... switch to HDG SEL for the remainder of the approach.

Conduct the LOCALIZER ONLY approach in the LOC mode of the AUTOPILOT.

Inside the FAF; USE THE V/S wheel.

USE OF LNAV:

The autopilot WILL track a localizer course; BUT it WILL NOT TRACK a VOR or an NDB course.

During a VOR or NDB approach using the FMC generated magenta line and the LNAV ... BE AWARE that the course may deviate slightly from the published VOR or NDB course. That's OK, if it is expected.

APPROPRIATE raw data information MUST BE MONITORED throughout the approach.

> *IF A DISAGREEMENT DEVELOPS BETWEEN THE TWO; USE OF THE LNAV MUST BE TERMINATED AND HDG SEL USED TO FOLLOW RAW DATA.*

For ALL non-precision approaches, stepdown fixes MUST be verified by reference to **RAW DATA**. *You are specifically denied the use of **ONLY** the FMC generated information.*

To display raw data:
TUNE appropriate navaid on the CDU **NAV RAD** page, and
if applicable use the **EFIS** VOR/ADF switches to display the appropriate information on the **ND** (Navigation Display).

VOR APPROACHES:

> *VOR APPROACHES REQUIRE that the PF MUST display the VOR on the ND no later than the FAF.*

At the Captain's discretion, the PNF may display the MAP MODE with the VOR bearing pointers displayed as a back-up to a *CONSTRUCTED* approach.

NDB APPROACH:

> *NDB APPROACHES REQUIRE that both pilots may display either MAP MODE or MAP CTR MODE with the ADF pointers displayed.*

LOCALIZER ONLY APPROACH:

During LOCALIZER ONLY APPROACHES, raw data is displayed on the PFD. *Note that it isn't necessary to switch to the ND APP mode.* Both pilots may stay in the ND MAP mode.

MORE NON PRECISION APPROACHES OP SPECS

APPROACH:

Use these **ROLL MODES**: **VOR** and **ADF** .... **LNAV OK**, if confirmed with **RAW DATA**, otherwise **HDG SEL** or **LOC** mode on **MCP** to intercept and track.

SET NEXT LOWER ALTITUDE ON THE MCP BEFORE DESCENDING IN V/S.
If you do not do this, the airplane will not level off, and the jet will fly into the ground and you will die !!!
HELLO ... DUH!!! This is the real BIGGIE!!!

FINAL APPROACH:

WHEN FMA GOES TO VNAV PATH, SET IN THE TDZE.

~about 5 miles before FAF:
   GEAR DOWN,
   FLAPS 20,
   20 FLAP speed + 10 KTS
   FINAL DESCENT CHECKLIST.

Just prior to FAF (~.3 miles)
   LANDING FLAPS
   reduce to TARGET SPEED.

Time/distance can be determined elsewhere; but the easiest place to look is the ND display.

DESCENT to MDA:

At .3 miles prior to FAF, **AGGRESSIVELY** ROLL THE V/S to calculated DESCENT RATE (probably around 800 fpm).

TRANSITION TO LANDING:

AT MDA (DH) with runway in sight:

STAY ON AUTOPILOT UNTIL 50 feet BELOW MDA.

USE V/S knob to adjust descent.
USE HDG SEL to tweak line-up
BRIEF PNF to call out 50 feet below!

SET-UP for the NON-ILS APPROACH

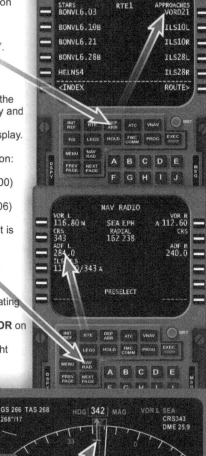

This is just a suggestion, but setting up the cockpit this way allows the crew to fulfill ALL the requirements for the approach ... as well as making all the required information readily available for quick visual reference.

Select the **DEP ARR** page and make sure that the desired **NON-ILS APPROACH** is "LINE SELECTABLE".

To tune the **VOR** and the **NDB (ADF)** approaches, use the **NAV RAD PAGE** of the **CDU**. Simply type the frequency and the heading into the scratchpad and line select the information into the appropriate position on the **CDU** display.

You can use the following formats to enter the information:
-**VOR NAVAID** identifier. (such as JFK, SEA, LAX)
-**VOR/DME** Frequencies: (such as 111.80, 115.85, 116.00)
-**IDENTIFIER/COURSE**: (such as SAN/215, LAX/090)
-**FREQUENCY/COURSE** (such as 111.80/215, 116.8/006)

This will also automatically tune the associated **DME** if it is applicable.
The display on the **CDU** page will include:
FREQUENCY/TUNING STATUS/IDENTIFIER (such as 116.80rLAX)
There are four tuning status modes:
-**A = AUTO**: Automatically tunes radios for position updating
-**M = MANUAL**: Pilot entered.
-**R = ROUTE**: Automatically selected by **FMC** of next **VOR** on active flight plan.
-**P = PROCEDURE: FMC** selected navaid for active flight plan procedure.

> The **PF MUST** display the **VOR** mode on the **ND** no later than the **FINAL APPROACH FIX**.

To do this you need to select the **VOR MODE** on the **ECU**.

Select **VOR L** on the **ECU** ... and observe the **GREEN VOR L** indication at the lower left of the **ND** and the "**THUMBTACK**" **ARROW** pointing to the **VOR**.

To be on the VOR COURSE inbound, the **MAGENTA** bar should be centered on the triangle. The 4 little donuts represent the 10 degree boundary of the **VOR FINAL APPROACH** corridor.

... and **PROCEDURES FOR STUDY and REVIEW** ONLY!

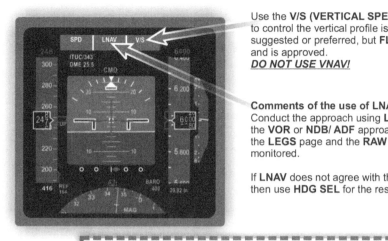

Use the **V/S (VERTICAL SPEED)** pitch mode to control the vertical profile is generally suggested or preferred, but **FLCH** works **OK** and is approved.
DO NOT USE VNAV!

Comments of the use of LNAV:
Conduct the approach using **LNAV** only when the **VOR** or **NDB/ ADF** approach is depicted on the **LEGS** page and the **RAW DATA** is monitored.

If **LNAV** does not agree with the **RAW DATA** ... then use **HDG SEL** for the rest of the approach.

For all non-precision approaches, **STEP-DOWN** fixes (**VOR** radials, **DME** distances, etc) **MUST** be verified by reference to appropriate **RAW DATA** information.
DO NOT RELY ON FMC GENERATED INFORMATION!

The technique is to set the **NEXT LOWER ALTITUDE** on the MCP *BEFORE* descending in **V/S**. This is **NECESSARY** because it is possible to "fly away" from an altitude without altitude capturing protection!

EXTRA CREDIT TECHNIQUE

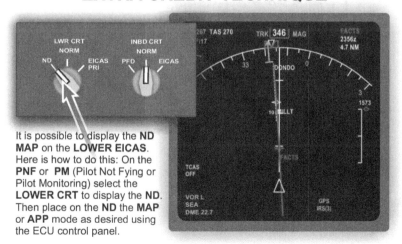

It is possible to display the **ND MAP** on the **LOWER EICAS**. Here is how to do this: On the **PNF** or **PM** (Pilot Not Fying or Pilot Monitoring) select the **LOWER CRT** to display the **ND**. Then place on the **ND** the **MAP** or **APP** mode as desired using the ECU control panel.

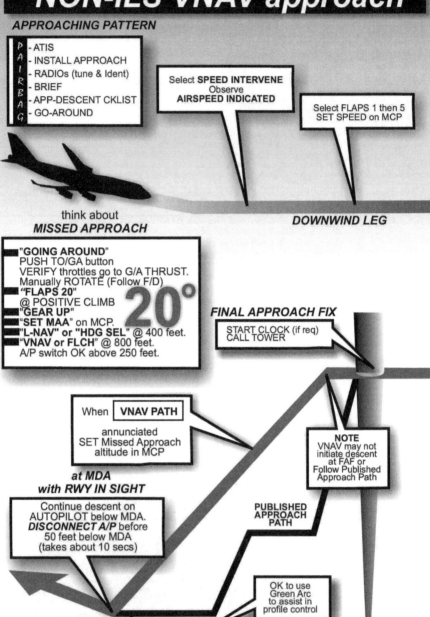

... and PROCEDURES FOR STUDY and REVIEW ONLY!

Select **VNAV**

Select **DA or MDA**

Select **SPEED INTERVENE** (push knob) Check for airspeed indication

Check **FMA** SPD/LNAV/VNAV PATH

Check **FLIGHT PATH DEVIATION INDICATOR** present on the ND

Check **VNAV PAGE** Observe **MCP SPEED** annuciated

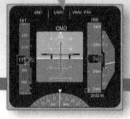

Check AUTO TUNE active by observing **DASHED GREEN COURSE LINE** present on the ND

About 2 miles prior to FAF:
GEAR DOWN,
Select **FLAPS 20**,
SET the **DA or MDA**,

CROSSWIND LEG
SELECT FLAPS 10

If on Engine failure; Terminate fuel transfer

Just prior to FAF:

select **LANDING FLAPS**,
Set **FINAL APPROACH SPEED**,
complete **LANDING CHECKLIST**,
Set **MINIMUMS (MDA or DA)** on MCP.

USE AUTOPILOT in VNAV/AUTOTHROTTLE

AT "ALT" SET NEXT "CLEARED" ALTITUDE on MCP.

USE FL CH for descents outside FAF

PILOTS MUST MONITOR ADF POINTERS on ND

USE ND MUST continuous monitor ADF NEEDLES during approach.

LNAV OK FOR TRACKING IF CRS deviates from RAW DATA; GO TO HDG SEL and fly the RAW DATA.

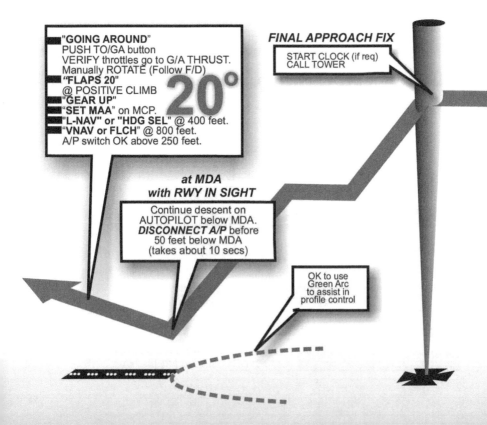

... and PROCEDURES FOR STUDY and REVIEW ONLY!

Check **PFD** and **FMA** or **APPROPRIATE INDICATIONS**

Check **VOR AUTO TUNE** active by observing **DASHED GREEN COURSE LINE** on the **ND**

Check "**HEAD**" of the **VOR /ADF POINTER** present on the **ND**

Check **FLIGHT PATH DEVIATION INDICATOR** present on the **ND**

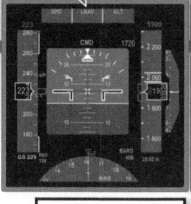

About 2 miles prior to FAF:

GEAR DOWN,
Select **FLAPS 20**,
SET the **DA or MDA**,

CROSSWIND LEG
SELECT FLAPS 10

If on Engine failure;
Terminate fuel transfer

Just prior to FAF:

select **LANDING FLAPS**,
Set **FINAL APPROACH SPEED**,
complete **LANDING CHECKLIST**,
Set **MINIMUMS (MDA or DA)** on MCP.

AT "ALT"

SET NEXT "CLEARED" ALTITUDE on MCP.

USE AUTOPILOT

USE FL CH for descents outside FAF

PILOTS MUST MONITOR ADF POINTERS on ND

USE ND MUST continuously monitor **ADF VOR NEEDLES** during approach.

LNAV OK FOR TRACKING IF **COURSE** deviates from **RAW DATA**;
GO TO **HDG SEL** and fly the **RAW DATA**.

© MIKE RAY 2014
WWW.UTEM.COM

NON-ILS CALL-OUTS

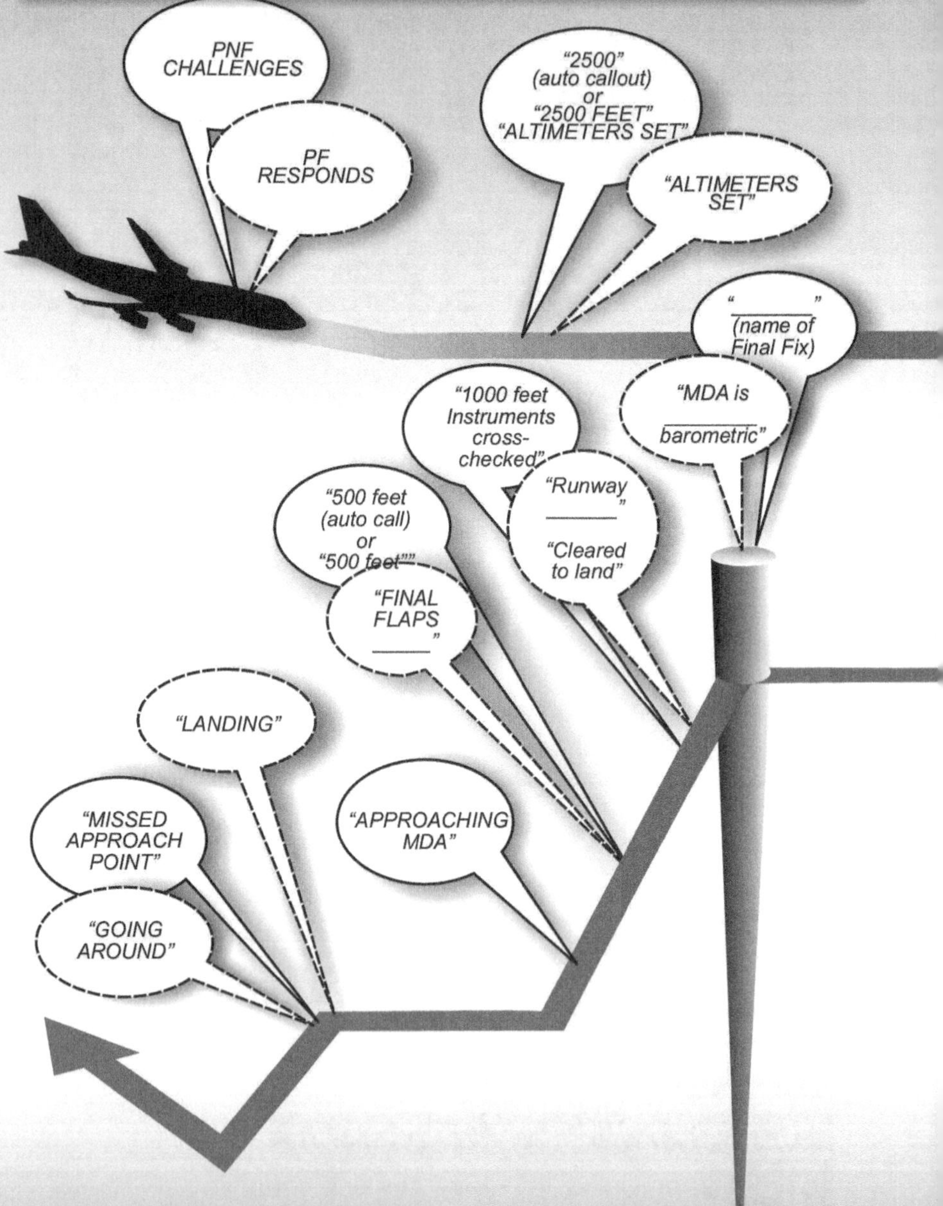

- PNF CHALLENGES
- PF RESPONDS

- "2500" (auto callout) or "2500 FEET" "ALTIMETERS SET"
- "ALTIMETERS SET"

- "_____" (name of Final Fix)
- "MDA is _____ barometric"

- "1000 feet Instruments cross-checked"
- "Runway _____" "Cleared to land"

- "500 feet (auto call) or "500 feet""
- "FINAL FLAPS _____"

- "LANDING"
- "APPROACHING MDA"
- "MISSED APPROACH POINT"
- "GOING AROUND"

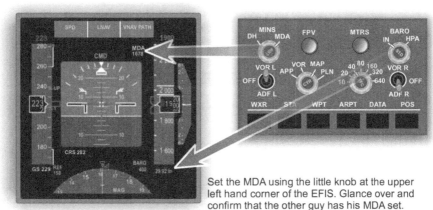

Set the MDA using the little knob at the upper left hand corner of the EFIS. Glance over and confirm that the other guy has his MDA set.

Confirm that the altimeter is set on inHg or Hp. You can't check this too many times, especially when operating in foreign places.

Ensure that the STANDBY ALTIMETER has the correct barometric setting. Specifically notice whether hectropascals or inches of mercury are applicable.

During the approach, monitor the CDU **LEGS PAGE** for specific distance, heading, and fix information.

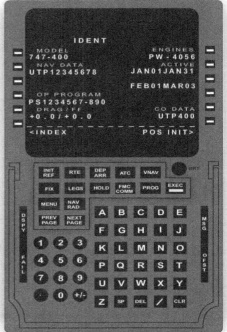

Descents outside of the FAF (Final Approach Fix) use the FLCH MODE or VNAV. When **VNAV PATH** is annunciated on the FMA, set Missed Approach altitude in the MCP.

At the MDA with the runway environment in sight suitable for landing, continue your descent below MDA. Stay on the autopilot until 50 feet below the MDA, then disconnect autopilot and make the landing manually.

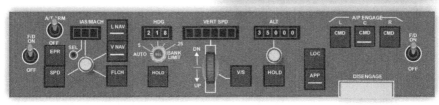

ADF (NDB) SETUP

HOW TO SET UP THE RADIOS and STUFF

It is assumed that you have completed the **A-I-R-B-A-G** steps which include:
- installing the approach on the **DEP ARR** page,
- resolving the *DISCONTINUITIES* on the **LEGS** page,
- and set the frequencies in the **NAV RAD** page.

Because I have shown that in detail earlier, I won't repeat those steps.

BOTH PILOTS must be is the MAP mode, either CTR or EXPANDED. I have depicted using the CTR mode on the outbound vector, because the airport information is behind the airplane and CTR allows more information from behind us without going to larger scale. We obtain the CTR or EXPANDED display by depressing the center of the EFIS selector knob labelled appropriately enough, CTR..

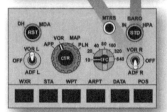

TECHNIQUE:
THE VOR AND THE ADF(NDB) NEEDLES "SHARE" THE SAME SYMBOLS ON THE ND YIPE!!!. If they are indicating the VOR; they are GREEN. If they are indicating the ADF, they are BLUE.

Big place for confusion: The indicator in the lower corner of the ND will be the same color as the "ARROW" symbols on the ND. Example: If the symbol is green and you can't remember which is which, just look at the lower corner.

> The #2 indicator is **BIG AND NICE**, the #1 indicator is a *puny, skinny thing* that you can hardly see. I prefer to use the **_ADF #2_** to fly the approach. The #1 nav radio can be used for a VOR/ADF that is required for approach definition, or turned **OFF**. That way, I always know what I am doing.

The approach may use LNAV, but BOTH pilots must monitor the MAP mode during the whole approach to continuosly monitor the ADF raw data signal.
Appropriate RAW DATA information MUST be monitored throughout the approach. If a disagreement between the MAP mode and the RAW DATA occurs, use of the LNAV must be terminated and RAW DATA followed using the HDG SEL knob on the MCP.

> **TECHNIQUE:**
> Throughout the approach, even while using the LNAV and AUTOPILOT; it is good technique to be constantly "Marrying the bugs" (setting the HDG SEL bug on the ND to the heading desired) so that should a transition to HDG SEL become necessary, the MCP will be set up and the airplane will not make a turn off course.

more HOW TO SET UP THE ADF RADIOS and STUFF

I think that using the ND in the EXPANDED MODE once you have turned inbound makes sense. That makes the top of the instrument larger and more easy to line up the indicators.

Outbound from the FAF (Final Approach Fix) which is (almost always) the ADF, here is the picture we want to see.

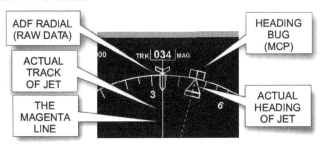

The MAGENTA LINE and the AIRCRAFT TRACK super-imposed, AND

The "BOMB" resting on the depicted RADIAL from the approach chart

ALL THREE MUST REMAIN LINED UP.

If these three item should not remain lined up,
and you are flying the approach in LNAV:
YOU MUST TERMINATE USING LNAV !

NOTE ON TECHNIQUE:

Even though it is not controlling the direction of the airplane if using LNAV, the Aircraft Heading Selector should be continually aligned with the Aircraft heading Selector using the heading selector knob on the MCP.
THEN ... If you need to transition to
MANUAL" control, when you push the heading slector knob, the airplane won't make some unexpected turn off course ... and csrew up the approach further.

When a heading is selected on the MCP and the HDG SEL knob is pushed in, the autopilot if selected and/or the flight director will fly the SAILBOAT (aircraft heading indicator) into BUCKTOOTH HARBOR (heading selector).
Knowing this, we can fly a FABULOUS ADF with the HDG KNOB. Here's how: If the Bomb slips off the ADF radial, tweak the heading selector so the jet will fly back to the appropriate heading to line up the **AIRCRAFT TRACK, THE ADF RADIAL AND THE BOMB.**

KEEP ALL THREE ALIGNED;
THEY MUST REMAIN LINED UP.

747-400 SIMULATOR TECHNIQUES ...

VOR SETUP

HOW TO SET UP THE RADIOS and STUFF

At this point in the approach set-up, we will assume that you have completed the **A-I-R-B-A-G** steps which include:
- installing the approach on the **DEP ARR** page,
- resolving the *DISCONTINUITIES* on the **LEGS** page,
- and set the frequencies in the **NAV RAD** page.

Since those items are shown that in detail earlier, we won't repeat those steps.

PNF may be is the MAP mode at the Captain's discretion. I have depicted using the CTR mode simply because on the outbound vector, the airport information largely is behind the airplane and CTR allows more information without going to larger scale. We obtain the CTR display by depressing the center of the EFIS selector knob.

PF may be use the MAP mode up to the FAF; after passing the FAF the PF MUST be in the VOR mode. *Appropriate RAW DATA information MUST be monitored throughout the approach. If a disagreement between the MAP mode and the RAW DATA occurs, use of the LNAV must be terminated and RAW DATA followed using the HDG SEL knob on the MCP.*

Throughout the approach, even while using the LNAV and AUTOPILOT; it is good technique to be constantly "Marrying the bugs" (setting the HDG SEL bug on the ND to the heading desired) so that should a transition to HDG SEL become necessary, the MCP will be set up and the airplane will not make a turn off course.

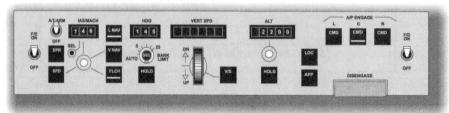

page 316

© MIKE RAY 2014
published by *UNIVERSITY of TEMECULA PRESS*

LOC ONLY SETUP

HOW TO SET UP THE RADIOS and STUFF

At this point in the approach set-up, we will assume that you have completed the **A-I-R-B-A-G** steps which include:
- installing the approach on the **DEP ARR** page,
- resolving the *DISCONTINUITIES* on the **LEGS** page,
- and set the frequencies in the **NAV RAD** page.

Because I have shown that in detail earlier, I won't repeat those steps.

SOME *specifically* LOC STUFF:

G/S INHIBIT should be pushed to prevent spurious warnings during the approach.

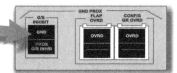

Once **CLEARED FOR THE APPROACH**, arm the **LOC** mode.

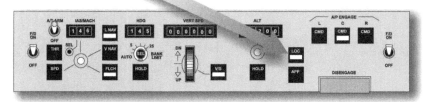

The **LOC ONLY** approach has guidance in the lateral mode only. All vertical guidance must be provided by the pilot.

VNAV is NOT AUTHORIZED due to complexity in the set-up.

Use **FLCH** outside the **FAF** and **V/S** wheel inside the **FAF**.
Leaving FAF altitude for MDA, use 1200 FPM.
Leaving MDA on descent to landing, use 800 FPM.

The ILS
"Instrument Landing System"

While modern technology has provided the aviation community with many advanced solutions to assist the airline-pilot in trying to find the ground and consistently land in the confines of the airport, there is nothing more heavily utilized and preferred by Air Traffic Controls around the whole world than the lowly and venerable ILS (Instrument Landing System). I think that you will find that most (99%) of your approaches will be ILS approaches. However, even this robust and highly accurate system has been the focus of the inevitable rise of technology. Application of advances in the instrumentation and on-site runway improvements have reached the point where it is actually a daily routine to be able to land the Boeing 747-400 precisely where it is intended, consistently and do it smoothly without the pilot visually acquiring the runway. This is referred to as the CAT IIIb Auto-land triple auto-pilot approach ... and can be used down to an amazing reported visibility minimum of RVR 300.

It is considered S.O.P. (Standard Operating Procedure) to use the ILS system whenever it is available, regardless of the weather. Even if you should elect NOT to actually "couple it up" it is recommended that the ILS (or the highest level of approach aid available) be tuned and set-up for the approach. It is intended that the instrument landing indicators be available and used to monitor the progress and quality of your flight profile.

Here is a BIG problem with the ILS. If during an ILS approach, once you have "coupled up" and you should subsequently desire to fly away from the selected ILS course, you will have only three options:
- *Disconnect the auto-pilot and hand-fly the airplane,*
  in which case the ILS signal remains tuned
  and frequency changes are inhibited, or
- *Depress one or both of the TOGA buttons,*
  in which case the airplane reverts aggressively to
  modes of the HDG SEL and V/S with appropriate
  massive application of engine power (WOWEE!), or
- *Disconnect the auto-pilot, then turn BOTH flight directors OFF*
  and back ON, Then re-engage the auto-pilot.

This is useful knowledge if you should get a runway re-assignment during the approach.

FLYING THE ... ILS APPROACHES

These used to be called the PRECISION APPROACHES and are the type that you will be flying MOST of the time out on the line. They are ILS related:

ILS CAT I
ILS CAT II / III

ILS CAT 1 CALL-OUTS

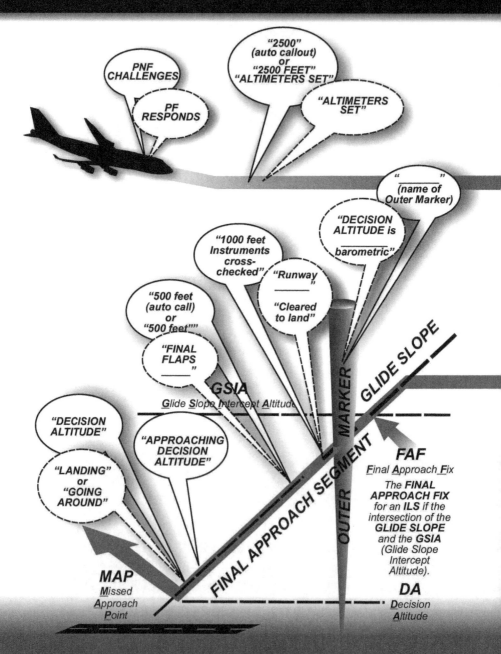

ILS APPROACHES — OP SPECS

SOME ILS DEFINITIONS:
There seems to be a lot of confusion about terminology regarding the ILS approach environment. Here are a few of the terms and their definitions.

GLIDE SLOPE: That is the pathway that the airplane flies to the runway. It is strictly defined by the ILS transmitter on the ground. The airplane MUST receive and fly the GLIDE SLOPE in order to be on the approach.

GSIA or GLIDE SLOPE INTERCEPT ALTITUDE: This is an altitude depicted on the approach chart that represents the ceiling of the Final Approach Segment.

FAF (Final Approach Fix): The place where the Final Approach Segment begins.

> **The point in space where the GSIA and the GLIDE SLOPE intersect IS the FAF.**

Technically, the approach has not begun until the airplane has passed the Final Approach Fix.

OM (OUTER MARKER): The "*Outer Marker "IS NOT*" the FAF. It is simply a point that is designated by reference to a ground based transmitter as the outer marker.

ALERT HEIGHT v. DECISION ALTITUDE: *Here the Training People got in their licks. AH and DA are the same thing, but designated differently because of some reason we don't have to know about. So, if you are flying an approach with a DA(Decision Altitude), treat it the same as an AH (Alert Height).*

COMMENT about ENGINE INOPERATIVE OPS:

Believe it or not, the airline training people had to admit that this airplane was so fabulous that they could make the regular all engines working approach the same as flying it with one engine shut down. Fantastic. It is incredible, not because the airplane can do that ... but because the Training Kingdom would actually admit that they didn't have a case for cluttering up the procedures. Mark up one for the pilots.

ENGINE INOPERATIVE ILS STUFF:

With an engine inoperative; Use **NORMAL APPROACH FLAP SETTINGS** (25 or 30, Captain's choice).

During landing, the **RUDDER TRIM** *may be zeroed at the outer marker or left in; but the training procedures don't suggest attempting to center the rudder on short final. I never liked doing that anyway and usually forgot.*

Autopilot and Flight Directors stuff is the same for both engine out and all engines running.

> **You CAN fly a CAT II or CAT III AUTOLAND with an engine inoperative.**

747-400 SIMULATOR TECHNIQUES ...

ILS CAT 1 APPROACH
RVR 1800 or BETTER

OP SPECS: All landings with 1800 RVR or less will be *AUTOLAND*.

POSBD NOTE: Must use AUTOLAND below 2400 RVR if the required lighting for RVR 1800 is inop.

APPROACHING PATTERN

- **A** - ATIS
- **I** - INSTALL APPROACH
- **R** - RADIOs (tune & Ident)
- **B** - BRIEF
- **A** - APP-DESCENT CKLIST
- **G** - GO-AROUND

think about
MISSED APPROACH

- **"GOING AROUND"**
  PUSH TO/GA button
  VERIFY throttles go to G/A THRUST.
  MANUALLY ROTATE (Follow F/D)
- **FLAPS 20"**
  @ POSITIVE CLIMB
- **"GEAR UP"**
- **"L-NAV" or "HDG SEL"** @ 400 feet.
- **"VNAV or FLCH"** @ 1000 feet.
  A/P switch above 1000 feet.

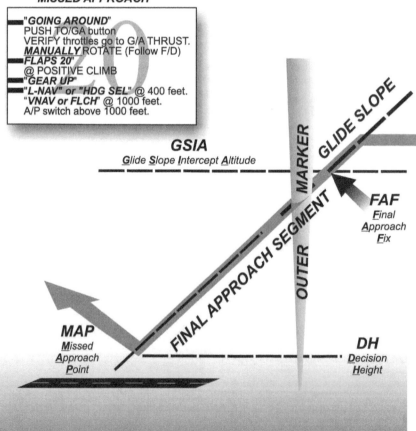

GSIA — Glide Slope Intercept Altitude

FAF — Final Approach Fix

FINAL APPROACH SEGMENT

OUTER MARKER

GLIDE SLOPE

MAP — Missed Approach Point

DH — Decision Height

... and **PROCEDURES FOR STUDY and REVIEW** ONLY!

NOTES:
- ALL LANDINGS BELOW CAT I MINIMUMS (less than 1800 RVR) WILL BE "AUTO-LAND."
- *THE AUTO-LAND IS A MULTIPLE AUTOPILOT MANEUVER.*
- *DO NOT try to AUTOLAND ON A SINGLE AUTOPILOT.*
- *DO NOT AUTOLAND ON A RUNWAY THAT IS NOT CAT II or CAT III CAPABLE.*
- FAF is (USUALLY) INTERSECTION OF GSIA and GLIDE SLOPE.
- WX MUST BE AT or ABOVE MINIMUMS TO CONTINUE PAST GSIA.
- IF WX GOES BELOW MINIMUMS ONCE PAST FAF; OK TO CONTINUE TO MAP; and LAND IF ALL OTHER CRITERIA ARE MET (for example: You see the runway)

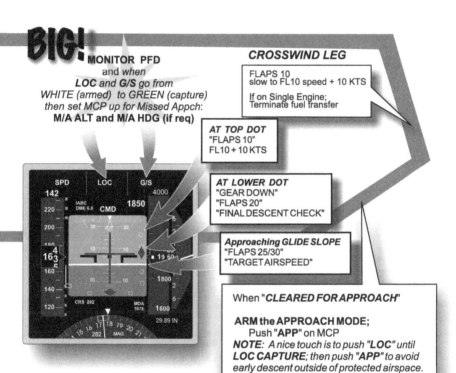

BIG!

MONITOR PFD
and *when*
LOC and **G/S** go from
WHITE (armed) to GREEN (capture)
then set MCP up for Missed Appch:
M/A ALT and M/A HDG (if req)

CROSSWIND LEG
FLAPS 10
slow to FL10 speed + 10 KTS
If on Single Engine;
Terminate fuel transfer

AT TOP DOT
"FLAPS 10"
FL10 + 10 KTS

AT LOWER DOT
"GEAR DOWN"
"FLAPS 20"
"FINAL DESCENT CHECK"

Approaching GLIDE SLOPE
"FLAPS 25/30"
"TARGET AIRSPEED"

When "*CLEARED FOR APPROACH*"
ARM the APPROACH MODE;
Push "**APP**" on MCP
NOTE: A nice touch is to push "LOC" until LOC CAPTURE; then push "APP" to avoid early descent outside of protected airspace.

USE ND IN EITHER APP or MAP MODE
PFD is considered the PRIMARY NAVIGATION INSTRUMENT.

CDU in LEGS PAGE for distance to waypoints

USE AUTOPILOT/AUTOTHROTTLE EVEN WITH AN ENGINE FAILURE, it is OK to USE AUTO-PILOT TO FLY APPROACH.

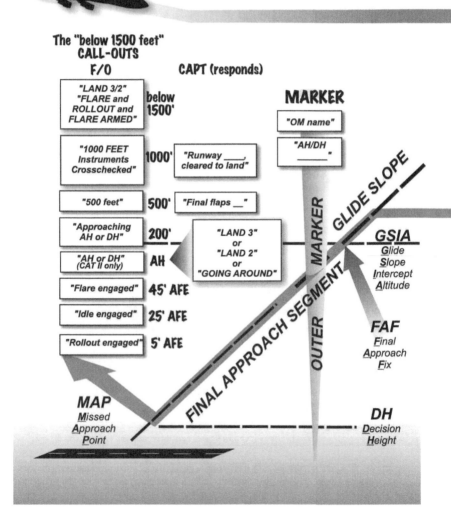

... and PROCEDURES FOR STUDY and REVIEW ONLY!

NOTES

- THE AUTO-LAND IS A MULTIPLE AUTOPILOT MANEUVER.
- DO NOT try to AUTOLAND ON A SINGLE AUTOPILOT.
- DO NOT AUTOLAND ON A RUNWAY THAT IS NOT CAT II or CAT III CAPABLE.
- FAF is (USUALLY) INTERSECTION OF GSIA and GLIDE SLOPE.
- WX MUST BE AT or ABOVE MINIMUMS TO CONTINUE PAST GSIA.
- IF WX GOES BELOW MINIMUMS ONCE PAST FAF; OK TO CONTINUE TO MAP;
  BUT, you CANNOT LAND CAT III UNLESS RVR LEGAL (Even if you see the runway).
- **OK to AUTOLAND with ONE ENGINE INOPERATIVE.**

BIG !

MONITOR PFD
and when
LOC and **G/S** go from
WHITE (armed) to GREEN (capture)
then set MCP up for Missed Appch:
M/A ALT and **M/A HDG (if req)**

CROSSWIND LEG
FLAPS 10
slow to FL10 speed + 10 KTS
If on Single Engine;
Terminate fuel transfer

AT TOP DOT
"FLAPS 10"
FL10 + 10 KTS

AT LOWER DOT
"GEAR DOWN"
"FLAPS 20"
"FINAL DESCENT CHECK"

Approaching GLIDE SLOPE
"FLAPS 25/30"
"TARGET AIRSPEED"

USE ND IN EITHER APP or MAP MODE
The PFD is considered
the PRIMARY NAVIGATION INSTRUMENT.

CDU in LEGS PAGE for distance to waypoints

USE AUTOPILOT/AUTOTHROTTLE
EVEN WITH AN ENGINE FAILURE,
it is OK to USE AUTO-PILOT TO FLY
APPROACH and AUTO-LAND.

"CLEARED FOR APPROACH"
you MUST do **2** things
1. **ARM the APPROACH MODE;**
   Push "**APP**" on MCP
   NOTE: A nice touch is to push "LOC" until
   LOC CAPTURE; then push "APP" to avoid
   early descent outside of protected airspace.
2. **ARM OTHER TWO AUTOPILOTS.**

For some inexplicable reason,
this is **EASY TO FORGET** !
You should NOT autoland
using single autopilot.

© MIKE RAY 2014
WWW.UTEM.COM

some thoughts about ...
Doing the
GO-AROUND

SOMETIMES, PILOTS PUSH THE TOGA BUTTON WHEN IT MAY NOT BE APPROPRIATE!

Just remember, there is a time to push the TO/GA button and then...
there is a time to simply add power and fly a missed approach profile.
On a light 747-400, especially, if you elect to push the TO/GA button the sudden burst of power will turn the jet into a rocket ship. I tell you this, because I have seen some pilots pushing the GO-AROUND BUTTON when it was probably the wrong thing to do ... and then get flustered and confused trying to regain control.

Let me explain. If you are close in at the MDA or MISSED APPROACH POINT on an approach, then of course, using the GO-AROUND capability of the airplane is entirely the right thing to do.
However, should you elect to abandon an approach while the airplane is well above the terrain, then *it may be a better idea to simply push up the power a bit and level off using other available tools*.
Let's talk about this a bit and using an example try and understand the situation in greater detail.

Probably one of the most abused events on the check-ride is the Go-around. The problem comes from pilots pushing **TO/GA** button inappropriately.

WOWWWW! The power rapidly comes on, the nose rapidly pitches up, and the airplane starts climbing and the ILS guidance shuts off and the pilot becomes very disoriented. Particularly if there is an assigned intermediate level-off, the airplane will be climbing such that it will likely "bust" the altitude. I have observed pilot candidates getting behind the airplane and not fully aware of how to regain control using the automated control systems.

For example: Descending on the 25L glide-slope into LAX, you are passing through 3000 feet. The Tower advises you to:
"*Execute missed approach, level off at 5000 feet and maintain runway heading.*"

If you push the **TO/GA** button, the power will come on big-time, the nose will pitch up. The next thing the pilot will instinctively do is, shut off the autopilot and push on the yoke to keep from exceeding the 5000 foot restriction.

WHEEEEEE!

Then the airspeed will be building so rapidly, that flap speeds will be exceeded. At that point the pilot candidate with pull the throttles back ... And if the auto-throttle is not disconnected, the throttles will attempt to go back to the high power setting. So the confused aviator is over-riding the thrust levers in idle while trying to hand-fly the speeding missile and interpret what is going on.

All the while, the terrified passengers are wondering if they are going to die ... and the pilot is wondering if she/he will pass their check-ride.

I think you get the idea.

> *Just because you hear the command, "Go-around" does not mean you have to push the TO/GA button.*

CONSIDER THIS TECHNIQUE:

1. **Disconnect the auto-pilot** and
2. **Re-set BOTH the flight directors**.

This is the easiest way to dis-engage the ILS. Then simply

3. **Re-engage the auto-pilot**, and
4. Roll the **altitude selector on the MCP to 5000 feet**.
5. **Depress FL/CH** and

...you are all set. The nose will come gently up and the power will keep the airspeed right where it was and if a turn is required, simple select and twiddle the heading selector to the desired heading.

You don't raise a sweat and it is all so smooth and simple.

747-400 SIMULATOR TECHNIQUES ...

Missed Approach
Go around or Overshoot

> Quote from the manual:
> *"IF A MISSED APPROACH IS REQUIRED FOLLOWING AN AUTOPILOT APPROACH, LEAVE THE AUTOPILOT ENGAGED."*

■ **"GOING AROUND"**
PUSH TO/GA button

■ **"FLAPS 20"**

VERIFY throttles go to **G/A THRUST**.
Monitor autopilot or Manually (Follow F/D)

■ **ROTATE**
Observe the autopilot Initially command about 15 degrees and then
pitch between COMMAND SPEED BUG and BUG +25.

@ POSITIVE CLIMB
■ **"GEAR UP"**

■ Set **MISSED APPROACH ALTITUDE on MCP** !!!!

20°

"VNAV or FLCH" @ 800 feet
FLCH is recommended in order to open **SPEED INTERVENE** and move the **COMMAND SPEED BUG** for **FLAP** retraction.

- - - - - 800 feet AGL - - - - -

"L-NAV" or "HDG SEL" @ 400 feet.
If you don't select another heading mode, the jet will remain on the heading that it was on when you pushed the **TO/GA** button.

- - - - - 400 feet AGL - - - - -

NOTE 1:
Below 400 feet, selecting **LNAV** or **VNAV** does not change the autopilot or flight director modes.
NOTE 2:
If you are one the go-around from an engine out, autoland, triple autopilot approach; the rudder trims for the assymetry. HOWEVER, when altitude capture occurs above **400 feet AGL**, the autopilot reverts to the original single autopilot operation and
REMOVES THE AUTOMATIC RUDDER INPUT.

- - - - - 250 feet AGL - - - - -

A/P *(if not already on)*
OK above 250 feet.

Expect about a 40 foot descent below the altitude where the go-around is initiated.

page 328

© MIKE RAY 2014
published by UNIVERSITY of TEMECULA PRESS

... and PROCEDURES FOR STUDY and REVIEW ONLY!

Complete the **"AFTER TAKE-OFF"** Checklist.

MAP (Missed Approach Altitude)

When ALT is annunciated, move the BUG speed up and begin to clean up the airplane.

"A-B-C"

A: At ALT,
B: BUG UP
C: CLEAN UP

It is Company SOP to remain configured at 20 degree flap and Approach Speed bug all the way to the missed approach altitude.

"BUG IN - BUG OUT"

BIG PROBLEM

If you have not set in the MISSED APPROACH ALTITUDE when you raised the gear, FLCH will not have an appropriate target altitude. It may attempt to push over and return to the altitude set in the MCP, in this case it could be the TDZE.

FLCH is preferred over VNAV SPEED INTERVENE because if VNAV ALT is displayed, a premature leveloff may occur requiring the selection of FLCH to correct. So, why not already be in FLCH?

1. At **ALT**, Move **"BUG UP"** to Speed greater than the minimum flaps maneuvering speed for the next flap setting.
   When **COMMAND SPEED BUG** is at **FLAP** marker on speed tape,
2. PF Request **FLAPS** ____ .
3. PNF sets **COMMAND SPEED BUG** on the **MCP**.
4. PNF moves **FLAP HANDLE**.
5. **RETRACT** flaps on schedule to desired setting.

FL CH TECHNIQUE RECOMMENDED

RECOMMENDATION

*Set the **BUG** to about **225 Knots or greater** if you intend to raise the flaps full up. This will assist in keeping the engines from chasing airspeed and surging back and forth. However, if you intend to return to the field for another landing, perhaps you should consider leaving the flaps 1 or 5 degrees and set speed accordingly.*

TO/GA SWITCHES

QUICKIE REVIEW OF THE GO-AROUND MODE OF THE MOST POWERFUL BUTTONS ON THE BOEING 747-400 !

Here is the SIMPLE description without all the details:

IF A/C ON THE GROUND; and
AIRSPEED BELOW 50 KTS; then

Pushing either **TO/GA** button causes the engines to start producing **LOTS** of thrust. *I am assuming that you have engines running, flaps selected, and the* Auto-throttles engaged ... of course. That is one reason why we delay arming the auto-throttles until we are taking the runway.

IF AIRBORNE; and EITHER
   GLIDE SLOPE IS CAPTURED, or
   FLAPS ARE SELECTED; then

Pushing either **TO/GA** button will cause the airplane to pitch up and the thrust levers to spool up.

AUTOTHROTTLE engages in the **THR** mode. This provides thrust **UP TO** the reference **LIMIT** in order the achieve a **2000 FPM climb**. Once the 2000 FPM rate of climb is established, the **AUTOTHROTTLE** controls thrust to maintain that 2000 FPM climb rate.

TARGET AIRSPEED: The **AFDS** will increase **PITCH** to hold the **GREATER of CURRENT AIRSPEED or MCP SELECTED AIRSPEED**. The **TOGA** algorithm will change the **TARGET AIRSPEED** after 5 seconds to a maximum of **IAS/MACH indication plus 25 Kts**.

DISCUSSION
There are two places where **TO/GA** can get you into trouble:

FIRST, on take-off.
If you inadvertently allow the airplane to attain a speed greater than 50 knots before you push the **TO/GA** button, then the **TO/GA** function will be **DEACTIVATED** and the throttles will *NOT MOVE TO TAKE-OFF SETTING*.

SECOND, during the approach.
If you should elect to abandon the procedure before the **MAP** (Missed Approach Point) and depart from the glide-slope; using the **TO/GA** button will result in a radical application of power and a departure from the **LNAV** or approach course. These things could cause you to bust an altitude or deviate from the assigned clearance. Generally speaking, the **TOGA** feature is used for **"CLOSE IN"** abandonment of the approach.

... and PROCEDURES FOR STUDY and REVIEW ONLY!

CAUTION

ENGINE FAILURE PROBLEM

Without getting bogged down in a complex system description, take it as true that when operating with "**MULTIPLE AUTO-PILOTS**" engaged the Boeing 747-400 can fly an extraordinarily accurate and fabulous automated approach ... even with an engine failed.
It will fly the whole approach, flare, touchdown and land and roll-out on centerline perfectly ... even with an engine shut-down. Now where it gets you into trouble is that during the **APPROACH** with the engines spooled back to approach setting, it will adjust the **RUDDER** automatically without actually moving the **RUDDER TRIM**.
And, if you initiate an automated go-around, the **AUTO-PILOT** will move the **RUDDER** to place the airplane in a **TRIMMED CONDITION** without moving the rudder trim.

When either:

- Above 400 feet, another roll mode is selected (LNAV or HDG SEL, e.g.), or
- At ALT annunciation, or
- The three autopilots revert to single autopilot operation,

THE RUDDER WILL RETURN TO THE ORIGINAL TRIMMED POSITION UNLESS THE PILOT EXERTS THE RUDDER PEDAL FORCE REQUIRED TO MAINTAIN THE RUDDER POSITION.

This means that if you are not holding the rudders with your feet or have it trimmed out; when the autopilot releases the **RUDDER BIAS** that it was holding and "lets go," the airplane could possibly

ROLL OVER AND YOU WILL DIE!!!

I have actually observed pilots flying this airplane in the approach environment with **THEIR FEET ON THE FLOOR**.
NOT A GOOD IDEA.

Here is the statement about this situation directly from an airline Pilot Handbook:
"If the autopilots are compensating for an asymmetric thrust condition when they revert to a single autopilot engaged configuration, the rudder will return to the trimmed position unless the pilot exerts the rudder pedal force required to maintain the rudder position."

OMIGOSH !!!

CAUTION
ENGINE FAILURE PROBLEM CONTINUED

While it is easy to get really excited about the power and capability of this airplane, even with an engine failure; it must be remembered that it is like any other airplane and only operates at its engineered capabilities when the airplane is in trim.

How do we know if the airplane is in trim: The "**SLIP INDICATOR**" referred to by pilots as the "**SAILBOAT**".

The technique is to
"FLY THE SAILBOAT WITH YOUR FEET."

Now, let me develop this because it is not frequently emphasized in training and can be very important. To keep the airplane in trim, you have to
"STEP ON THE RUDDER and TRIM OUT THE PRESSURE"
This is critical in situations where the airplane is undergoing extreme asymmetric thrust. You need to push on the appropriate rudder to keep the "**BOX**" centered under the "**TRIANGLE**".

For example; During go-around with an outboard engine failure, without appropriate rudder input from the pilot, the airplane is virtually uncontrollable. It will take massive amounts of yoke movement and the heading will still be slewing into the bad engine. You can trim out the pressure by pushing the **RUDDER** under the lower **YOKE HORN**.

Even on spool-up during a manual go-around. There simply **MUST BE RUDDER INPUT** to control the airplane, not only during **NON-AUTO-PILOT** approaches, but also during **MULTIPLE PILOT AUTO-LAND** approaches.

During normal all engines operating operations, particularly with protracted turns, it may not be necessary or desirable to displace the rudder by pushing or trimming in order to keep the rudder/sailboat centered. Indeed, without some rudder, this big hog actually flies a sloppy turn and if you notice, the sailboat will be 1/3 or so out of alignment. In normal operations, I don't think it is necessary to chase that correction, just accept the fact that this is one of the airplanes inherent tendencies.

HOWEVER, during engine out operations ... keep the boat under the sail!.

Example

HOW TO **LAND** the Boeing 747-400

It may be useful to define some of our terms when talking about the landing.

REF - This is the speed for a specific flap configuration (For example Flaps 30 = 30 REF).

TARGET SPEED - This is the **REF** speed corrected for the wind component. It is the speed that we use to fly the approach. It is equal to:

> **REF** + 1/2 the **STEADY WIND** component
> + the **FULL GUST** value.
> *The maximum correction for the wind is not to exceed **20 Kts**.*

TOUCHDOWN SPEED - This is the speed at which we LAND the airplane. That is, after the flare and approaching the runway. It is equal to:

> **REF** + the **FULL GUST** value.
> *The maximum gust correction not to exceed 20 Kts.*

NEVER - NEVER - NEVER LAND THE JET WITH THE SPEED BRAKES OUT!!!!

747-400 SIMULATOR TECHNIQUES ...

LANDING
... including ILS CAT 1 and NON-ILS approaches.

There is a problem and it is a **BIG PROBLEM**.
Pilots that have mastered the art of flying the Non-ILS approach are frequently screwing the whole thing up when they get to MDA and decide to make a landing.
There are two problems:

LINE-UP and DESCENT.

Any of the suite of NON-ILS approaches available have the inherent problem of line up. The VOR, for example, can have a really big alignment problem and still be within parameters. With an airplane the size of an apartment house, making late course adjustments, even though the jet is very responsive, can be tricky.
With the ADF, even if it is flown EXACTLY PERFECT, is **_NEVER_** going to be aligned with the runway ... Period. It is just an inherent part of the ADF system.
So, I said all this to say that alignment **IS ALWAYS A PROBLEM**.

The decent problem can be directly traced to the habit of the pilot flying turning **OFF** the auto-pilot **TOO SOON**. There is no requirement to turn off the auto-pilot at the point where ground contact is made.

REMEMBER:
While the **MDA** is the point where the **GO-AROUND/CONTINUE** decision must be made, you may still keep the auto-pilot operating up to a point **50 feet below the MDA**. This 50 extra feet takes several seconds and gives valuable time to continue operating the airplane using the automation. Pilots tend to turn off the autopilot **AT MDA** and, thereby, attempt to hand-fly the final descent too early.
This is **_NOT GOOD_** !

The best technique seems to be to remain on instruments almost totally and as those last seconds pass, gradually transition more and more to an outside sight picture until you arrive at the runway with about 80% of your attention outside.

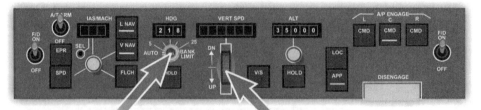

The auto-pilot can be used by selecting V/S (vertical speed) and HDG SEL (Heading Select); making small, teensy changes to assist in the line-up process.
When the V/S is selected, it will default to the existing rate of descent, and if you have constantly kept the heading buckteeth aligned with the heading boat, the Heading select will default to the airplane heading.
Then if everything works out, and you don't drag the trucks in the lights, or land in a skid, or bounce down the runway ....

SIGHT PICTURE

The concern, of course is that you will not drag the trucks in the light array.

GOING LOW IS NO-NO!

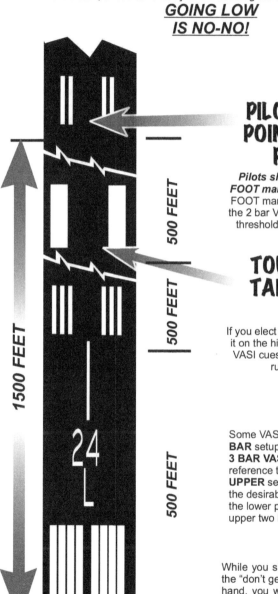

PILOT AIM POINT 1500 FEET

Pilots should aim for the 1500 FOOT marker. Aiming for the 1000 FOOT marker (equivalent to having the 2 bar VASI centered) results in a threshold clearance of 2.6 FEET! YIPE!!!

TOUCHDOWN TARGET 1000 FEET

If you elect to fly a 2 BAR VASI ... Fly it on the high side and abandon the VASI cues at **300 FEET** above the runway threshold!

VASI

Some VASI installations have a **3 BAR** setup. Normal procedure for a **3 BAR VASI APPROACH** is to reference the **MIDDLE** and the **UPPER** sets of lights to determine the desirable **VASI** path. Disregard the lower par and fly using the upper two bars.

While you should be spring loaded to the "don't get low" profile, on the other hand, you want to get it on. It is an awareness problem. Don't say I didn't warn you. Try to land this bird beyond 1000 FEET but don't float either. Tough assignment.

Here is the underlined number

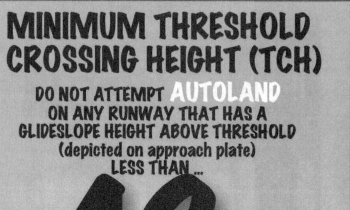

MINIMUM THRESHOLD CROSSING HEIGHT (TCH)

DO NOT ATTEMPT **AUTOLAND** ON ANY RUNWAY THAT HAS A GLIDESLOPE HEIGHT ABOVE THRESHOLD (depicted on approach plate) LESS THAN...

42 FEET

The key word here is ... **AUTOLAND**.

BUT ... Here are some other landing stuff you will be required to know.

If the height shown is LESS THAN 42 FEET, then the GLIDESLOPE must be abandoned no later than **200 FEET AFE** and visual cues used to land the jet.

The RUNWAY THRESHOLD should disappear under the nose when approximately **75 FEET** is displayed on the Radio Altimeter and "PRIOR" to the 50 foot callout.

FYI: Pilots eyes are 45 FEET above the MAIN LANDING GEAR during the landing evolution.

LIMIT CROSSWINDS

reported BRAKING action	RUNWAY conditions	limit CROSSWIND takeoff	limit CROSSWIND landing
DRY	DRY	36	36
GOOD	WET	25	32
FAIR / MEDIUM	DRY SNOW	15	25
FAIR / MEDIUM to POOR	STANDING WATER / SLUSH	15	20
POOR	ICE - NO MELTING	15	15

It is important to emphasize that these are **GUIDELINES** and **NOT LIMITS**.

NOTE 1: Touchdown with bank angles greater than 6-8 degrees may result in engine nacelle contact with the runway. With crosswinds greater than approximately 16 knots, this bank angle may be exceeded using the sideslip only technique. Use a combination of crab and sideslip to avoid exceeding 8 degrees of bank at touchdown.

NOTE 2: ATIS will include the term *"BRAKING ACTION ADVISORIES ARE IN EFFECT"* when braking action is POOR to NIL.

FOM NOTE: In the FOM there is this note: *"TAXI, TAKEOFF, and LANDING ARE NOT RECOMMENDED WHEN BRAKING ACTION IS REPORTED AS NIL."*

NEVER EXCEED 8° ROLL ON LANDING

PITCH and ROLL limits

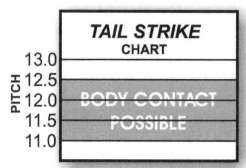

Applies to all weights and speeds.

MAX PITCH 11°

MAX ROLL 6°

Depends on the amount of pitch involved. It can vary from 6 degrees to 8 degrees ... SO, we will assume the maximum roll allowed is 6 degrees and that will cover you in all situations.

The part of the jet that contacts the earth is the outboard engine nacelle.

When you are doing your walk-around, it is a good idea that you check that area on the outboard engines for ground contact. It is possible that the crew could be unaware that they have "drug a pod."

TWO ENGINE LANDING

Maneuvering for a landing with two engines inoperative requires "PRECISE CONTROL" and "DELIBERATE PLANNING". Consideration must be given to FUEL DUMPING to reduce the weight of the airplane. Maneuvering the airplane can be accomplished CLEAN; since any loss of airspeed can place the airplane in an unrecoverable situation. You have to keep in mind that the jet will quickly slow down, but that it is very difficult to regain airspeed. The recommended maneuvering speed is NOT LESS than the flap maneuvering speed plus 10 Kts.

CAUTION

V_{MCA} with two engines inoperative on the same side is **161 Kts**. *DO NOT* reduce airspeed below **161 Kts** or extend the landing gear until landing is assured.

Below 161 Kts or after the landing gear has been extended... **DO NOT** attempt a **MISSED APPROACH!**

 161 Kts

When a speed reduction is required, select the next flap setting and allow the drag to reduce the speed. Do not extend the landing gear until the landing is assured. Plan landing with flaps 25 and plan the approach so as to extend flaps 25 no later than 500 feet AFE.

Once the gear has been extended or the airspeed is below 161 Kts :

DO NOT ATTEMPT A MISSED APPROACH.

FLARE and LANDING

The jet is so big, that the same techniques apply to all landings including crosswind and slippery runway conditions.

DO NOT input sudden, abrupt, or violent movements to the controls UNLESS a EMERGENCY or EXTREMIS situation develops.

Begin with a stabilized approach on glide path, on speed, and in trim.

At about 30 feet above the runway, increase pitch attitude approximately 2 - 3 degrees. This is all you need to slow the rate of descent without setting up a float situation.

After the flare is set, retard the throttles slowly to idle, and make small pitch attitude adjustments. Ideally, the touchdown should occur about the same time that the throttles reach the idle position.

NOTE:
The nose will have a tendency to pitch nose down as the thrust is reduced. As the engines wind down, you will have to increase the pull force on the yoke to overcome the pitch down tendency and apply the 2 - 3 degree additional pitch up.

The airplane should touch down at approximately VREF plus any gust correction.

NOTE:
DO NOT TRIM DURING FLARE OR AFTER TOUCHDOWN.

During the landing, the airplane will have a tendency to pitch up. It is right here that a potential exists for a tail-strike.

Let's go over the yoke profile again;
Pull - touchdown - relax some of the pull
- allow nose gear to touchdown - relax back pressure.

DO NOT hold the jet off, fly it on to the runway.
Smoothly roll on nose-gear onto the runway.

LOWER NOSE WHEEL to the runway **PROMPTLY**.
Direct quote from a Pilots Flight Handbook:
"However, applying excessive nose down elevator during landing can result in substantial forward fuselage damage."

This means that the pilot MUST get the nose gear on the runway as quickly as possible but **NOT BANG IT DOWN TOO HARD OR IT WILL BEND THE JET**.

AVOID
Try not to touchdown with thrust above idle.
This may establish an airplane pitch-up and increased landing roll.

AFTER TOUCHDOWN and LANDING ROLL PROCEDURE

SPOILER ALERT
Both pilots should be alert to the possibility of spoilers NOT deploying. The braking effectiveness is reduced by 60% if the spoilers do not deploy.

In the event the spoiler do not deploy automatically, **MANUALLY EXTEND** them immediately. There are no adverse effects if the spoilers are raised while the nosegear is still being lowered to the runway.

SPOILER DEPLOYMENT IS IMPORTANT !

PNF NOTE *From the TRAINING MANUAL ...*
"The position of the spoilers should be announced during the landing phase by the PNF."

NOTE: Reverse thrust and spoiler drag are *MOST EFFECTIVE* during the high speed portion. Deploy spoilers and activate reversers as quickly as possible after landing.

Mr. Boeing recommends: *"Immediate initiation of reverse thrust at touchdown, and FULL REVERSE."*

REDUCE REVERSE THRUST LEVERS and move REVERSE LEVERS to forward so as to ensure IDLE THRUST is achieved by 60 Kts. Good technique is to begin moving the thrust levers to idle about 80 Kts.

The technique for releasing autobrakes is to gradually increase the brake pedal force until the autobrake systems disarms. A **NORMAL** braking condition would have the autobrakes releasing about 60 knots.

Do not use the NOSE WHEEL STEERING TILLER until reaching taxi speed.

BRAKES STUFF

NOTE: *"Autobrakes 2 or greater results in continuous brake application, which can increase carbon brake life."*

CAUTION
Some braking comments

Maximum braking effectiveness is achieved either the selection of **MAX** position on the **AUTOBRAKES** selector or **FULL MANUAL** brake pressure.
Once touchdown and wheel spin-up is achieved, it takes about *2 seconds for the AUTOBRAKES* to apply the brake pressure. This is a critical time because it may seem to the pilot that the **AUTOBRAKES** have failed to operate and there is the impulse to initiate **MANUAL BRAKING** prematurely. Even with full steady brake application, early in the roll-out detection of braking is difficult and may be interpreted as brake failure. this airplane, the pilot can't feel the wheel brake releases when the brakes are fully depressed.

The situation is exacerbated when some rudder has been displaced to adjust for crosswind. If manual braking is used, there is an almost certainty that "differential" pressure" will inadvertently be applied to the pedals. This increases the potential for control problems during the initial roll-out. Use of the **AUTOBRAKES** will eliminate this problem.

So, as good pilot technique, it is suggested that you allow the **AUTOBRAKE** to supply the initial braking during the touchdown and rollout. **AVOID** "pumping" the brakes as this causes the **ANTI-SKID** system to become ineffective. The use of **AUTOBRAKES**, combined with **AUTO-SPOILER** and the timely and effective use of **REVERSE THRUST**, is the most cost effective way to bring the jet to a speed slow enough to exit the runway.

**UNLESS HERE IS SOME COMPELLING REASON,
AVOID USING MANUAL BRAKING AT HIGH SPEEDS.**

BE ALERT
Pilots should be alert to **EICAS** messages that relate to slowing the airplane. Messages such as
- ANTISKID
- ANTISKID OFF
- BRAKE LIMITER
- HYD PRESS SYS 4

indicate that **MANUAL BRAKING** should be initiated if required.

---

*IF A SKID OR LOSS OF DIRECTIONAL CONTROL
DEVELOPS with AUTOBRAKES applied,
IMMEDIATELY DISARM the AUTOBRAKE SYSTEM.*

SURPRISE!!!
One comment about the **CATIIIb TRIPLE AUTO-PILOT** approach. After landing with the auto-pilot still engaged and the airplane tracking the runway centerline, when turn-off is desired (because of the extremely high force required to manually overpower the **AUTO-PILOT RUDDER CONTROL**) it <u>**WILL BE NECESSARY**</u> to turn off the **AUTO-PILOT** in order to use the tiller bar to re-gain control of the airplane and execute the turnoff.

CAPTAIN and FIRST OFFICER PROCEDURES

- **TAXI-IN**
- **PARKING**
- **SECURE**

DO NOT START ANY CLEAN-UP or SOPs UNTIL "CLEAR OF THE RUNWAY".
Making absolutely certain that the tail of the airplane is **"CLEAR"** of the active runway is the **NUMBER ONE PRIORITY** after landing!

MAXIMUM RECOMMENDED TAXI SPEEDS

- STRAIGHT AHEAD .... 25 Knots
- 45 DEGREE TURN .... 15 Knots
- 90 DEGREE TURNS .... 10 Knots

747-400 SIMULATOR TECHNIQUES ...

CAPTAIN TAXI-IN STUFF

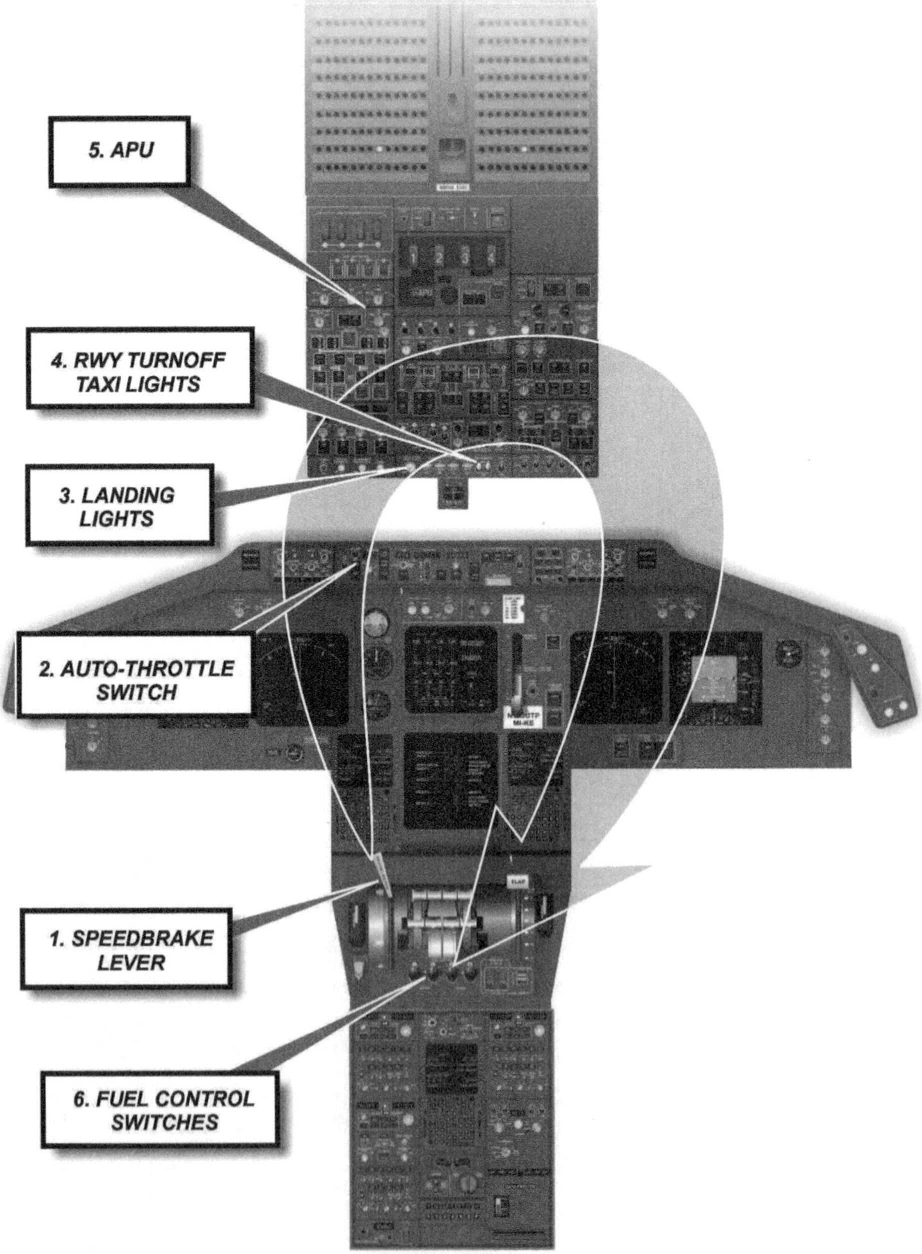

- 5. APU
- 4. RWY TURNOFF TAXI LIGHTS
- 3. LANDING LIGHTS
- 2. AUTO-THROTTLE SWITCH
- 1. SPEEDBRAKE LEVER
- 6. FUEL CONTROL SWITCHES

1. SPEED BRAKE LEVER .... RETRACT

When throttles #1 or #3 are advanced on the ground, the SPEED BRAKE/SPOILER lever moves automatically to the DN position and the GROUND SPOILERS retract.

2. AUTO-THROTTLE SWITCH .... OFF

Since we know that the **TOGA** is armed on the ground when the **FIRST** flight director is selected and the **AUTO-THROTTLE** is armed; So it makes good sense to dis-arm the auto-throttles so that any inadvertent actuation of the **TOGA** buttons will not result in undesired thrust application.
POTENTIAL FOR PILOT SCREW UP ... the airplane goes to THR REF mode at the gate and the nose of the jet imbeds itself into the terminal.

DO NOT ARM THE AUTO-THROTTLES UNLESS YOU ARE ON THE RUNWAY READY FOR TAKE-OFF!

3. LANDING LIGHTS .... FF

It is always annoying to on-coming airplanes on the adjacent taxiway to be greeted to the full blinding power of your landing lights. Definitely a safety item.

4. RUNWAY TURNOFF and TAXI LIGHTS .... AS NEEDED

5. START APU (or not) .... START

If it is to be used at the gate, start it up. Guidelines for use of the **APU** (Auxiliary Power Unit) are predicated on fuel use. The sucker burns a lotta gas. So, even if you are anticipating the use of the **APU** after engine shutdown at the gate, if the taxi is a prolonged one, consider delaying the **APU** start until needed.

The start cycle is simple: Switch right turn to the **START** position, release and it is spring loaded to the ON position.

6. FUEL CONTROL SWITCHES 2 and 3 .... CUTOFF
(for two engine taxi)

Two engine taxi is a matter of opinion and circumstance. If you have determined that it is a good idea, then you MAY (at Captain's discretion) shut down the inboard engines in order to save fuel and engine wear. NOT MANDATORY ... strictly your call.
POTENTIAL PILOT SCREW UP: Now, here is the rub. If you elect to use the two engine taxi and subsequently have to exceed the maximum taxi thrust level (about 40% N1) ... then you will find yourself at the long green table trying to explain why you blew the baggage carts onto the runway.
So, if there are impediments to taxi such as a heavy airplane, up-slope segment of taxiway, clutter on the taxiway, strong surface winds, etc. ... **BEWARE!**

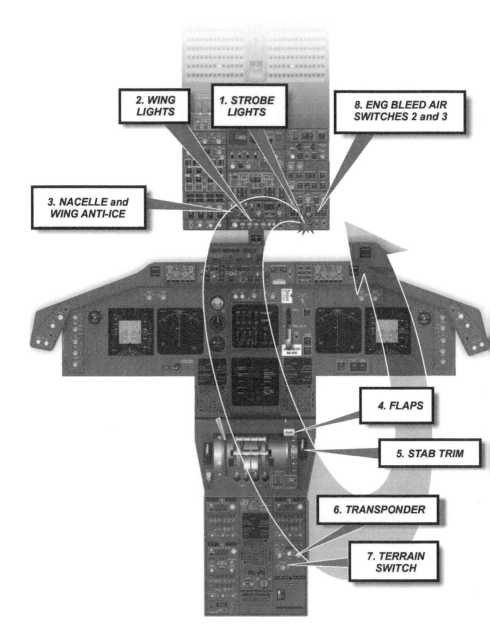

1. STROBE LIGHTS .... **OFF**

2. WINGS LIGHTS .... **OFF**

3. NACELLE/WING ANTI-ICE .... **OFF**

when the **NACELLE ANTI-ICE** is **ON**, the engine thrust (EPR) is automatically adjusted to maintain a minimum N1 setting, giving additional thrust when not needed. The associated **CONTINUOUS IGNITION** is also energized.

4. FLAPS .... **RETRACT**

Carefully allow the flap handle to pause briefly in the gates and ensure that it is then placed positively in the next detent before proceeding. Failure to do this could result in a **FLAP FAULT** as one or more of the **FCU** (Flap Control Units) may disconnect. If this occurs, it can usually be corrected by cycling the **ALTN FLAPS ARM** switch to **ALTN** and then back to **OFF**.

5. STAB TRIM .... **6 UNITS**

If you set the STABILIZER TRIM with the yoke switch, remember to use "both" switches at the same time.

6. TRANSPONDER .... **STANDBY**

Moving the switch to removes the radar return clutter from controllers screen. One problem that I have heard about is pilots resetting the TRANSPONDER codes on the ground while they are in transmit mode ... particularly if they pass through codes such as 7700 they will transmit undesirable information to the controlling radar.

7. (PF/PNF) TERRAIN SWITCH .... **OFF**

8. BLEED AIR SWITCHES (2 and 3) .... **IF 2 ENG TAXI**

Co-ordination with the Captain is required. If you shut down the engines with the Bleed Air Switches selected ON, then when you select them OFF, they may not close fully and you may receive erroneous **EICAS** messages.

747-400 SIMULATOR TECHNIQUES ...

CAPTAIN PARKING STUFF

- 3. ELECTRICAL POWER
- 8. EMER LIGHTS SWITCH
- 4. HYD DEMAND PUMP 1, 2, and 3
- 6. FUEL PUMP SWITCHES
- 5. HYD DEMAND PUMP 4
- 7. EXTERIOR LIGHTS
- 9. LEFT FLT DIR SWITCH
- START HERE
- 1. PARKING BRAKE
- 10. MIC SELECTOR
- 2. SEAT BELT SIGN

© MIKE RAY 2014
published by UNIVERSITY of TEMECULA PRESS

1. PARKING BRAKE LEVER .... SET

Depress the toe brakes, and while holding the brakes, pull the paking brake lever "spoon" up and while holding the lever up, release the brakes. Then, observe the EICAS message PARK BRAKE SET displayed.

2. SEATBELT SIGN .... OFF

3. ESTABLISH ELECTRICAL POWER
*There are two sources for the power, **EXTERNAL POWER** or **APU POWER**.*
   ***EXTERNAL POWER 1 or 2 AVAIL lights**: Check if the external power **AVAIL** light are on.*
*If they are on, Push the switches **ON** and **VERIFY** that the switch **ON** lights are illuminated.*

   ***APU POWER 1 or 2 AVAIL lights**: If the **APU AVAIL** lights are on, then push the switches one at a time and **VERIFY** the switch **ON** is illuminated.*

4. HYDRAULIC DEMAND PUMP SELECTORS ..... 1, 2, 3 OFF
*Select pumps 1, 2, and 3 **OFF**.*

---

5. HYDRAULIC DEMAND PUMP selector 4 .... AUX
*Selecting Hydraulic Demand Pump 4 to the AUXILIARY (AUX) position allows pressurized hydraulic fluid to the **ACCUMULATOR** to prevent decay of the parking brake pressure.. The SOP is to keep it in AUX for a minimum of*
THREE (3) MINUTES.

---

6. EXTERIOR LIGHTS .... AS REQUIRED
It is required that the NAVIGATION LIGHTS remain on at all times.

7. FUEL PUMP SWITCHES .... ALL OFF

8. EMERGENCY LIGHTS SWITCH .... OFF

9. LEFT FLIGHT DIRECTOR SELECTOR .... OFF

10. MICROPHONE SELECTORS .... FLT
*The Captain is responsible for selecting **FLT** on both his and the **FIRST OBSERVER'S** audio control panels.*

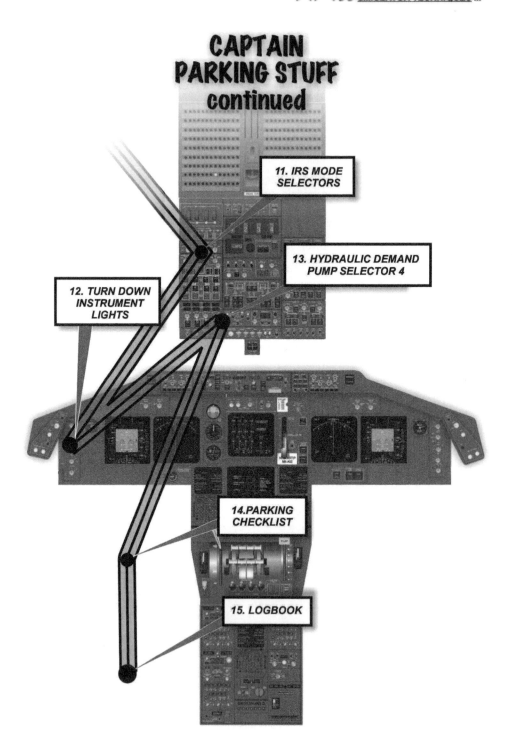

11. IRS MODE SELECTORS .... **OFF**
Ensure that the FIRST OFFICER has completed the IRS LOGBOOK DATA before turning OFF the IRS units.

12. SEATBELT SIGN .... **OFF**

13. INSTRUMENT LIGHTS .... **TURN DOWN**
Turn down the intensity of the CRT (LCD) of the displays and instruments.

14. HYDRAULIC DEMAND PUMP SELECTORS 4 .... **OFF**
*Select pumps 1, 2, and 3 **OFF**.*

AFTER COMPLETING THE PARKING CHECKLIST

15. LOGBOOK .... **COMPLETE**
It is a FEDERAL REQUIREMENT that the Captain to ensure that all the entries in the LOGBOOK are completed. Any additional COMPANY REPORTS must be completed inn accordance with company policy.

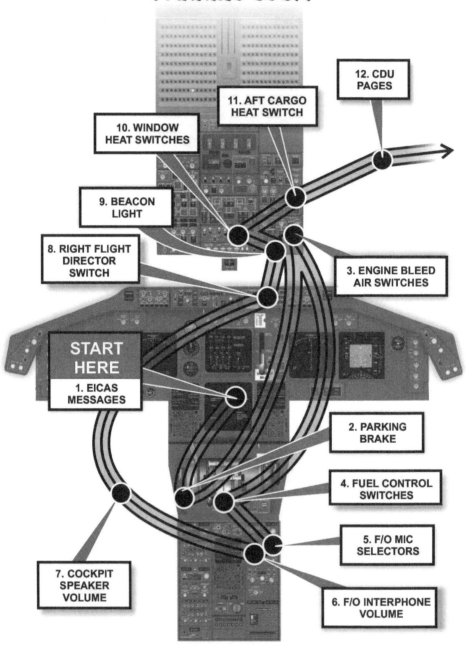

1. EICAS MESSAGE .... **CANCEL**
Subsequent pushes of the RCL switch may be required to cancel ALL of the EICAS messages..

2. PARKING BRAKE .... **VERIFY SET**
Observe the EICAS message PARK BRAKE SET displayed on the UPPER DISPLAY.

3. ENGINE BLEED AIR SWITCHES................................. **OFF**

4. FUEL CONTROL SWITCHES .... **CUTOFF (on order)**
You should VERIFY ENGINE cutoff by observing
- DECREASING FUEL FLOW
- DECREASING N1
- DECREASING EGT

WARNING
Ground personnel expect the RIGHT SIDE ENGINES
to be shut down upon arrival at the gate.

NOTE:
All engines except #1 should be shut down and the PARKING BRAKE set ASAP so the ground personnel can plug in the GROUND POWER. Keep the #1 engine running until the alternate power (APU or EXTERNAL) has been connected and is powering the airplane systems.

5. MIC SELECTOR .... **FLT**

6. F/O INTERPHONE .... **SET**
Adjust the FLT volume FULL UP.

7. COCKPIT SPEAKER CONTROL KNOB .... **SET**
Adjust the volume to the midposition.

8. RIGHT FLIGHT DIRECTOR SWITCH .... **OFF**

9. BEACON LIGHT SWITCH .... **OFF**

10. WINDOW HEAT SWITCHES .... **OFF**

11. AFT CARGO HEAT SWITCH .... **OFF**

12. CDU PAGE continued on next page

FIRST OFFICER PARKING STUFF
continued

12. CDU NAV RAD page
Delete the manually tuned stations.

13. CDU NAV DATA page
Push the **INIT REF** key.
Line Select **6L (INDEX)**.
Line Select **1R (NAV DATA)**.
Delete any **INHIBITS**.

14. CDU ATC log
DELETE all messages.

15. FMC COMM
CHECK any inhibits.

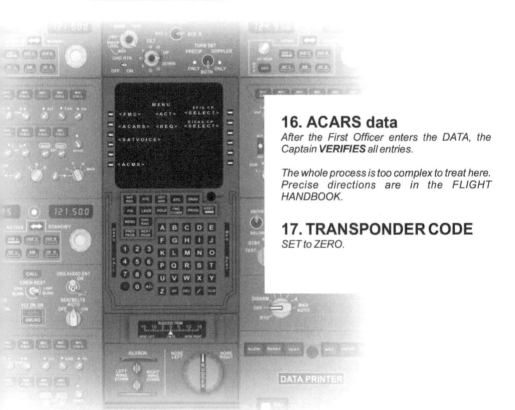

16. ACARS data
After the First Officer enters the DATA, the Captain **VERIFIES** all entries.

The whole process is too complex to treat here. Precise directions are in the FLIGHT HANDBOOK.

17. TRANSPONDER CODE
SET to ZERO.

... and PROCEDURES FOR STUDY and REVIEW ONLY!

18. IRS LOGBOOK DATA .... RECORD

NOTE:
This data **MUST** be recorded at the end of every flight,
whether it is a DOMESTIC or OVER-WATER flight.
It is essential that these records be kept so that
maintenance can conduct IRS/INS accuracy checks.

The following information is to be recorded on the IRS/INS REPORT
part of the AIRPLANE FLIGHT LOG and SERVICE FORM.

- **LEFT, CENTER, and RIGHT IRS POSITIONS** from the **CDU POS REF** page 3/3

- **RESIDUAL GROUNDSPEED** for the **L,C,** and **R** IRSs from the **CDU** page **POS REF** 3/3. Record it using a "/" after the **IRS** positions.

- **BLOCK-TO-BLOCK TIME.**
This is considered to be the operating time in the **NAV MODE** for the **IRS**s.

Here is an example:

L N3355.3 W11824.2/02
C N3356.2 W11823.7/11
R N3356.5 W11824.5/05
12:09

MUST DO THIS EVERY FLIGHT!

19. INDICATOR, DISPLAY, and INSTRUMENT LIGHTS

It is considered standard operating procedure to turn all the CRT (LCDs) to black on the panel and CDUs to avoid "burn-in". As a result if you should someday get into a cold-dark cockpit, remember that the displays may all be turned down ... and they appear to be inoperative, but it just that they are turned down to black.

THE PARKING CHECKLIST
PARKING CHECKLIST in APPENDIX (Page 304)

LAST ITEM .... PARKING CHECKLIST
Both Captain and First Officer participate in completing the PARKING CHECKLIST. It is customary for the First Officer to READ the list and the Captain and First Officer to respond to the items.

© MIKE RAY 2014
WWW.UTEM.COM

747-400 SIMULATOR TECHNIQUES ...

SECURE CHECKLIST

It is necessary to use this check-list only if you do not anticipate that there will be a crew change or that the airplane will not continue in service soon. It is intended for those lay-overs that are overnight or for prolonged "storage" or removal from service for some reason ... such as maintenance.

1. (F/O) PACK SELECTORS .... OFF

2. (F/O) RECIRCULATION FANS .... OFF

3. (C) APU/EXTERNAL POWER SWITCHES .... OFF

4. (C) APU SELECTOR .... OFF
Observe the EICAS message PARK BRAKE SET displayed on the UPPER DISPLAY.

5. (C) STANDBY POWER SELECTOR .... OFF

NOTE:

NOTE:

Some discussion needs to take place to describe something about the APU. You will have to wait two minutes after the APU selector if placed off BEFORE you place the BATTERY SWITCH OFF because there will not be any fire protection for the APU during the shutdown.

6. BATTERY SWITCH.. OFF
Only after waiting for two minutes after shutting down the APU

7. WHEELS.. CHOCKED

CAUTION:
As you leave the airplane, take a moment to ensure that the jet is either attached to something (like a push-back tug) or that the wheels are securely chocked. There is the very real chance that if the airplane is left without something to restrain it, when the brake accumulator bleed off it may take a stroll down the ramp and stop only when it hits something.

... and PROCEDURES FOR STUDY and REVIEW ONLY!

When I was a young pilot, (in the really old days) It always seemed to me that memorizing the emergency checklists was a pretty stupid concept. In the first place, even when things were going good, I couldn't remember the 34 steps to the DC-6 engine fire procedure ... and when everything started going bonkers and we were having a real emergency, I would have a brain dump and generally couldn't even remember my own name. So, one day, some really smart person decided to create a checklist for emergency procedures. The smartest people on earth are Albert Einstein, Isaac Newton, and this guy (probably a retired airline pilot).

So, here is his creation ... the Quick Reference Checklist or as some pilots like to refer to it: **the QRC, QRH, QUICKLIST, etc.**
Each airline seems to be too proud to submit to a common nomenclature and refuses to use the others designation. We will use QRC, simply because I get to decide which airline's designation to use.

What is expected is:

    1. You know which event IS a **QRC** event, and
    2. If there are **MEMORY ITEMS** and what they are.

There are 15 QRC events and 5 of them have memory items:

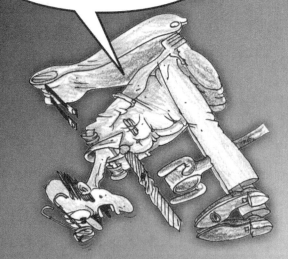

... and PROCEDURES FOR STUDY and REVIEW ONLY!

EMERGENCIES

QRC with MEMORY ITEMS

Don't be confused by the title ... very airline calls their **EMERGENCY CHECKLIST** a different thing: **QRC** (Quick Reference Checklist), **QRH** (Quick Reference Handbook), **Emergency Checklist**, Hail Mary Handbook, Immediate Action Items, and so forth.

You will be expected to know 3 things about the QRC:

1 The First three items of QRC which are:

> **FLY THE AIRPLANE**
> **SILENCE THE WARNING**
> **CONFIRM THE EMERGENCY**

2 Which "**15 EMERGENCIES**" are on the QRC Checklist

3 What "**MEMORY ITEMS**" (if any) are on the QRC emergency checklist for each situation.

The 15 QRC EMERGENCIES *

* the 5 with memory items

- Abnormal Engine Start*
- Starter Cutout
- Multiple Engine Flameout or Stall
- High Engine EGT/Compressor Stall
- Fire Eng/Severe Damage/separation
- Fire APU
- Fire Cargo FWD (AFT)
- Galley Fire/Smoke
- Fire Wheel Well
- Smoke/Fumes/Odor*
- IAS Disagree/Airspeed/Mach Unreliable
- Stab Trim Unschd*
- Cabin Altitude/Rapid Decompression*
- Emergency Descent
- Evacuation*

© MIKE RAY 2014
WWW.UTEM.COM

ABNORMAL ENGINE START

THIS PROCEDURE HAS IMMEDIATE ACTION STEPS ...

WHAT IS AN ABNORMAL START ?

- NO EGT rise within .... 20 seconds

- HI FUEL FLOW .... > 700 PPH
- EGT approaching .... 535 C
- EGT exceeds .... 535 C

- NO N1 by 40% N2
- NO OIL PRESS by 40% N2

- N2 NOT at ~62% N2 by .... 2 MIN

- PNEU .... FAILS
- ELECT .... FAILS

IF ANY OF THESE HAPPEN DO THIS

IMMEDIATE ACTION MEMORY STEP

1 FUEL CONTROL SWITCH ... CUTOFF

Do step one from memory, and then

GO TO THE QRC ...

REFERENCE ACTION SEZ ...

VERIFY ENGINE START SWITCH ILLUMINATED

IF:

IF Start switch is NOT ILLUMINATED and N2 is BELOW 15% ... THEN:

PULL ENGINE START SWITCH
and
MOTOR ENGINE for 30 SECONDS or more

HOWEVER:

IF PROBLEM WAS:

NO EGT RISE within 20 SECs or
HIGH INITIAL FUEL FLOW or
EGT rapidly APPROACHING 535 C

AUTO IGNITION SWITCH ..... BOTH
FUEL CONTROL SWITCH .... RUN

BUT IF:

IF PROBLEM WAS:

NEITHER OF THOSE TWO or
RE-OCCURS ..... then:

ENGNE START SWITCH .... PUSH

and when all else fails ... CALL SAM (maintenance)

STARTER CUTOUT ( )

Displayed on the upper EICAS
THERE ARE NO IMMEDIATE ACTION STEPS - GO TO QRC.

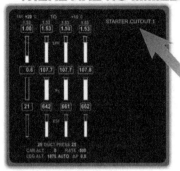

WHAT IS THE PROBLEM?
The engines are started by using pneumatic air, usually from the APU, from an air cart or another running engine (BLEED AIR). There are two (2) separate valves that have to open for the pneumatic air to reach the starter motor:
  The START VALVE, and
  The ENGINE BLEED AIR VALVE.
They BOTH open when the START SWITCH is PULLED. During start the switch is "normally" held open by a solenoid that automatically closes at 50% N2.

The **EICAS MSG: STARTER CUTOFF ( )** means the starter valve did not close when it was supposed to. This is normally a "start" problem on the ground, but could occur also in the air.

IMHO discussion: I personally can't believe the light would just come on by itself for no reason, so I am interpreting the procedure to mean that this might occur when you are doing an in-flight re-start. The question is, why would you be doing an in-flight re-start unless you have something else going on.

THE QRC sez:

1. ENGINE START SWITCH .... PUSH IN

2. CHECK EICAS
  If STARTER CUTOUT message goes away ...
    Checklist is considered complete ... end of problem.
  If STARTER CUTOUT message still displayed ...
3. ENGINE BLEED AIR SWITCH .... OFF
Push the switch to select the **OFF** position, check the light in the switch.

THE REFERENCE sez:

ON GROUND:
Shut down the engine. Place fuel selector to the **CUTOFF** position.

IN AIR:
AVOID icing. The engine will **NOT** have any nacelle anti-ice.

... and PROCEDURES FOR STUDY and REVIEW ONLY!

One of the catastrophic failures that pilots have have nightmares about is ...

ALL ENGINES FLAMING OUT SIMULTANEOUSLY?

We probably need to have a discussion about this event ... because it is a high probability that you will have this presented to you as a problem during your training or check-ride. So, here are my thoughts regarding this event.

HOW COULD ALL 4 ENGINES FLAME OUT AT THE SAME TIME?

OMIGOSH!!!! Believe it! It has happened!!! Here are some potential ways this could (and has) occurred. By no means can we discuss every possible way this could happen, but these few comments may get your mind thinking about this potential (real) disaster. The **NUMBER ONE** way this can (and already has) happen is ...

1. VOLCANIC ASH CLOUD ENCOUNTER...

It is not likely that you would encounter volcanic ash at higher cruising altitudes. It is during climbs and descents that the greatest threat exists. My recommendation is that if you are transiting an area of reported or potential volcanic ash, even at cruise altitude; but particularly when in a descent or climb:

> **CONTINUOUS IGNITION SWITCH ....ON.**

Here is the problem. The chemical make-up of the volcanic ash cloud can cause "coking" in the burner cans, with the potential for interruption of the combustion flame ... and subsequent multiple flame-out. If the check-guy or ATC reports that there is the potential for volcanic ash, be proactive and **AVOID** the situation if possible by re-directing your path of flight. Potential encounters during a night or **IMC** descent makes it extremely difficult, or maybe even impossible to detect the **ASH CLOUD** visually.

> **VOLCANIC ASH DOES NOT DISPLAY ON WEATHER RADAR**

There may be no PIREP or NOTAM or any pre-flight or warning possible if the volcanic cloud is developing extremely rapidly without prior warning.

I will mention the next item only because of a recent "high profile" event that caused the "loss of all engines" to occur.

2. MASSIVE BIRD STRIKE ...

It is not likely that you would encounter a flock of birds of such number that they would cause the loss of all four engines; however, it is possible that in the simulator, a huge flock of migratory flamingos or swans could be ingested while climbing out of Namibia, Africa and all the engines could flame out at a low altitude. Any event at low altitude would require all your skill in finding and landing dead-stick ... especially at night-time or in IMC.

© MIKE RAY 2014
WWW.UTEM.COM

3. FUEL CONTAMINATION. YIPE!!!! How could we "possibly" determine that we have contaminated fuel on board **BEFORE TAKE-OFF** and avoid the problem. Here are some things to consider.

    A. Fuel "floats" on water. If the fuel is contaminated with water, since the fuel tank feed line is in the "**LOWEST**" part of the tank the chances are that the engines would not start or operate. However, it is conceivable, that during rotation after take-off, the fuel pumps could be exposed to contaminated fuel not initially sampled by the engine feed system.

    B. Fuel contaminated with Unknown particulate matter or the wrong "grade" of fuel is loaded. Tip-off could be "hot" start or no start. erratic or inaccurate fuel gauge indications, excessive EGT or exhaust emissions (smoke) on start, and so forth. It is essential that we include the potential for FUEL CONTAMINATION PROBLEMS" when evaluating unusual start situations.

If there is "erratic" or unexpected readings on the fuel tank indicators, then contaminated fuel load could be suspect.

4. FUEL ICING and sunsequent **CLOGGED FUEL FILTERS**. There is the chance that the fuel could get so cold, it could form ICY FUEL PARTICLES that will restrict the flow of fuel through the filters. Fuel temperature guidelines are established. The Flight Operation Manual states:

> "During cruise, particularly in areas with **AIR TEMPERATURE -65 DEGREES C OR LESS**, pilots must remain alert to ensure that the **OPERATIONAL FUEL TEMPERATURE LIMIT** is not exceeded."

The limit is defined as: "If the measured freezing point of the loaded fuel is known, the <u>**OPERATIONAL LIMIT is 3 DEGREES C ABOVE THAT LIMIT**</u>."

<u>**"MINIMUM FUEL TEMPERATURE for TAKE-OFF is -43 DEGREES C.**</u>"

It is also worthy of comment, that while the potential for an engine shutdown exists; there is this quote from the Flight Operations Manual of a major airline:

> "There have been no known civil jet aviation engine shutdowns due to operations with fuel temperatures below the fuel freeze specifications limit."

RVSM - Revised Vertical Separation Minima means that coming right at you, a mere 1000 feet below, is a string of other airplanes right on the magenta line. Also, remember that your airspeed is probably decreasing and a stall is imminent. So, consider:
1. **GET OFF THE AIRWAYS, and**
2. **START YOUR DRIFT-DOWN.**

Obviously it is not possible to cover everything in this simple discussion, so "WHEN" the event occurs. keep you wits about you and handle the situation with as much professionalism as you can.

ALWAYS STEP NUMBER ONE:

PF DOES ONE THING
FLY THE AIRPLANE !!!

... and PROCEDURES FOR STUDY and REVIEW ONLY!

MULTIPLE ENGINE FLAMEOUT or STALL ( )

THERE ARE NO IMMEDIATE ACTION STEPS - GO TO QRC.

THINK THIS COULDN'T HAPPEN? THINK AGAIN!
DISCUSSION:
(1) There actually was a -400 that lost **ALL** it's engines while transiting a volcanic ash cloud on descent into Alaska. They got 'em started again (Thank God!) and were able to make an emergency landing.
(2) In another incident, a -400 crew lost an engine during the enroute phase. The airplane was pretty heavy and was right at that point where it was struggling to maintain altitude. The crew kept the airplane at altitude. They did not initiate a timely descent and eventually the airspeed decreased and the jet stalled. They did a 22,000 foot high-dive. The airplane entered denser air and they were able to recover. This tells us some things.
First, sometimes we can get complacent because the -400 has so many motors and so much reliable power that a significant loss of that power seems unthinkable.
Second, while at altitude, it seems so unbelievable that the airplane cannot stay there indefinitely. However, in the loss of power problem, we simply MUST consider a DRIFT-DOWN procedure if applicable.

DO NOT BE CONFUSED!!!

If the airplane is in a descent, the cockpit indications **LOOK LIKE** an
ELECTRICAL POWER FAILURE!

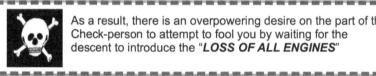

As a result, there is an overpowering desire on the part of the Check-person to attempt to fool you by waiting for the descent to introduce the "**LOSS OF ALL ENGINES**"

THE DISCUSSION CONTINUES:
Assuming the worse case (**LOSS OF ALL ENGINES**); when the airplane is without engine generators (and this assumes that the "windmilling" effect doesn't drive the generators sufficiently) there will be only battery power available.
This means there will be:
          NO EICAS for engine EGT indications, and
          ENGINE IGNITION
                will be supplied from the MAIN STANDBY BUS
                when the FUEL CONTROL switches are in RUN, and
          NO AUTOPILOTs, and
          AUTO-THROTTLES, and
          BUNCHES OF OTHER STUFF GOES AWAY.

In other words, as the motors wind down and stuff starts shutting down, it is going to be confusing and virtually impossible to determine the nature of the problem. It will look a lot like a total electrical power failure of some sort.

REMEMBER: **FLY THE AIRPLANE !!**

If some engines are working and the autopilot is available, use it, If not
Grab the yoke, get your feet on the rudders and start looking at the standby instruments, because "YOU GOT IT."
Prepare to maintain 200 plus knots and turn the airplane off airways.

MULTIPLE ENGINE FLAMEOUT
or STALL ( ) continued

THERE ARE NO IMMEDIATE ACTION STEPS - GO TO QRC.

THE QRC sez:

1 CONTINUOUS IGNITION SWITCH .... ON

DEPRESS the light switch,
OBSERVE the ON light illuminates.

2 FUEL CONTROL SWITCH(es) .... *CONFIRM* engine with PF, then CUTOFF

Keep your fingers on the switch and When EGT starts to decrease:

If the EICAS' are blank, use this technique. After a suitable delay, ASSUME that the temperature is decreasing and proceed to the next step.

The idea is to shut down the fuel control lever for all the engines that are flamed out, even if it is all four. Don't do one at a time. We are initially trying to start ANY engine, in the event that all four are flamed-out.

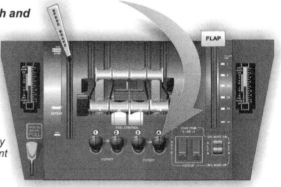

3 FUEL CONTROL SWITCH(es) .... RUN

NOTE: If unable to get engines running, go to the:

GO TO THE REFERENCE STEPS:
REMINDER

PF FLIES THE AIRPLANE ... ONLY !!!

NOTE:
If you were thinking about starting the APU ...
here is some things to consider.

APU STUFF:
The APU **CANNOT** be started in flight.
If already started, it can run in the air up to 20,000 feet.
In this case, BLEED AIR can be used to power 1 (ONE) pack up to 15,000 feet.
The APU **CANNOT** supply electrical power IF the airplane thinks it is airborne.

So much for the APU idea.

MULTIPLE ENGINE FLAMEOUT
or STALL ( ) continued on next page

MULTIPLE ENGINE FLAMEOUT or STALL ( ) continued
REFERENCE STEPS

4 AUTO IGNITION SELECTOR .... BOTH

IF AIRSPEED IS LESS THAN 200 KTS:

5 ENGINE START SWITCH .... PULL

IF ENGINE(S) FAILS TO RESTART:
NOTE: Restart is indicated by EGT RISE within 30 seconds:

6 FUEL CONTROL SWITCH .... CUTOFF

7 THROTTLE .... IDLE

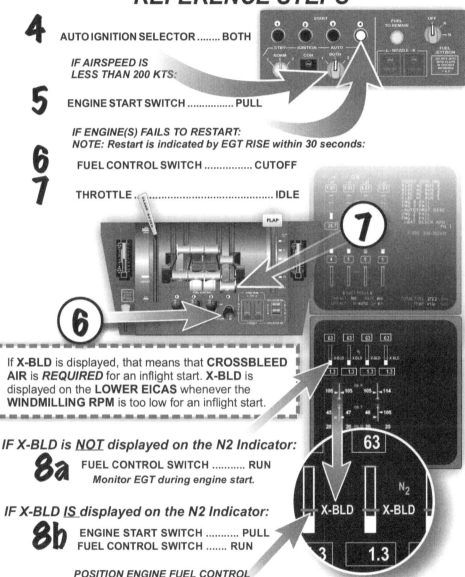

If **X-BLD** is displayed, that means that **CROSSBLEED AIR** is *REQUIRED* for an inflight start. **X-BLD** is displayed on the **LOWER EICAS** whenever the **WINDMILLING RPM** is too low for an inflight start.

IF X-BLD is NOT displayed on the N2 Indicator:

8a FUEL CONTROL SWITCH .... RUN
   Monitor EGT during engine start.

IF X-BLD IS displayed on the N2 Indicator:

8b ENGINE START SWITCH .... PULL
   FUEL CONTROL SWITCH .... RUN

   POSITION ENGINE FUEL CONTROL TO RUN when N2 exceeds FUEL-ON indicator.
   Monitor EGT during engine start.

IF ANY ENGINE FAILS RESTART ATTEMPTS:
   DO THE ENGINE ( ) FAIL IRREGULAR PROCEDURE

747-400 SIMULATOR TECHNIQUES ...

HIGH ENGINE EGT / COMPRESSOR STALL

THERE ARE NO IMMEDIATE ACTION STEPS - GO TO QRC.
THE QRC sez:

1 THROTTLE .... CONFIRM, IDLE

*The procedure calls for placing the throttle lever **ALL THE WAY TO IDLE**.*

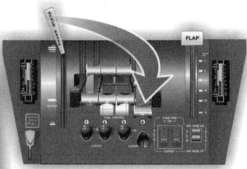

If the engine indications are stabilized or EGT decreases.

2 THROTTLE .... ADVANCE

Advance the throttle slowly and observe that RPM and EGT follow throttle movement. Note that under certain circumstances, rpm may increase very slowly.

3 ENGINE OPERATION .... MONITOR

Go ahead and operate engine normally or at a reduced thrust level where the indications are stable and EGT is within limits.

If the engine indications are abnormal or
if EGT continues to increase or exceeds the EGT limit:

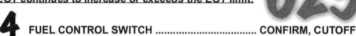

4 FUEL CONTROL SWITCH .... CONFIRM, CUTOFF

If there is no apparent damage to the engine:

5 ENG INFLIGHT START IRREG PROCEDURE (7-31) .... ACCOMPLISH

NOTE:
If you exceeded the EGT limit, it doesn't preclude you from attempting a restart. However, obviously, the engine should continue to be closely monitored.

IF ENG DOES NOT RESTART:

6 TRANSPONDER .... TA ONLY.

NOTE:
Taking the Transponder out of RA and placing it in TA prevents climb commands that would exceed the performance capability.

7 *NOTE: Complete and submit the "FLAMEOUT / COMPRESSOR STALL / THRUST LOSS report.*
Complete and submit a "CAPTAIN'S REPORT."

... and PROCEDURES FOR STUDY and REVIEW **ONLY!**

EVACUATION

THERE IS AN IMMEDIATE ACTION STEP

1 **ADVISE ATC** (*Tower, ground, whomever is in control*).

2 set **PARKING BRAKE**
Setting the parking brake with the airplane in motion denies the use of the anti-skid and could lead to blown tires.

3 retract **SPEED BRAKES**
IMPORTANT: If not retracted, the use of the wings as potential escape route is compromised.

4 **FUEL CONTROL switches** (ALL) CUTOFF.
Unless APUs are operating (not likely after a landing) there will be a reversion to battery power.

5 **PRESS. OUTFLOW VALVES ... OPEN.**
The PRESSURIZATION should have relieved automatically. We back it up because with residual pressure, the doors will NOT open ... YIPE!

6 **EVACUATION INITIATE**
Once you initiate the EVAC ALARM; it is OK to silence the warning horn in the cockpit. The warning will continue to sound at each individual station until extinguished at that site.

7 **ENGINE and APU FIRE HANDLES ... OVERRIDE, PULL.**
There are four ways to release the handles:
  If a FIRE WARNING exists,
  If FUEL CONTROL LEVER in CUTOFF,
  FIRE/OVHT test switch is pushed,
  Manually push release button under handle.

8 **IF FIRE HANDLES ILLUMINATED ... ROTATE**

CREW MEMBER'S RESPONSIBILITIES

CAPTAIN: *IMMEDIATELY AFTER AIRPLANE STOPS*:
Accomplish shutdown, Go to cabin and exercise **OVERALL COMMAND INSIDE THE AIRPLANE.**

FIRST OFFICER: *IMMEDIATELY AFTER AIRPLANE STOPS*:
Accomplish cockpit shutdown, Go to cabin and determine that all usable exits are open. After you have helped as much as possible, **LEAVE THE AIRPLANE.**

If you can't get to the main cabin, leave the airplane using the escape slides and inertia reels.
Once on the ground; **ASSEMBLE THE PASSENGERS** upwind a safe distance from the airplane and approaching rescue vehicles.

747-400 SIMULATOR TECHNIQUES ...

ENGINE FIRE ( ) or SEVERE DAMAGE or SEPARATION

PF: FLY THE AIRPLANE

MASTER WARNING LIGHTS and BELL, ENGINE FIRE SHUTOFF HANDLE, and FUEL CONTROL SWITCH

THERE ARE NO IMMEDIATE ACTION STEPS - GO TO QRC.

Depressing the warning light on the glare shield will extinguish the bell and reset the warning system.

WHAT RELEASES the handle:
1. FIRE WARNING
2. FUEL CONTROL in CUTOFF
3. FIRE/OVHT test switch
4. MANUALLY using button under handle

THE QRC sez:

1 THROTTLE .... CONFIRM, And then pull to IDLE

CONFIRM-CONFIRM-CONFIRM.

SMOOTHLY retard the throttle so that the auto-pilot is able to control any yaw that may be induced.

2 FUEL CONTROL SWITCH(es): .... CONFIRM, and then Pull up and back to CUTOFF

3 ENGINE FIRE HANDLE: ..... CONFIRM, PULL

Pulling the handle **CLOSES**:
**FUEL,
AIR ,
ELECTRICAL,
HYDRAULICS,**
and **ARMS EXTINGUISHER**

4 IF FIRE MESSAGE DISPLAYED: FIRE HANDLE..... ROTATE

5 IF FIRE MESSAGE REMAINS After 30 seconds: FIRE HANDLE..... ROTATE IN OPPOSITE DIRECTION

GO TO THE REFERENCE STEPS:

page 370

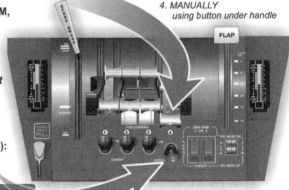

© MIKE RAY 2014
published by UNIVERSITY of TEMECULA PRESS

... and PROCEDURES FOR STUDY and REVIEW ONLY!

continuing ENGINE FIRE ( ) or SEVERE DAMAGE or SEPARATION

HERE ARE THE REFERENCE STEPS:

IF "HIGH" AIRFRAME VIBRATION OCCURS:

6 AIRSPEED and/or ALTITUDE .... REDUCE.

Vibration may be reduced by changing airspeed / altitude. If, however, reducing didn't work ... consider increasing airspeed.

NOTE:
It may seem that the "**SEVERE**" vibration is about to shake the airplane to pieces, but the Boeing guys tell us that it is "**UNLIKELY**" that the vibration will damage the airplane or critical systems.
Of course, they were sitting at their engineer's desk when they made that statement.

7 TRANSPONDER .... TA ONLY.

DISCUSSION:

The idea here is to avoid some spurious RA. I **DISAGREE** with that for two reasons.

FIRST: It is common for the 747-400 to be cruising at it's highest performance altitude commensurate with fuel considerations. This would preclude any "CLIMB-CLIMB" RAs from being responded to anyway. Further, if the airplane is cruising in RSVM, it becomes **CRITICAL** to avoid contact with another airplane as the drift-down starts.

SECOND: The simulator seldom takes into account the fact that an airplane in distress will most likely be transiting both tracks and altitudes assigned to other airplanes. In reality, if in MNPS, or NCA, or other Oceanic airspace there will be no ATC radar support and it is up to the pilots of the airplane in distress to avoid other airplanes.

My argument is for the pilots to assess the situation BEFORE shutting off the RA on the transponder.

8 IF FIRE CONTINUES:

> ## LAND AT NEAREST SUITABLE AIRPORT !
> **DUH !** I guess they put that in there for brain surgeons who are disguised as pilots.

9 THREE/TWO ENGINE
APPROACH AND LANDING CHECKLIST (Flight Handbook 7-92)
.... COMPLETE.

... and, of course, the MOST IMPORTANT step in this procedure:

10 Complete a FLAMEOUT/COMPRESSOR STALL/THRUST LOSS REPORT.

747-400 SIMULATOR TECHNIQUES ...

FIRE APU

THERE ARE NO IMMEDIATE ACTION STEPS - GO TO QRC.

THE QRC sez:

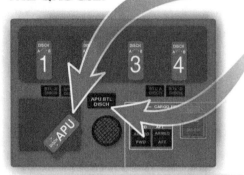

1 APU FIRE HANDLE:
..... PULL and ROTATE

2 VERIFY:
APU BOTTLE DISCHARGE LIGHT is ON.

NOTE:
Rotating the handle the other way to discharge the "other" bottle just serves to show that you don't know what you are doing. There is **ONLY ONE BOTTLE** for the APU(s). Turning the handle either way works.

DISCUSSION:
I cannot think of a reason why the APU(s) would ever be operating with the airplane airborne. The APU CANNOT be started in the air and CANNOT supply electrical energy once airborne. if the APU was left running at take-off, it can only supply enough air to operate ONE PACK up to 15,000 feet.

So ... the situation addressed in this procedure seemingly would apply to
ON THE GROUND operations ONLY!

THEREFORE ... I SUGGEST:

 GET THE PASSENGERS OFF THE AIRPLANE.
Coordination with the Passenger Agent and/or the Flight Attendants and use the Jetway if available. Use the slides if appropriate, but stop the loading process and minimize the exposure of the passengers to a potential fire on board the aIrplane.

4 NOTIFY THE FIRE FIGHTERS
Use the radios to notify the Tower or the Company so that they can coordinate with the Fire Department.

PULLING THE APU FIRE HANDLE does:

SHUTS DOWN APU, bypassing the 60 second cool-down.

CLOSES APU FUEL VALVE.

CLOSES APU BLEED AIR VALVE

TRIPS APU GEN FIELD AND BREAKER

ARMS APU FIRE EXTINGUISHER

SILENCES HORN in WHEEL WELL.

... and PROCEDURES FOR STUDY and REVIEW ONLY!

STAB TRIM UNSCHD

Uncommanded stabilizer motion is detected and automatic cutout does not occur.

THERE IS AN IMMEDIATE ACTION STEP

1 CONTROL COLUMN .... MOVE TO OPPOSE TRIM

SCENARIO: You are cruizin' along at altitude and you perceive that the trim wheel is turning and the trim is running away. You also see that the **YOKE IS MOVING TOWARDS YOUR LAP**.

Question: QUICK, which way do you move the yoke to "oppose the trim."

Here's the problem. The intuitive reflex is to grab the yoke and "OPPOSE" it's *motion*, that is to push *IT* towards the instrument panel.

"*Is that your final answer?*"

Let's think about this for a moment. The autopilot is tied to the yoke, that is, when the autopilot commands a turn or climb, the yoke moves in response to the autopilot input. So, in this case, with the trim motor running away, the autopilot (controls) are moving to oppose the pressures imposed by the trim.

So, here is the REAL answer, **YOU MAY HAVE TO MOVE THE YOKE IN THE SAME DIRECTION IT WAS MOVING IN ORDER TO "HELP" THE AUTOPILOT OPPOSE THE TRIM MOVEMENT**.

The problem is, by the time you figure out that you have been pushing the yoke the wrong way, the airplane may be outside of the flight envelope where recovery is possible.

2 STAB TRIM SWITCHES .... CUTOUT

HERE ARE THE REFERENCE ITEMS

3 AUTOPILOT .... DISENGAGE

It is UNDER STATEMENT to say that there will be "HIGHER THAN NORMAL" column forces when the autopilot is disengaged. Particularly if you were pushing in the wrong direction, you will have a handful of yoke pressure.

 STAB TRIM SWITCHES (one at a time) .... AUTO
  Check for correct stabilizer movement.
  Trim is available after a short period.
  Trim rate is one half normal rate.

 IF UNSKED STAB TRIM reoccurs .... CUTOUT
  Repeat procedure on other switch.

IF unable to re-establish stabilizer trim:

Treat the situation as a JAMMED STABILIZER LANDING AND IRREGULAR PROCEDURE (7-70) situation.

FIRE CARGO FORWARD (AFT)

THERE ARE NO IMMEDIATE ACTION STEPS - GO TO QRC.

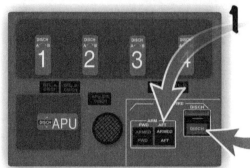

1 CARGO FIRE FORWARD/ AFT ARM SWITCH:.......... PUSH

2 CARGO FIRE DISCHARGE SWITCH: .... PUSH

NOTE:
The cargo fire warning lights will remain on until the fire is extinguished and the smoke has dissipated.

THE REFERENCE ACTION ITEMS:

3 PACK CONTROL SELECTORS: .... PACK 1 or PACK 2 OFF.

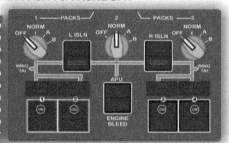

DISCUSSION:
The idea is to shut off all but 1 pack to reduce airflow to the cargo compartments. Here is a **747-400 MYSTERY.**
When you have all three packs running, and you shut down either Pack #1 or Pack #2, ... Pack # 3 will shut down automatically.

4 LANDING ALTITUDE SWITCH: .... PULL OUT (MANUAL).

5 LANDING ALTITUDE CONTROL: .... SET 8000-8500 feet.

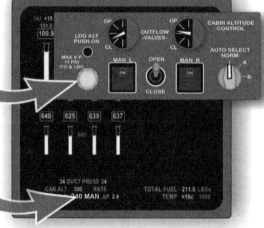

When you pull out the switch, it activates the manual mode. You twist it to set the landing altitude which is shown on the EICAS followed by a "**MAN**."

MORE REFERENCE ACTION ITEMS:

FIRE CARGO FORWARD (AFT) cont.

LAND ASAP
LAND AT NEAREST SUITABLE AIRPORT:

NOTE:
The decision as to where to park the airplane is a difficult one and I SUGGEST STRONGLY that you get on the radio and get DISPATCH involved as early as possible. They can do wonderful things for you. ATC is good ... but DISPATCH is WONDERFUL !

PRIOR TO STARTING DESCENT:

 LANDING ALTITUDE SWITCH: .... PUSH IN (AUTO).

DISCUSSION:
In auto, while in the descent, this will maintain a differential pressure for equipment cooling and fire fighting agent concentration.

ON THE GROUND:

Next time you are out over the Pacific Ocean and *R-E-A-L-L-Y* bored, take the logbook with the metal cover and turn to that little red book in the back.

> **ICAO publication DOC 9481-AN/928 "Emergency Response Guidance for Aircraft Incidents involving Dangerous Goods"**
> *(That red booklet stuck in the back of the Logbook)*

IT WILL TELL YOU:

FIRST: If you have a cargo compartment fire, the Fire personnel are trained to ask for the Hazardous Materials Report BEFORE they open the compartment and fight the fire.

SECOND: The Firefighters expect you to evacuate the aircraft BEFORE they open the Compartment.

> **WARNING**: Even though all indications of fire may be negative and the fire may actually be out, they are concerned that the sudden influx of fresh air may ignite residual combustible material and cause the airplane to blow up.

So, (and these are my interpretations) Keep these things in mind during your checkride:

1. Have hazardous materials report in hand when you evacuate.
2. Park airplane at a "remote safe location," **NOT AT THE GATE**!
3. *Always* Evacuate the airplane even though the fire seems out.
4. Anticipate fire/explosion after compartment is opened;
   i.e. get everybody away from the airplane.

GALLEY FIRE/SMOKE

THERE ARE NO IMMEDIATE ACTION STEPS - GO TO QRC.
THE QRC sez:

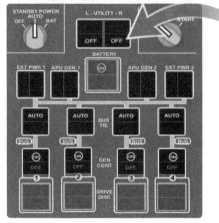

1 UTILITY BUS SWITCHES: .... OFF

HERE ARE THE REFERENCE ACTIONS:

IF THE FIRE/SMOKE IS EXTINGUISHED:

2 UTILITY BUS SWITCHES (one at a time): .... ON

DISCUSSION:
Right here is a step that I would be only marginally motivated to perform. Depending on the location of the airplane and the distance to the closest suitable airport, I may consider cancelling the inflight service and giving away food vouchers. Unless I can **POSITIVELY** determine the location and cause of the fire, I cannot think of any reason to troubleshoot a galley fire just so I can restore electrical power to the cabin for food services.

IF THE FIRE/SMOKE CONDITION RECURS AND IT CAN BE DETERMINED WHICH UTILITY BUS IS THE BAD ONE:

3 ASSOCIATED UTILITY BUS SWITCHES: .... LEAVE OFF.

IF THE FIRE/SMOKE BECOMES UNCONTROLLABLE.

4 VACATE THE AREA AND *CONSIDER* AN IMMEDIATE LANDING OR DITCHING.

It is my opinion that this situation warrants **SERIOUS** treatment. I consider troubleshooting which bus is the cause is **EXCESSIVE**. While proceeding to destination may certainly be an option, but if the fire/smoke is significant enough to warrant an emergency response, I would at least talk with dispatch and SAMC and consider a divert to intermediate airport.

FIRE WHEEL WELL

THERE ARE NO IMMEDIATE ACTION STEPS - GO TO QRC.

THIS IS AN AUTOPILOT/AUTO THROTTLE ON PROCEDURE.

1. PUSH ON THE MASTER WARNING LIGHTS.
   This will silence the warnings and re-set the warning system.

2. DECLARE AN EMERGENCY AS SOON AS ABLE.
   This will allow deviation from airspeed requirements from ATC or below 10,000 feet.

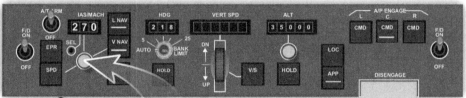

3. SLOW TO 270 KIAS/.82 M
   STEP 1: PUSH the airspeed control knob on the MCP.
   STEP 2: SET 270 knots/.82 Mach

4. LANDING GEAR .... DOWN

> Slow the jet expeditiously but smoothly.
> I suggest this thought process:
> 1. SET 270 on MCP SPEED WINDOW
> 2. SPEED BRAKE TO FLIGHT DETENT
> 3. At 270 KTS, GEAR DOWN
> 4. SPEED BRAKE UP
> 5. confirm that THROTTLES ADJUST POWER.

270 knots/.82 Mach is the maximum airspeed limitation for the extention and the retraction of the landing gear. Once the gear has been extended, the airspeed limitation is 320 Klts/.82 MACH

HERE ARE THE REFERENCE ACTIONS:

5. <u>LAND AT THE NEAREST SUITABLE AIRPORT:</u>

> **CAUTION:**
> 1. LEAVE THE LANDING GEAR DOWN FOR A MINIMUM OF 20 MINUTES AFTER THE EICAS GOES OUT.
>
> 2. CONSIDER RETRACTING THE GEAR ONLY IF NECESSARY TO REACH THE NEAREST SUITABLE AIRPORT.
>
> 3. IF A GO-AROUND OCCURS, CONSIDER LEAVING THE GEAR DOWN TO PREVENT FURTHER DAMAGE.
>
> 4. TERMINATION OF THE WHEEL WELL FIRE WARNING DOES NOT NECESSARILY MEAN THAT THE FIRE IS OUT.

SMOKE/FUMES/ODOR

THERE ARE 2 IMMEDIATE ACTION STEPS

1 OXYGEN MASK and REGULATORS .... ON, 100%.

2 CREW COMMUNICATIONS .... ESTABLISH.

3 SMOKE GOGGLES (IF REQUIRED) .... ON.

> **WARNING:**
> Leave cockpit door
> **CLOSED**
> unless required by procedure or greater emergency.

HERE ARE THE REFERENCE ACTIONS:

4 SMOKE/FUMES/ODOR source .... DETERMINE.

If smoke source CAN be determined:

5 ACCOMPLISH appropriate IRREGULAR PROCEDURE

- Air Conditioning Smoke (7-1)
- Electrical Smoke or Fire (7-24)
- Lavatory Smoke or Fire (7-72)
- Smoke/Fumes/Odor Removal (7-87)

> **NOTE:**
> The procedure referred to as "**SMOKE/FUMES/ODOR**" in the QRC is different from "**SMOKE/FUMES/ODOR REMOVAL**" in the IRREGULARS.

If smoke source CANNOT be determined:

6 **LAND AT THE NEAREST SUITABLE AIRPORT:**

... and PROCEDURES FOR STUDY and REVIEW ONLY!

IAS DISAGREE/ AIRSPEED/MACH UNRELIABLE

THERE ARE NO IMMEDIATE ACTION STEPS ... GO TO QRC.

QUICK CHECK TECHNIQUE:
CROSSCHECK between Captain and First Officers and Standby instruments. Use the three to resolve ambiguity about which instruments are unreliable.

---

1 DISENGAGE AUTOPILOT
   Use the YOKE SWITCH

2 AUTOTHROTTLES OFF
   Use the MCP panel OFF switch.

3 FLIGHT DIRECTORS OFF
   Use the MCP panel OFF switch.

4 ATTITUDE and THRUST adjust.
   Set the power using the **PITCH-EPR** chart.

If the airspeed indications are unreliable ...

5 LAND AT THE NEAREST SUITABLE AIRPORT:
   Regarding the concept of what is a **SUITABLE AIRPORT**. With the situation that might require shooting an approach without any indication of airspeed, it might be useful to select a **VMC** location with a **LONG RUNWAY**, etc. Selecting a destination airport that had an **ILS** would be extremely useful.

DISCUSSION:

The most challenging part of this problem may be the APPROACH and LANDING. While it is assumed that the loss of airspeed indication occurs in isolation, the pilots may also be faced with other pressing airframe and powerplant issues that have occurred. So, in this problem, I think it is useful to select a diversion airport that has a long runway and an ILS. If weather is a factor, then it my be useful to include that as a factor and determine the best choice for a positive outcome.

Bring the ATC controller and the Airline Dispatcher into the problem. Once clearly defined, the solution to the larger problem of getting the airplane safely on the ground may include the use of outside resources. This will definitely earn you points with the Check-Guy.

747-400 SIMULATOR TECHNIQUES ...

IAS DISAGREE/ AIRSPEED/MACH UNRELIABLE

THERE ARE NO IMMEDIATE ACTION STEPS ... GO TO QRC.

DISCUSSION:

This failure is an extremely **CRITICAL** and potential **DEADLY** situation. In the history of aviation, there have actually been some instances where an otherwise perfectly good airplane crashed! Early detection, proper recognition, and the timely exercise of a solution by the crew is essential. If you can determine that you have failures that will affect this issue during the pre-flight checks and **_BEFORE_** starting the take-ff evolution, this is the obviously best solution. While not the only potential source of a problem, the most common oversight by the crew is a failure to ensure that the **PITOT-STATIC PROBE** heat system is powered and operating properly. This "could" potentially become a MAJOR PROBLEM during an otherwise routine flight.

There are two venues where **PITOT STATIC** failures are a problem:
    FIRST, right after take-off, and
    SECOND, Icing or system failure in the flight phase.

Let's take a quick look at the "**PROBE HEAT**" system (Called lovingly the "PETER HEATER" by the crews).
On the 747-400, the operation of the **PROBE HEAT SYSTEM** is **_AUTOMATIC_**.
The **PITOT-STATIC, PT2, TT2, EPR** probes, and the **ANGLE OF ATTACK** sensors are **_ELECTRICALLY HEATED WHENEVER ANY ENGINE IS OPERATING_**.
The **TAT** probes are electrically heated **ONLY IN FLIGHT**.

Failures of the systems that could result in an **IAS** disagreement or an **AIRSPEED/MACH UNRELIABLE** problem are indicated by a **YELLOW EICAS** warning message!
Here is a list of the possible warning **EICAS** messages:

```
AILERON LOCKOUT
>AIRSPEED LOW
HEAT L AUX
HEAT P/S CAPT
HEAT P/S F/O
>OVERSPEED
RUD RATIO DUAL
RUD RATIO SNGL
```

The cause of a problem "may be" associated with the pitot static system being taped shut. This could occur during painting or washing the fuselage when the personnel will place tape over the outlets to keep out the paint or cleaning materials. it is particularly difficult to see on the walk-around if the tape is painted over.
BE READY ON THE CHECK-RIDE FOR THIS!

The airborne situation is more difficult to discern. However, **PITOT-STATIC** blockage (usually icing) while airborne is generally associated with icing conditions.

OTHER ISSUES

- **AIRSPEED/MACH** failure flags visible.
- **AIRSPEED** indicator blanking or fluxuating.
- **PFD AIRSPEED** indicator turns **AMBER**.
- **CAPTAIN** and **F/O AIRSPEED INDICATORS** don't match within **_5 KNOTS_**.
- **OVERSPEED** indications.
- **AMBER LINE** displayed "through" **PFD FLIGHT MODE ANNUNCIATIONS**
- simultaneous **OVERSPEED** and **STALL** warnings.
- **RADOME DAMAGE** (Lightning strike?)

If **AIRSPEED** and **ALTITUDE** information not consistent with the pitch attitude and thrust settings. Consult the "**PITCH/EPR**" chart for that information. I have put an abbreviated "**PITCH/EPR**" chart on the next page to give a feel for what is being referred to.

IAS DISAGREE/ AIRSPEED/MACH UNRELIABLE

GREATLY SIMPLIFIED
AIRSPEED UNRELIABLE CHART
(Called the "PITCH/EPR" chart)

	PITCH	EPR	TARGET SPEED
CLIMB	5° to 9° UP	MAX CONT	290 Kts
LEVEL FLIGHT	2° to 3° UP	1.2 to 1.4	.84 M
DESCENT	1° UP to 2° DN	IDLE	300 Kts
MANEUVERING	9° UP	1.08	150 Kts
APPROACH	1.3° UP	1.07	140 Kts

DISCUSSION:
The idea is that we can fly an airplane like the 747-400 without any reference to the **AIRSPEED** indicator by using the **PITCH** and **POWER SETTINGS** (Either **EPR** or **N2**). These numbers are ballpark targets, just to give you something to shoot for until you can get the actual numbers from the more complex **"REAL" PITCH/EPR** chart in your Pilots Handbook.

If you should experience an airspeed indicator failure, chances are that doing "nothing" to the power or pitch will likely maintain the airplane in controlled flight. Looking at the numerical values, you can see that the values are actually quite small ... indicating that rapid or LARGE changes in **PITCH** or **THRUST** settings are not required to maintain control the airplane.

CONSIDER THIS:
Be aware that the Air Traffic Controller has the capability of monitoring your airspeed ... and is more than willing to share that information with you. This is particularly useful during the approach when you are uncertain of the flaps speeds.

CABIN ALTITUDE/ RAPID DEPRESSURIZATION

THERE ARE 2 IMMEDIATE ACTION STEPS

1 OXYGEN MASK and REGULATORS .... ON, 100%.

2 CREW COMMUNICATIONS .... ESTABLISH.

3 ALL 3 PACK SELECTORS .... NORMAL

4 ISOLATION VALVES (BOTH) .... CLOSE

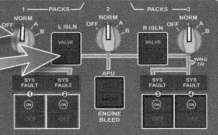

5 PRESSURIZATION OUTFLOW VALVES .... CLOSE

A: FIRST PUSH BOTH LIGHT SWITCHES.
B: THEN HOLD SWITCH TO CLOSE.
C: OBSERVE VALVE IND CLOSED.

IF CABIN ALTITUDE IS UNCONTROLLABLE:
DO AN EMERGENCY DESCENT

6 CABIN SIGNS .... ON

IF CABIN ALTITUDE IS CONTROLLABLE BUT above 14,000 feet.
PASSENGER OXYGEN SWITCH .... ON

IF CABIN ALTITUDE IS CONTROLLABLE BUT above 10,000 feet.
CREW OXYGEN REGULATORS .... NORMAL
   Place both regulator levers to the N position to conserve oxygen.
PURSER .... ADVISE.
   Inform the Purser of what has happened and what may be expected.

   After cabin has stabilized at 14,000 feet or below, advise the Puser of the cabin altitude, when they may leave their seats, and when oxygen masks may be removed. Consider using the PA.

AFTER AIRPLANE LEVELS OFF

CAPTAIN MAKE ANNOUNCEMENT:

NOTE on ANNOUNCEMENT:
The MOST important thing is that you modulate your voice, speak slowly and distinctly, and project an air of confidence.
Tell them what happened and explain what is going on.

AFTER cabin altitude descends below 9500 feet

PASSENGER OXYGEN SWITCH .... RESET.

NOTE:
Since this configuration isolates the CENTER BLEED DUCT, an EICAS msg **TEMP ZONE** is displayed.

ALL PASSENGER ZONES will be maintained at **75 degrees F**.

EICAS msg CARGO DET AIR because of insufficient airflow for cargo smoke detector.

IF cabin altitude CAN BE controlled and BOTH DUCT PRESSURES are normal:

PACK 2 CONTROL SELECTOR .... OFF.

IF cabin altitude CAN BE controlled but ONE DUCT PRESSURE remains low:

ENG BLEED AIR (affected side)
................OFF

ISOL VALVE (normal side)
................OPEN

PACK SEL (affected side)
................OFF

BOTH HYD DEMAND PUMP (affected side)
................OFF

WING ANTI-ICE
....................................OFF

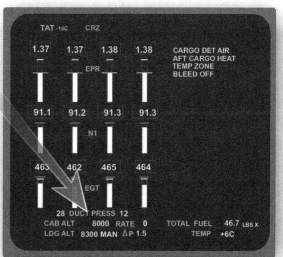

NOTE:
DO NOT use WING ANTI-ICE.
Sufficient bleed air may not be available.

EMERGENCY DESCENT

LEAVE AUTOPILOT ON !

The reasoning goes like this: Should the crew become incapacitated at some point during the procedure, the airplane will continue and level off on it's own.

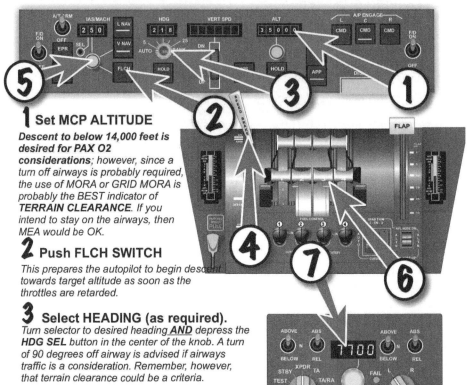

1 Set MCP ALTITUDE
Descent to below 14,000 feet is desired for PAX O2 considerations; however, since a turn off airways is probably required, the use of MORA or GRID MORA is probably the BEST indicator of **TERRAIN CLEARANCE**. If you intend to stay on the airways, then MEA would be OK.

2 Push FLCH SWITCH
This prepares the autopilot to begin descent towards target altitude as soon as the throttles are retarded.

3 Select HEADING (as required).
Turn selector to desired heading **AND** depress the **HDG SEL** button in the center of the knob. A turn of 90 degrees off airway is advised if airways traffic is a consideration. Remember, however, that terrain clearance could be a criteria.

4 Extend SPEED BRAKES
Expeditiously - **BUT SMOOTHLY** - pull speed brake lever to the flight detent. Rapid jerking of the handle could de-stabilize the airplane and result in upset. The idea is to descend to an altitude where pressurization considerations are no longer a factor.

5 Set AIRSPEED ON MCP to Vmo.

SELECT Vmo. The jet will likely be using **MACH** number, so depress the little black **SEL** button on the **MCP** and set the highest allowable airspeed on the **PFD**, just below the high speed stall indicators (red bricks). If structural integrity is in doubt ... limit airspeed and avoid high maneuvering loads.

6 THROTTLES to IDLE (*USE CAUTION*)
Move the throttles to IDLE in a FIRM AND AGGRESSIVE manner; **BUT DO NOT JERK THE THROTTLES AT ALTITUDE**.

7 TRANSPONDER 7700

8 ATC advise

MAYDAY-MAYDAY! This is an **URGENT** situation. Obtain heading information for terrain and traffic avoidance and divert information if necessary. Treat this event as an **EMERGENCY**!

EMERGENCY DESCENT continued

REFERENCE ACTION

9 CONT IGNITION switch ... ON

10 CABIN SIGNS ... ON

IF CABIN ALTITUDE above 14,000 feet:

11 PASSENGER OXYGEN switch ... ON

> **OXYGEN SYSTEM REVIEW DISCUSSION:**
> With the system in NORMAL, when the cabin altitude reaches 14,000 feet, the masks are automatically deployed. Manually switching the system to ON will ensure activation of the mask deployment and if below 14,000 feet will result in a reduced rate of O2 flow proportional with altitude.
> All normal flight plans take into account the fuel requirements for a decompression. However, that fuel is predicated on:
> > **ALL ENGINES** operating
> > **14,000 feet**
> > **Long Range Cruise**
> > **EQUI-TIME** point (most extreme position) to overhead designated diversion airport with **NO RESERVES**.
> > **Predicated on 18,000 feet winds.**
>
> Where fuel considerations are a factor, descent should be limited to that altitude where a minimum 14,000 foot **CABIN ALTITUDE** can be maintained.

12 ALTIMETERS (at trans level) SET BAROMETRIC PRESS.

APPROACHING LEVEL/OFF ALTITUDE:

ABOUT 3000 ABOVE DESIRED LEVEL/OFF.

13 monitor DESCENT RATE

EXTRA CREDIT: Use the **MCP** tools, such as using the **V/S** knob on the **MCP** to reduce the Vertical Speed. Generally speaking, this is NOT a problem area and strictly a suggestion to reduce the level off "**G**" forces.

14 retract SPEED BRAKES **DON'T FORGET!**

BE SMOOTH! Slamming the handle to the forward stop "could" produce pitch moments that if structural considerations exist could result in a potential upset ... or worse.

15 set AIRSPEED as desired.

This will allow the nose of the airplane to rise slowly, making a smooth transition and a very efficient leve-off.

747-400 SIMULATOR TECHNIQUES ...

comments and techniques regarding the ...

EMERGENCY DESCENT

RULE NUMBER ONE: LEAVE AUTOPILOT ON !

It is V-E-R-Y common first reaction for the student pilot to shut off the auto-pilot. Remember, that even though using the auto-pilot is actually a REQUIREMENT, the procedure actually works better using the auto-flight tools. Using the auto-pilot is much better than thrashing around with that annoying oxygen mask on, trying to hand-fly and communicate with the auto-pilot off.

OXYGEN MASK considerations:

You WILL "probably" have your **OXYGEN** mask on your face **BEFORE** you start the descent. Ordinarily, the Check-person will treat you to a **"CABIN ALTITUDE WARNING"** or a **"RAPID DECOMPRESSION"** just prior to the **"HIGH DIVE"**. With the mask in place, it is very difficult to be constantly changing the comm panel so as to talk with ATC controller, other airplanes, and then back with the other pilot. So here is what I suggest. It is MUCH EASIER if you simply talk with the other pilot, ATC controller, and other airplanes all on the same frequency. Trying to setup and switch the comm panel back and forth is check-ride suicide..

V_{mo} versus M_{mo}

Here is a simple description of the relationship between **MACH** and **IAS**. If you operate the airplane at a constant **MACH** number, as the airplane descends the **IAS** (indicated airspeed) will be increasing. When the **IAS** reaches about **310 Kts**, the **MCP** is designed to switch automatically to **IAS**. The reason is that if you remain in the **MACH** mode below about 29,000 feet, the **IAS** commanded will exceed the structural limits of the airplane. This leaves us with the thought that if we start the descent in the **MACH** mode, we will probably level off at about **310 Kts IAS**. If we start the descent in the **IAS** mode, we will most likely level-off with about **250 Kts IAS** on the **MCP**.

Here are the maximum allowable operating speeds:

The maximum **MACH (M_{mo})** is **.92 MACH** at altitudes above FL 290.

The maximum **IAS (V_{mo})** varies between **365 Kts** (at Sea Level) to about **254 Kts** (at FL 41.0).

The Check-pilot may want to know the definition of some terms. Here are the ones he will be interested in.

MEA: MINIMUM ENROUTE ALTITUDE: This is the altitude on the airways that ensures that you will be able to receive the NAVAIDS suitable for navigation and ATC communication so long as you remain within 4 NM of the airway centerline.
MOCA: MINIMUM OBSTACLE CLEARANCE ALTITUDE: Ensures at least 1000 feet about the terrain and in the mountainous terrain at least 2000 feet clearance as long as you remain on the airway. MOCA is for emergency use.
GRID MORA: MINIMUM OFF AIRWAYS ALTITUDE: Some navigation charts are divided into grids. There is an altitude listed that provides terrain clearance of 1000 feet up to 5000 foot elevation and after that clearance by 2000 feet. Used for off airways descents.

LIMITS and SPECIFICATIONS
APR 15

The thing about the "LIMITATIONS and SPECIFICATIONS" part of the check-ride is that nobody warns you that you gotta know them. The training people don't specifically talk about them, they just assume you know them ... and then when you are in your oral, the check-guy usually spends an inordinate amount of time dwelling on those tiny, unknown parts of the Limits section, asking embarrassing questions about stuff you can't even remember.

So ... in addition to the <u>UNDERLINED ITEMS</u> (You gotta know those definitely) I have include be some of the other items in this section that may be part of the ORAL EXAM.

FLYING NON-PRECISION APPROACHES in LNAV
... some really picky technical details.

First, we have to understand some of the jargon associated with this evolution. They have invented a special language and every phrase and word is loaded with meaning.

CDU LINE SELECTED APPROACHES: These are approaches that can be selected from the list on the DEP/ARR page.

CONSTRUCTED APPROACHES: These are approaches for which there are available procedures and approach plates, but for some reason are not on the list in the **DEP/ARR** page. Generally, these are **_NOT_** allowed.

SUPPLEMENTAL NAVIGATION: This is the buzz-word which means that LNAV may be used to steer the airplane during the approach.

DISCUSSION:

*Here is what this is all about, when you are flying a **NON-ILS** approach,*

> **YOU WILL BE REQUIRED TO MONITOR RAW DATA VOR/ADF INDICATION FROM THE FINAL APPROACH FIX INBOUND.**

Even if this may be an engine out approach or some other irregularity, still, anytime you are flying the NON-ILS, remember that at least **ONE PILOT** has gotta be monitoring **RAW DATA (VOR/ADF MODE)** from the **FAF** inbound.

ALSO: On some departure or arrival or Go-around procedures, it is required to track outbound on a certain **VOR "RADIAL"** (Such as the SFO RWY 28 L/R). Remember, You **MUST** have the **VOR** selected on the **ND**.
AND: There are departures that **REQUIRE** climbing out on a **VOR "RADIAL."**

CRITICAL NOTES:

■ WE DO NOT DO BCRS APPROACHES IN THIS AIRPLANE.

■ The AFDS will track a LOCALIZER COURSE,
BUT
The AFDS Autoflight system WILL NOT track a VOR or NDB (ADF) COURSE!

■ Appropriate RAW DATA MUST BE MONITORED throughout the approach!

■ The PILOT FLYING (PF) MUST display the VOR MODE on the ND no later than the FINAL APPROACH FIX!

continued ...

MAP MODE OK as reference if desired ONLY IF these requirements are met for the approaches indicated.

If you are flying this approach:

Then: At least one of the pilots must be monitoring these sources.

LOC	PFD OK
-DME	DME
NDB	ADF bomb on ND
VOR	VOR pointer on ND

from the FAF inbound.

TECHNIQUE:

Select the ND to be displayed on the LOWER (supplemental) EICAS in the MAP mode. Select the PF ND to the VOR mode. That way, the PF can monitor the MAP mode and the VOR.

ADJUSTMENTS:

>70 degrees add 3 psi/1 degree above 70.
<70 degrees subtract 3 psi/1 degree below 70.

$+3_{psi} / 1_{degree\ F} > 70°F < -3_{psi} / 1_{degree\ F}$

some non-underlined but nice (IMPORTANT) to know

ESCAPE SLIDE STUFF

EXCEPT IN EMERGENCY (of course):

- DOORS 3L/R SHOULD NOT BE OPENED WITH MORE THAN 60,000 POUNDS FUEL ON BOARD.

- ON UPPER DECK, YELLOW KNOB SHOULD BE VISIBLE IN THE AUTOMATIC VIEWPORT WHEN PASSENGERS ARE UP THERE.

- FERRY FLIGHTS: (see checklist in additionals) CABIN DOORS 2L/R should be ARMED.

... and PROCEDURES FOR STUDY and REVIEW ONLY!

OIL MINIMUM QUANTITY

18 QUARTS BEFORE ENGINE START

AFTER ENGINE START 15 to 21 QUARTS

Here is the **EXCEPTION**: After engine start, it is not uncommon for the engine oil quantity indication to decrease to about ...

CLOGGED OIL FILTER problem

Let me take this opportunity to make a comment about responding to such things as this. They may seem to be minor annoyances, but on the check-ride, they are opportunities to get some major brownie points ... and look like you know what you are doing.

CLOGGED OIL FILTER discussion. Heads up!!! This is a favorite Check-guy situation, usually setup in preparation for the impending **ENGINE SHUT-DOWN** procedure. Go to your pilot manual ... and the scriptures will say something like "**MONITOR OIL PRESSURE**" and if it goes to **REDLINE** consider reference to "***ENGINE SHUTDOWN PROCEDURES***".

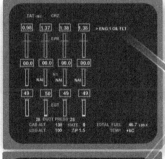

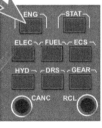

Here are some of my thoughts about this. This is a clue that the Check-Airman is getting ready for you to experience an Engine shutdown. This gives us some time to prepare ...
DON'T SIT THERE LIKE A DUMMY ...

They are going to want to see you do some things.

1. Ask some questions such as: Is the airplane flying at an altitude that would be unsustainable on three engines. We have charts and the FMC/CDU to give us some clues about that. Perhaps we could **DESCEND TO A LOWER ALTITUDE** and monitor the situation. *DUH!*

2. Will it be necessary to divert for an off schedule landing or can we continue to destination. We might want to **COMMUNICATE WITH THE DISPATCHER** to help us resolve the situation. I don't think, for example, that I would want to start out across the North Atlantic with this situation, but sometimes continuing to destination may be an OK alternative.

3. Get out the materials necessary for a shutdown ... find the stuff in your "PILOT BAG" and get them out and brief for the event. Know what to do ahead of time.

4. Use your imagination. I don't know what the flight situation will be ... but use your head.
THINK LIKE A CAPTAIN!

© MIKE RAY 2014
WWW.UTEM.COM

747-400 *SIMULATOR TECHNIQUES ...*

ENG STUFF

some **NOT** underlined ... but still important

BUT ... you are definitely required to know this engine parameter COLD!

MAX EGT FOR ENGINE START

535 °C

Of course, the **"RED INDICATION"** is always a good answer, but how do you know when you are heading for a **HOT START** if you don't know the number?

SOME REVERSE LIMITS:

- ☠ ■ NOT ALLOWED IN FLIGHT !
- ■ POWERBACK PROHIBITED !

■ AFTER LANDING: START REMOVING REVERSE AT 60 kts so as to be in idle reverse by TAXI SPEED.

... and PROCEDURES FOR STUDY and REVIEW ONLY!

FUEL STUFF

FLEET PLANNED LANDING FUEL (FPL)

19,000 pounds

Of course, the Checkguy will be interested in knowing just what that fuel number means:

> **"ENSURES:**
> **OVER ONE HOUR OF USABLE FUEL**
> **AT MAX ZFW PLUS FPL (535,000 + 19,000)**
> **AT 10,000 feet MSL**
> **AT HOLDING SPEED."**

FYI: The amount of fuel required to execute a go-around at runway threshold to 1000 feet AGL, fly a VFR pattern, intercept a 3 degree glideslope approximately 2 1/2 miles from the runway, and continue to land is: **5600** pounds

FYI: The MINIMUM DISPATCH fuel: **22,000** pounds

FYI: FUEL JETTISON rate: **4,650** pounds per minute

FYI: TAXI PER MINUTE: **130** pounds per minute

FYI: MINIMUM ALTERNATE: **6,500** pounds

FYI: MINIMUM DESIRED LANDING FUEL: **11,100** pounds

> *"ENSURES SUFFICIENT FUEL ON BOARD AT THE THRESHOLD IN A WORST CASE CONDITION WITH MAXIMUM FUEL QUANTITY INDICATOR ERROR (indicator reads too high)."*

HYDRAULICS

747-400 SIMULATOR TECHNIQUES ...

MINIMUM BRAKE ACCUMULATOR PRECHARGE PRESSURE

750 psi

Some things about the **HYDRAULIC BRAKE / ACCUMULATOR** pressure indicator that you will be asked about on the oral.

1. This gauge indicates **HYDRAULIC PRESSURE AVAILABLE** from the **NORMAL BRAKE SYSTEM**.

2. When **NORMAL BRAKE SYSTEM is *NOT* PRESSURIZED** ... This gauge indicates **ACCUMULATOR PRECHARGE**.

3. **ACCUMULATOR** can *ONLY* be pressurized by the **#4 HYDRAULIC SYSTEM**.

4. The accumulator provides sufficient pressure to (1) **SET** and (2) **HOLD** the **PARKING BRAKES** when all other sources of brake pressure fail.

THE ACCUMULATOR WILL NOT STOP THE AIRPLANE!

This light indicates **LOW PRESSURE** in the active brake source. It comes on well before 750 psi.

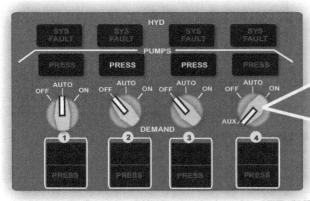

THIS COULD BE IMPORTANT: During the Captain's **"FINAL COCKPIT PREPARATION"** set-up, turn the **#1** hydraulic pump switch to **AUTO** and the **#4** to **AUX**. It is IMPORTANT that the **#4 AUX** be selected *"BEFORE"* the #1 switch is turned to **AUTO**.

REASON: The parking brake accumulator can **ONLY** be replenished by **HYD SYS #4**; If **HYD SYS #1** is pressurized (without pressurizing **HYD SYS #4**) and **HYD SYS #1** subsequently fails; there *may* be **NO PARKING BRAKE AVAILABLE** (depending on the residual status of the accumulator which has not been replenished). If, however, **HYD SYS #4** is pressurized, should **HYD SYS #1** subsequently fail, parking brake capability will be available.

... and PROCEDURES FOR STUDY and REVIEW ONLY!

When to USE
NACELLE ANTI-ICE

On the GROUND and IN THE AIR WHEN ...

10°C OR BELOW and VISIBLE MOISTURE

What can pilots screw up.
10 degrees C is about 50 degrees F. So, you could be planning a departure on a 50 degree day and think that just because it isn't 32 degrees F that it is NOT

50°F = 10°C

Visible moisture is restricted visibilities of 1 mile or less, and on the ground there is the additional consideration about ground clutter. it could be CAVU to the moon, but if there is crap on the taxiways or runways that could freeze on

OFFICIAL DEFINITION OF ICING CONDITIONS

The -40°C RESTRICTION:
NACELLE ANTI-ICE must be **ON** during **ALL** ground and flight operations when icing conditions are present or expected
... *EXCEPT WHEN TEMPERATURE BELOW -40°C SAT.*

HOWEVER ...
If the airplane is in a **_DESCENT_** and icing conditions are present or anticipated, you **MUST** have the **NACELLE ANTI-IC ON**
.... **EVEN THOUGH THE TEMPERATURE IS BELOW -40°C SAT.**

DO NOT USE
WING ANTI-ICE

■ *WHEN TAT EXCEEDS 10 degrees C.*

■ *WHEN FLAPS ARE EXTENDED.*

747-400 SIMULATOR TECHNIQUES...

How fast can you ...
ROCK and ROLL?

TURBULENT AIRSPEEDS
Or ... Airspeeds to fly during turbulent air.

290 / .82

NOT UNDERLINED, but nice to know:

340 / .84 — STD CLIMB above 10K (FMC INOP)

340 / .86 — STD CRUISE above 10K (FMC INOP)

290 / .84 — STD DESCENT above 10K (FMC INOP)

... and PROCEDURES FOR STUDY and REVIEW ONLY!

MAX STRUCTURAL WEIGHTS (pounds)

Checkguys L-O-V-E this chart; and I can guarantee that they will ask you numbers from this matrix on your oral.

878,000	MAX TAXI
875,000	MAX T/O
630,000	MAX LAND
535,000	MAX ZFW

OK ... practice ... practice ... practice... practice

MAX TAXI	
MAX T/O	
MAX LAND	
MAX ZFW	

HOW HIGH CAN IT GO?

45,100
MAXIMUM OPERATING PRESSURE ALTITUDE
(feet msl)

The question about just how high the airplane can actually fly is not a single answer. Oh sure, the **MAXIMUM OPERATING PRESSURE ALTITUDE** is that 45,001 foot MSL figure we have cited, but as you intuitively know, a heavy airplane is restricted by the actual ability of the engines and wing to reach that altitude ... and even further, the airplane can probably, under certain circumstances actually fly higher that so-called maximum.

To derive a more useful and representative maximum altitude figure, we have to take into account the weight of the airplane, the density and temperature of the air and whole bunches of other stuff. And fortunately for us mere mortal airplane drivers, our airplanes computer heart can figure out just what altitude is our maximum.

To find that number:
STEP 1: Select the **VNAV** page on the **CDU**.
STEP 2: Select the **NEXT PAGE** on the keyboard and that should give display **M.XXX CRZ ..... PAGE 2/3**.

Across the bottom of that page are two airspeeds **OPT** (Optimum) and **MAX** (Maximum). The optimum altitude is the suggestion from the computer as to what is the most efficient altitude to fly at the airspeed you have placed in the database. The maximum is what it calculates as the highest performance altitude predicated on the information that it has (**ECON**, weight, temperature, air density, etc). It also adds a safety factor of about 30% into the result to accommodate turbulence induced stall.

However, none of these are the correct answer ... because the **MOST DESIRABLE OPERATING ALTITUDE**, while it takes all this into account, comes from the flight plan. It uses the winds and engine performance so as to plan for the most efficient fuel burnout profile consistent with the operational needs of the schedule. This is SO complex that mere human pilots cannot figure it out.

Even the old adage ..."Higher uses less fuel" ... simply will not work. The calculation actually used is predicated on the **NAM/1000#** chart and the wind matrix. It turns out, however, that the Boeing 747-400 likes to fly at lower altitudes when it is heavy and then climb as the burnout progresses, eventually reaching its highest cruising altitude near the end of the flight segment.

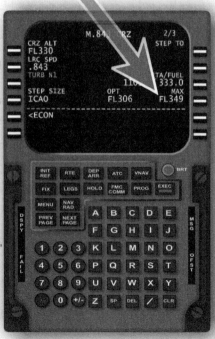

This is generally a useful feature of this airplane, since many other airplanes are trying to get the higher altitudes initially and this allows us to get clearances that otherwise would be taken by other airplanes.

The final word on whether the airplane "can" climb or not should consider the thought about whether it "should" climb or not.

THERE EXISTS A SERIOUS PROBLEM !!

Statement of PROBLEM: The higher an airplane flies, the less "STALL MARGIN" there is for the wing. This wing limitation is well understood and depends on factors that are constantly changing; such as weight of the airplane, density and temperature of the airmass, turbulence, and control input forces. The appropriate information can be described by a "V-n" diagram which illustrates the various speed/altitude capabilities of the airplane. Characteristically, there is a portion of every portion of the flight envelope for every airplane that is called the "COFFIN CORNER" and where there is a diminishing limit to the available airspeed that the wing can fly without either a HIGH SPEED STALL or a LOW SPEED STALL. There is a chart in your pile of manuals that will define that limitation. Usually, the ABSOLUTE MAXIMUM ALTITUDE suggested leaves about a 1.3G spread.

I don't know if you understand what that says, so here it is in plain English. If you try to fly "outside" the limits of the airplanes capability, the airplane will stall ... and it appears that aviators of big airplanes continue to attempt to fly their airplanes at altitudes that do not provide the STALL MARGIN required ... and they stall the wing and the airplanes sometimes cannot recover from the stall. Even though the engineers have created wing geometries that account for the span-wise flow of stall migration, and also have built in STALL BUFFET characteristics for the wing root to mitigate the situation; sometimes, it seems, pilots are either unaware or ignorant of the situation as it develops.

Description of how the problem manifests itself:

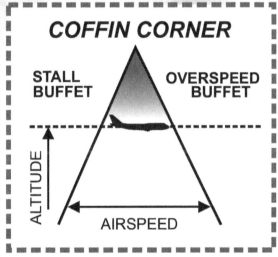

How could a typical garden variety airline crew get in such a situation? Simple. Here is an example situation. When using the airways around the world, there are frequently conflicts with the allocation of available altitude assignments. The center will offer available altitude alternatives. The decision as to whether to accept an available altitude that is marginally acceptable for the weight should "ALWAYS" be very carefully considered. The engines are capable of pushing the airplane well above the altitude where there is an acceptable "STALL" limit for the wing. On densely traveled route structures such as the North Atlantic Track System (NATS), frequently desired altitudes are simply not available and the acceptance of lower altitudes are less desirable due to wind components or increased fuel consumption. It is possible that crews may accept higher than optimum altitudes that present less than acceptable stall margins just to "get" the altitudes they want.

BE ALERT!

THE END

... of this pile of material, but not the end of the training and review that must always go on
FOREVER!.

AD SECTION
Where you can get other books and stuff by Captain Mike

If you enjoyed the 747-400 pictures that appeared in this book, you can get them in enlarged and professionally framed editions on Captain Mike Ray's *"FINE ART AMERICA"* website.

MIKE RAY

This guy loves airplanes ... and loves to create and look at pictures of airplanes. He also wants to share his stuff, so if you want to see some of the screen shots that he has stolen from his flight simulation world of aviation ... as well as a whole bunch of other images ... you can view them at his spectacular *"Fine Art America"* website:

www.mike-ray.artistwebsites.com

http://mike-ray.artistwebsites.com/

STOOFDRIVER
FLYING THE GRUMMAN S2 TRACKER
by
MIKE RAY

The USS Hornet Far East cruise of 1965 was a great adventure ... and Lieutenant Mike Ray was a part of the company of Sailors and Marines that made this historical journey. While it may be considered a "typical" mission carried out by the U.S. Navy for the time, it was also representative of the experiences of thousands of others who made up the vast cold war armada that swept the seas. Ride with them now as we revisit their legacy in this part of history.

This beautiful book has 200 pages crammed with photographs and explanatory text that can be found nowhere else. Hardbound with slipcover, the 8 X 10 inch landscape format is something that will become a treasured part of your library. It makes a terrific gift item for the veteran in your family.

VIEW THE BOOK IN ITS ENTIRETY
@
www.utem.com

8" X 10"
200 full color pages

UNOFFICIAL Airliner Training and Checkride Survival Guides

"The Unofficial Boeing 747-400 Simulator and Checkride Procedures Manual."

Referred to as "the best 747-400 training manual in the world" by airline professionals. It is a veritable treasure trove of beautiful graphics and illustrations. Written for pilots by a pilot, it serves to make simple what is arguably one of the most complex flying machine in existence. Must have for Professional airline 747-400 pilots and serious simmers alike.

"The Unofficial Boeing 757/767 Simulator and Checkride Procedures Manual."

This book goes well beyond merely the fabulous Boeing 757 and 767 cockpit ... and develops some of the most in depth presentations for Op Specs. Even for you guys that don't fly this airplane, there is much interesting and useful information about flying the approaches included. For professional airline 757/7676 pilots it is an indispensable guide book ... and for the serious simmer.

"Sim-flying the Airbus A300 Series Flight Simulations."

Here is a book developed SPECIFICALLY for the flight simmer and basic Airbus neophyte. If you just bought that terrific new Airbus A320 simulation ... and are staring blankly at the complicated images on the video screen, then this is the book for you. Full color and lavishly illustrated. It is written to be read, understood, and enjoyed. Must read.

"The Unofficial Airbus A320 Series Simulator and Checkride Procedures Manual."

TWO BOOKS in ONE! Brand new and a training TOUR-DE-FORCE. This massive 400+ page manual is simply beautiful. The full color pages are not only written in an interesting and readable style; but this book is also **THE** training tool-set for discovering and mastering the finer points of Airbus flying. A terrific bargain. Written for the professional airline pilot ... and the serious simmer.

"The Unofficial Boeing 737-300,400,500 Simulator and Checkride Procedures Manual."

This book has been the tutorial of choice for literally hundreds of airline pilots wanting to "get ready" for their initial and recurrent and check-ride. Profusely populated with hundreds of illustrations and diagrams and written in that easy-to-read Mike Ray style. Great resource for professional airline pilots and serious simmers alike.

"Flying the Boeing 700 Series Flight Simulators."

If you are "brand new to simming". If you have finally made the plunge and purchased one the terrific Boeing flight simulations that are appearing in the marketplace ... you gotta include this book in your list of "**MUST HAVE**" items. The mysteries of flying the Boeing "glass" are revealed in a humorous, interesting, and easy to read format. This book is written both beginners and veteran simmers alike in mind.

AIRPLANE STUFF

www.utem.com
created by
Captain Mike Ray

CHANGES TO WEBSITE ... SOME NEW STUFF
NEW AIRPLANE STUFF

- **AIRPLANE BOOKS**
- **COCKPIT PANEL** posters
- **AIRPLANE STUFF** book
- **E-BOOK DOWNLOADS**
- **AVIATION ARTWORK**
- **STOOFDRIVER** section
- **DOLLAR DOWNLOADS**
- **SPANISH ARTICULOS**

BAD ATTITUDE CLUB
IRES PERICULOSUS

VISIT FOR **FREE STUFF**

Come on over and view my website.
You can order your stuff there.
www.utem.com

Printed in France by Amazon
Brétigny-sur-Orge, FR

21252638R00231